ALGEBRAIC GEOMETRY MODELING IN INFORMATION THEORY

Series on Coding Theory and Cryptology

Editors: Harald Niederreiter *(National University of Singapore, Singapore)* and
San Ling *(Nanyang Technological University, Singapore)*

Published

Vol. 1 Basics of Contemporary Cryptography for IT Practitioners
by B. Ryabko and A. Fionov

Vol. 2 Codes for Error Detection
by T. Kløve

Vol. 3 Advances in Coding Theory and Cryptography
eds. T. Shaska et al.

Vol. 4 Coding and Cryptology
eds. Yongqing Li et al.

Vol. 5 Advances in Algebraic Geometry Codes
eds. E. Martínez-Moro, C. Munuera and D. Ruano

Vol. 6 Codes over Rings
ed. P. Solé

Vol. 7 Selected Topics in Information and Coding Theory
eds. I. Woungang, S. Misra and S. Chandra Misra

Vol. 8 Algebraic Geometry Modeling in Information Theory
ed. E. Martínez-Moro

Series on Coding Theory and Cryptology – Vol. 8

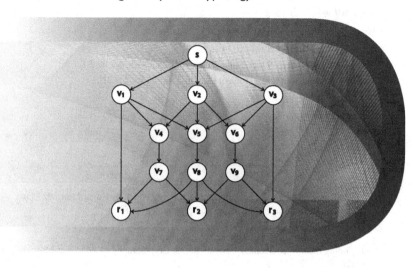

ALGEBRAIC GEOMETRY MODELING IN INFORMATION THEORY

Editor

Edgar Martínez Moro
University of Valladolid, Spain

NEW JERSEY • LONDON • SINGAPORE • BEIJING • SHANGHAI • HONG KONG • TAIPEI • CHENNAI

Published by

World Scientific Publishing Co. Pte. Ltd.
5 Toh Tuck Link, Singapore 596224
USA office: 27 Warren Street, Suite 401-402, Hackensack, NJ 07601
UK office: 57 Shelton Street, Covent Garden, London WC2H 9HE

Library of Congress Cataloging-in-Publication Data
Algebraic geometry modeling in information theory / edited by Edgar Martínez Moro.
  p. cm. -- (Series on coding theory and cryptology ; v. 8)
  Includes bibliographical references.
  ISBN 978-981-4335-75-1 (hard cover : alk. paper)
  1. Coding theory. 2. Geometry, Algebraic. 3. Cryptography. I. Martínez-Moro, Edgar.
  QA268.A415 2012
  003'.54--dc23
                                                    2012032887

British Library Cataloguing-in-Publication Data
A catalogue record for this book is available from the British Library.

Copyright © 2013 by World Scientific Publishing Co. Pte. Ltd.

All rights reserved. This book, or parts thereof, may not be reproduced in any form or by any means, electronic or mechanical, including photocopying, recording or any information storage and retrieval system now known or to be invented, without written permission from the Publisher.

For photocopying of material in this volume, please pay a copying fee through the Copyright Clearance Center, Inc., 222 Rosewood Drive, Danvers, MA 01923, USA. In this case permission to photocopy is not required from the publisher.

Printed in Singapore.

Preface

In this book some of the latest research developments in the area of algebraic modelling in a broad sense applied to coding theory and information theory are presented. This book is a revised version of the lectures held at the postgraduate school S3CM 2010. The S3CM series stands for 'Soria Summer School on Computational Mathematics' and it is a series of annual international schools mostly intended for master, PhD students and Postdocs working on computational mathematics. On 2010 the meeting was devoted to the topic 'Algebraic Geometric Modelling in Information Theory' and took place in the Campus of Soria of the Universidad de Valladolid and hosted by SINGACOM reseach group. The AGMINT aim was to fill in the gap between the theoretical part of algebraic geometry and the applications to problem solving and computational modelling in engineering in signal processing and information theory which involve non-trivial knowledge of algebra and geometry.

In this book the fundamentals and the 'state of the art' of each of the topics covered are also presented. I hope that this book will be useful for students and researchers both in applied algebraic geometry as well as information theory and coding theory.

I want to thank all the authors for their contributions to this volume, indeed their efforts made this book possible. I also want to thank the Spanish Research Council (MICINN), the 'Junta Castilla y León' and the Lifelong Learning Programme of the European Commission for their support that made possible the celebration of S3CM 2010. Finally I am really grateful to World Scientific and E.H. Chionh for their continuous support and excellent editorial work.

E. Martínez-Moro
Institute of Mathematics
University of Valladolid

Contents

Preface   v

Sage: A Basic Overview for Coding Theory and Cryptography
D. Joyner   1

Aspects of Random Network Coding
O. Geil and C. Thomsen   47

Steganography from a Coding Theory Point of View
C. Munuera   83

An Introduction to LDPC Codes
I. Márquez-Corbella and E. Martínez-Moro   129

Numerical Semigroups and Codes
M. Bras-Amorós   167

Codes, Arrangements and Matroids
R. Jurrius and R. Pellikaan   219

Chapter 0

Sage: A Basic Overview for Coding Theory and Cryptography

David Joyner

Mathematics Dept.
USNA, Annapolis, MD, 21402, USA.
wdjoyner@gmail.com

This brief paper surveys recent work in Sage, a free and open-source mathematics package, on implementing algorithms to compute with linear block codes and cryptography. The paper ends with a "wish-list" of coding-theoretic functionality that the author hopes will be added to Sage in the not-too-distant future.

Contents

0.1	Introduction	2
0.2	What is Sage?	2
	0.2.1 Functionality of selected components of Sage	3
	0.2.2 History	4
	0.2.3 Why Python?	4
	0.2.4 The CLI	6
	0.2.5 The GUI	8
	0.2.6 Open source philosophy	11
0.3	Coding theory functionality in Sage	11
	0.3.1 General constructions	12
	0.3.2 Coding theory functions	13
	0.3.3 Weight enumerator polynomial	16
	0.3.4 More code constructions	18
	0.3.5 Automorphism group of a code	19
	0.3.6 Even more code constructions	21
	0.3.7 Block designs and codes	22
	0.3.8 Special constructions	23
	0.3.9 Coding theory bounds	24
	0.3.10 Asymptotic bounds	26
0.4	Cryptography in Sage	28
	0.4.1 Classical cryptography	28
	0.4.2 Algebraic cryptosystems	29
	0.4.3 RSA	29

    0.4.4 Discrete logs 31
    0.4.5 Diffie-Hellman 31
    0.4.6 Linear feedback shift registers 32
    0.4.7 BBS streamcipher 33
    0.4.8 Blum-Goldwasser cryptosystem 34
0.5 Miscellaneous topics 35
    0.5.1 Duursma zeta functions 35
    0.5.2 Self-dual codes 38
    0.5.3 Cool example (on self-dual codes) 41
0.6 Coding theory not implemented in Sage 42
References 44

0.1. Introduction

This chapter gives a brief survey of Sage's functionality in coding theory and cryptography, with no pretention towards completeness. The emphasis is on examples but the level is aimed towards a mathematician or graduate student in mathematics.

First, we give a general introduction to Sage (and Python), though taylored to the subject at hand. Second, we give a selection of examples of Sage's functionality in coding theory. Next, we give a selection of examples of Sage's functionality in cryptography. Following that, we discuss a few particular topics in more detail (Duursma zeta functions, self-dual codes, and invariant weight enumerator polynomials). Finally, we end with a "wish-list" of coding-theoretic functionality that the author hopes will be added to Sage in the not-too-distant future.

0.2. What is Sage?

Sage is open-source general-purpose mathematical software available for download from http://www.sagemath.org. Sage is free to download, and will always remain so, and no registration forms have to be filled out in order to download Sage onto your machine, install it and use it. The Sage project is headed by the mathematician William Stein, who is at the University of Washington, in Seattle.

Sage includes Maxima and pynac and SymPy (for calculus and other symbolic computation), Singular and GAP (for algebra), R (for statistics), Pari (for number theory), SciPy (for numerical computation), libcrypt for cryptography, and over 60 more.

Other packages available in Sage:

Basic Arithmetic	GMP, NTL, flint
Command Line	IPython
Graphical Interface	Sage Notebook
Graphics	jmol, Matplotlib, ...
Graph theory	NetworkX
Interpreted programming language	Python
Networking	Twisted
Applied Math.	SciPy, GSL, GLPK, etc.
Source control system	Mercurial
Symbolic computation, calculus	SymPy, pynac

Sage now has a *huge* range of functionality.

To be a component of Sage, the software must be: free, open source, robust, high quality, and portable.

0.2.1. *Functionality of selected components of* Sage

This shall be a very selective and very abbreviated survey. We restrict to topics which have a tangential connection with algebraic block codes (in particular, algebraic-geometric codes). Sage provides significant functionality in the following areas.

- Commutative Algebra:
  - Clean, structured, object-oriented multivariate polynomial rings, coordinate rings of varieties, and ideals
  - Uses Singular as backend when possible for arithmetic speed and certain algorithms
  - Groebner Basis computations
- Algebraic geometry:
  - Varieties and Schemes
  - Genus 2 curves and their Jacobians (including fast p-adic point counting algorithms of Kedlaya and Harvey)
  - Implicit plotting of curves and surfaces
- Linear algebra:
  - Sparse and dense linear algebra over many rings
  - Highly optimized in many cases
  - In somes cases, possibly the fastest money can buy

- Group theory:
    - Sage includes GAP
    - Weyl groups and Coxeter groups,
    - Sage includes some "native" permutation group functions
    - Sage includes "native" abelian group functions
    - Sage includes a matrix group class, abelian group class and a permutation group class
    - Sage has some native group cohomology functions

Unfortunately, at the time of this writing, Sage lacks a free group class.

0.2.2. *History*

Here is a very abbreviated history of Sage.

- Nov 2004: William Stein developed Manin, a precursor to Sage.
- Feb 2005: Sage *0.1*. This included `Pari`.
- Oct 2005, Sage *0.8*: `GAP` and `Singular` included as standard.
- Feb 2006: Sage Days 1 workshop, UCSD – Sage 1.0
- May-July, 2006 (Sage 1.3.*) GUI Notebook developed by William Stein, Alex Clemsha and Tom Boothby.
- Sage Days Workshops at UCLA, UW, Cambridge, Bristol, Austin, France, San Diego, Seattle, MSRI, Barcelona, ...
- Sage won first prize in the Trophees du Libre (November 2007)
- Sage Days 23.5 – Kaiserslautern, Germany on "Singular and Sage integration," ends July 9, 2010.

See `http://wiki.sagemath.org/` for more details (lectures in pdf, lists of papers and books using Sage, mp3's of talks, dates of future Sage-days, and so on).

0.2.3. *Why Python?*

Sage is based on the mainstream programming language Python. If you wish to constribute to Sage, you should become very familiar with this great language.

Python is a powerful modern interpreted general programming language, which happens to be very well-suited for scientific programming.

- "Python is fast enough for our site and allows us to **produce maintainable features in record times**, with a minimum of developers."
  - Cuong Do, Software Architect, `YouTube.com`.
- "Google has made no secret of the fact that they use Python a lot for a number of internal projects. Even knowing that, once **I was an employee, I was amazed at how much Python code there actually is in the Google source code system.**"
  - Guido van Rossum, `Google`, creator of Python.
- "Python plays a key role in our production pipeline. Without it a project the size of **Star Wars: Episode II** would have been very difficult to pull off. From crowd rendering to batch processing to compositing, Python binds all things together."
  - Tommy Burnette, Senior Technical Director, `Industrial Light & Magic`.

Other great advantages of Python:

- Easy for you to define your own data types and methods on it. Symbolic expressions, graphics types, vector spaces, special functions, whatever.
- Very clean language that results in easy to *read* code.
- Easy to learn:
  - Free: Dive into Python http://www.diveintopython.org/
  - Free: Python Tutorial http://docs.python.org/tut/
- A *huge* number of libraries: statistics, networking, databases, bioinformatic, physics, video games, 3d graphics, ...
- Easy to use any C/C++ libraries from Python.
- Excellent support for string manipulation and file manipulation.
- `Cython` – a Python compiler (http://www.cython.org) which allows you to achieve speeds of corresponding C programs.

For all these reasons, Python was chosen as the scripting language for Sage. Other programs, such as Maple or Mathematica, have their own specialized language. If you use Sage, you don't have to learn a scripting language which you won't use anywhere else.

0.2.4. *The CLI*

Sage has a command line interface (CLI). When you start Sage in a terminal window you will get a small Sage banner and then the Sage command-line prompt sage:, after which you type your command. Most examples in this chapter are given in terms of the CLI[a]. If you wish to reproduce them yourself, just remember that sage: is a prompt and you should only type in the command *after* the prompt. If you want to switch to the graphical user interface (discussed below), type notebook() at the prompt and hit return.

If you are happy to work at the command line, here is an example of what a short Sage session could look like:

```
─────────────────────────── Sage ───────────────────────────
sage: 2^3
8
sage: t = var("t")
sage: integrate(t*sin(t^2),t)
-cos(t^2)/2
sage: plot[TAB]
plot                    plot_slope_field        plotkin_bound_asymp
plot3d                  plot_vector_field       plotkin_upper_bound
```

Tab-completion helps you select the command you want with less effort.

The Sage function below is written in Python but uses Sage classes, so can be used on the Sage command line.

```
─────────────────────────── Sage ───────────────────────────
def Hexacode():
    """
    This function returns the [6,3,4] hexacode over GF(4).
    It is an extremal (Hermitian) self-dual Type IV code.

    EXAMPLES:
        sage: C = Hexacode()
        sage: C.minimum_distance()
        4
    """
```

---

[a]Some minor edits are made in the appearance of the *output* you see in the Sage examples below, for typographical reason. I hope they cause no confusion.

```
F = GF(4,"z")
z = F.gen()
MS = MatrixSpace(F, 3, 6)
G = MS([[1, 0, 0, 1, z, z ], [0, 1, 0, z, 1, z ], [0, 0, 1, z, z, 1 ]])
return LinearCode(G)
```

The i-th **binomial moment** of the $[n,k,d]_q$-code C is

$$B_i(C) = \sum_{S, |S|=i} \frac{q^{k_S} - 1}{q-1}$$

where k_S is the dimension of the shortened code C_{J-S}, where $J = [1, 2, ..., n]$. Let's examine Sage's docstring and source code for the binomial_moment method using ? and ??.

─────────── Sage ───────────

```
sage: C = HammingCode(3,GF(2))
sage: C.binomial_moment?          # ? gives you the docstring
Type:              instancemethod
String Form:       <bound method LinearCode.binomial_moment
  of Linear code of length 7,
  dimension 4 over Finite Field of size 2>
File:    .../local/lib/python2.6/site-packages/sage/coding/linear_code.py
Definition:    C.binomial_moment(self, i)
Docstring:
       Returns the i-th binomial moment of the [n,k,d]_q-code C:

           B_i(C) = sum_{S, |S|=i} frac{q^{k_S}-1}{q-1}

       where k_S is the dimension of the shortened code C_{J-S},
       J=[1,2,...,n]. (The normalized binomial moment is b_i(C) =
       binom(n,d+i)^{-1}B_{d+i}(C).) In other words, C_{J-S} is
       isomorphic to the subcode of C of codewords supported on S.

<snip>
sage: C.binomial_moment??         # ?? gives you source code listing
  <snip>
           n = self.length()
           k = self.dimension()
           d = self.minimum_distance()
           F = self.base_ring()
           q = F.order()
           J = range(1,n+1)
           Cp = self.dual_code()
           dp = Cp.minimum_distance()
           if i<d:
               return 0
```

```
        if i>n-dp and i<=n:
            return binomial(n,i)*(q**(i+k-n) -1)/(q-1)
    P = SetPartitions(J,2).list()
    b = QQ(0)
    for p in P:
        p = list(p)
        S = p[0]
        if len(S)==n-i:
            C_S = self.shortened(S)
            k_S = C_S.dimension()
            b = b + (q**(k_S) -1)/(q-1)
    return b
```

0.2.5. *The GUI*

The graphical user interface, or Sage Notebook, is the most popular interface. It has no prompt but instead uses "cells" to input the commands. You can input one command per cell, or multiple commands in a cell. However, usually it is only the output from the last command in the cell which is printed on the screen. The Sage Notebook can be tried out for free by anyone with an internet connection and a good browser at http://www.sagenb.org.

Advantages to the GUI:

- Connect to Sage running locally *or elsewhere* (via internet).
- Create embedded graphics (in 2- and 3-d).
- Typeset mathematical expressions using LaTeX.
- Add and delete input, re-executing entire block of commands at once.
- Start and interrupt multiple calculations at once.
- The notebook also works with Maxima, Python, R, Singular, LaTeX, html, etc.!

The following screenshot illustrates a Notebook worksheet (solving a quadratic and a cubic algebraically).

Sage: *A Basic Overview for Coding Theory and Cryptography* 9

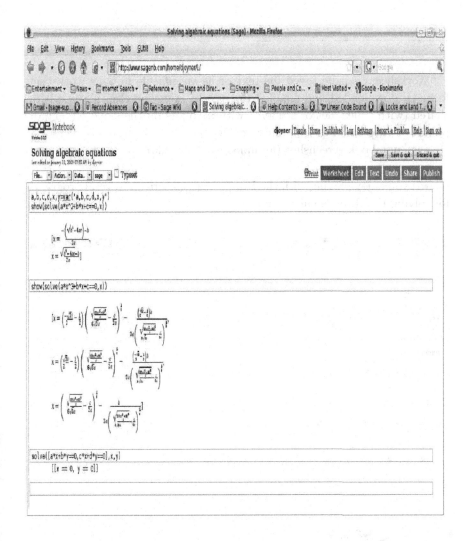

Here are the commands used to create the output in the Notebook session in the above screenshot:

```
                        Sage Notebook
a,b,c,d,x,y=var('a,b,c,d,x,y')
show(solve(a*x^2+b*x+c==0,x))
show(solve(a*x^3+b*x+c==0,x))
solve(a*x+b*y==0,c*x+d*y==0,x,y)
```

Worksheets can be saved (as text or as an sws file in Sage worksheet format), downloaded and emailed (for use by someone else, if you wish), shared (with colleagues or students with accounts on the same server), or published (if created on a public Sage server).

Sage notebook screenshot (an uploaded *.sws file)

- If you enjoy playing with the Rubik's cube, there are several programs for solving the Rubik's cube in Sage:

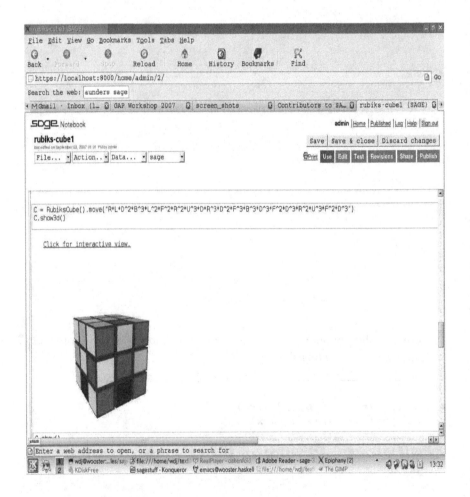

You can rotate the Rubik's cube interactively with your mouse.

0.2.6. Open source philosophy

- Sage is free software. You can check the algorithms yourself in the source code.
- You can legally serve all its functionality over the web (unlike Magma, Maple, Mathematica, and Matlab).
- Everything in Sage is 100% GPL-compatible,
- Sometimes we reimplement major algorithms from the ground up because of license problems (for example, McKay's Nauty was rewritten as NICE by Robert Miller [12]).
- You can change absolutely anything in Sage or any of its dependencies and definitely rebuild or publicly redistribute the result.

Why is open source relevant for mathematics?

> I think we need a symbolic standard to make computer manipulations easier to document and verify. And with all due respect to the free market, perhaps we should not be dependent on commercial software here. An open source project could, perhaps, find better answers to the obvious problems such as availability, bugs, backward compatibility, platform independence, standard libraries, etc. One can learn from the success of TeX and more specialized software like Macaulay2. I do hope that funding agencies are looking into this.
>
> Andrei Okounkov, *2006 Fields Medalist*

Why is open source relevant for industry?

> Open source software is part of the integrated network fabric which connects and enables our command and control system to work effectively, as people's lives depend on it.
> Open source software is all about "playing nice with others." It is all about "citizenship." We need more software collaboration in the DoD. My challenge to you: Become a citizen of the OSS community.
>
> Brig. Gen. N. G. Justice, *U. S. Army*

0.3. Coding theory functionality in Sage

A code is a linear block code over a finite field $\mathbb{F} = GF(q)$, i.e., a subspace of $\mathbb{F}^n$ with a fixed basis. In the exact sequence

$$0 \to \mathbb{F}^k \xrightarrow{G} \mathbb{F}^n \xrightarrow{H} \mathbb{F}^{n-k} \to 0, \qquad (0.1)$$

- G represents a generating matrix,
- H represents a check matrix,
- $C = Image(G) = Kernel(H)$ is the code.

The dual code $C^\perp$ satisfies $C = Image(H) = Kernel(G)$.

Excellent general references for error correcting codes include [9], [19], [20], [21].

0.3.1. *General constructions*

Sage contains the following functions which can be used to construct general linear block codes over a finite field.

General constructions	LinearCode, LinearCodeFromCheckMatrix LinearCodeFromVectorSpace, RandomLinearCode

Next is an example of how to construct a code via its generator matrix, find its parameters n, k, d, and then compute its spectrum.

```
Sage
sage: MS = MatrixSpace(GF(2),4,7)
sage: G  = MS([[1,1,1,0,0,0,0], [1,0,0,1,1,0,0],
               [0,1,0,1,0,1,0], [1,1,0,1,0,0,1]])
sage: C  = LinearCode(G) ; C
Linear code of length 7, dimension 4 over Finite Field of size 2
sage: C.base_ring()
Finite Field of size 2
sage: C.length(); C.dimension(); C.minimum_distance()
7
4
3
sage: C.weight_distribution()
[1, 0, 0, 7, 7, 0, 0, 1]
```

The following example illustrates the dual code of C.

Sage: *A Basic Overview for Coding Theory and Cryptography*

```
─────────────────────────────── Sage ───────────────────────────────
sage: MS = MatrixSpace(GF(2),4,7)
sage: G  = MS([[1,1,1,0,0,0,0], [1,0,0,1,1,0,0],
   [0,1,0,1,0,1,0], [1,1,0,1,0,0,1]])
sage: C = LinearCodeFromCheckMatrix(G); C
Linear code of length 7, dimension 3 over Finite Field of size 2
sage: C.length(); C.dimension(); C.minimum_distance()
7
3
4
sage: C.weight_distribution()
[1, 0, 0, 0, 7, 0, 0, 0]
```

The following example shows how you can "coerce" a subspace of $GF(q)^n$ into a linear code.

```
─────────────────────────────── Sage ───────────────────────────────
sage: V = GF(2)^7
sage: S = V.subspace([[1,1,1,0,0,0,0], [1,0,0,1,1,0,0], [0,1,0,1,0,1,0]])
sage: S.dimension()
3
sage: C = LinearCodeFromVectorSpace(S); C
Linear code of length 7, dimension 3 over Finite Field of size 2
sage: C.length(); C.dimension(); C.minimum_distance()
7
3
3
```

0.3.2. *Coding theory functions*

Recall the following definitions:

- Hamming metric is the function $d : \mathbb{F}^n \times \mathbb{F}^n \to \mathbb{R}$,

$$d(\mathbf{v}, \mathbf{w}) = |\{i \mid v_i \neq w_i\}| = d(\mathbf{v} - \mathbf{w}, \mathbf{0}).$$

- The weight is $wt(\mathbf{c}) = d(\mathbf{c}, \mathbf{0})$.
- minimum distance of C is $d(C) = \min_{\mathbf{c} \neq \mathbf{0}} wt(\mathbf{c})$.
- The Singleton bound is the inequality $d + k \leq n + 1$, satisfied by the parameters of a linear block code of length n, dimension k and minimum distance d. A code is maximum distance separable (MDS), if equality holds for the Singleton bound (i.e., $d+k = n+1$). In some sense, MDS is the term given to codes which are "best possible" in the sense described above.

- The weight distribution (or spectrum) of C is spec(C) = $(A_0, A_1, ..., A_n)$, where

$$A_i = |\{\mathbf{c} \in C \mid wt(\mathbf{c}) = i\}|.$$

- The support of a codeword is the set of coordinate indices which correspond to non-zero coordinates of the codeword.
- A code C having $k \times n$ generator matrix G is said to be in standard form if G is in row-reduced echelon form and if it is in block form $G = (I, A)$ where I is the $k \times k$ identity matrix. If C is not in standard form then there exists a code C' equivalent to C which is in standard form. Such a code C' is called a standard form of C.

Sage implements the following functions:

coding theory functions	spectrum, minimum_distance characteristic_function, binomial_moment gen_mat, check_mat, support, decode standard_form, divisor, genus random_element, redundancy_matrix weight_enumerator, chinen_polynomial zeta_polynomial, zeta_function

Here is an example of how **gen_mat** and **check_mat** are used:

──────────── Sage ────────────
```
sage: C = HammingCode(3,GF(2))
sage: C.gen_mat()
[1 0 0 1 0 1 0]
[0 1 0 1 0 1 1]
[0 0 1 1 0 0 1]
[0 0 0 0 1 1 1]
```

```
sage: C.check_mat()
[1 0 0 1 1 0 1]
[0 1 0 1 0 1 1]
[0 0 1 1 1 1 0]
sage: C.support()
[0, 3, 4, 7]
sage: Cd = C.dual_code()
sage: Cd.support()
[0, 4]
```

The following example shows the format Sage uses for the standard form of a code.

─────────────── Sage ───────────────
```
sage: C = HammingCode(3,GF(2)); C.gen_mat()
[1 0 0 1 0 1 0]
[0 1 0 1 0 1 1]
[0 0 1 1 0 0 1]
[0 0 0 0 1 1 1]
sage: Cs, p = C.standard_form()
sage: Cs
Linear code of length 7, dimension 4 over Finite Field of size 2
sage: p; p in SymmetricGroup(7)
(4,5)
True
sage: Cs.gen_mat()
[1 0 0 0 1 1 0]
[0 1 0 0 1 1 1]
[0 0 1 0 1 0 1]
[0 0 0 1 0 1 1]
```

The following example shows how to examine the *function* decode (e.g., decode(C,v)), as opposed to the *method* (e.g., C.decode(v)).

─────────────── Sage ───────────────
```
sage: decode??
Object 'decode' not found.
sage: from sage.coding.decoder import decode
sage: decode??
<snip>
File: ... /local/lib/python2.6/site-packages/sage/coding/decoder.py
Definition:    decode(C, v, method='syndrome')
Source:
```

```
def decode(C, v, method="syndrome"):
    """
    The vector v represents a received word, so should
    be in the same ambient space V as C. Returns an
    element in C which is closest to v in the Hamming
    metric.

    Methods implemented include "nearest neighbor" (essentially
    a brute force search) and "syndrome".
<snip>
```

Unfortunately, `decode` is not only slow but also a few decoding algoritms have been implemented.

Sage
```
sage: C = HammingCode(3,GF(2)); V = GF(2)^7
sage: v = V([1,1,0,1,1,0,1])
sage: v in V; v in C
True
False
sage: c = C.decode(v); c; c in C
(1, 0, 0, 1, 1, 0, 1)
True
```

This used syndrome decoding.

0.3.3. Weight enumerator polynomial

Let

$$A_C(x,y) = \sum_{i=0}^{n} A_i x^{n-i} y^i = x^n + A_d x^{n-d} y^d + \cdots + A_n y^n,$$

where

$$A_i = |\{c \in C \mid \operatorname{wt}(c) = i\}| = \# \text{ of codewds wt } i$$

denotes the number of codewords of weight i. A_C is called the weight enumerator of C.

Examples:

- $W_5(x,y) = x^8 + 14x^4y^4 + y^8$ is the weight enumerator of Type II [8, 4, 4] code C constructed by extending the binary [7, 4, 3] Hamming code by a check bit. This is the smallest Type II code.
- $W_6(x,y) = x^{24} + 759x^{16}y^8 + 2576x^{12}y^{12} + 759x^8y^{16} + y^{24}$ is the weight enumerator of the extended binary Golay code with parameters [24, 12, 8].

Sage can verify these facts.

──────────────── Sage ────────────────
```
sage: C = HammingCode(3,GF(2))
sage: Cx = C.extended_code()
sage: Cx.weight_enumerator()
x^8 + 14*x^4*y^4 + y^8
sage: C = ExtendedBinaryGolayCode()
sage: C.weight_enumerator()
x^24 + 759*x^16*y^8 + 2576*x^12*y^12 + 759*x^8*y^16 + y^24
```

More on these weight enumerators later.

The Duursma zeta function is implemented.

──────────────── Sage ────────────────
```
sage: C = HammingCode(3,GF(2))
sage: C.genus() # n+1-k-d
1
sage: C.weight_enumerator()
x^7 + 7*x^4*y^3 + 7*x^3*y^4 + y^7
sage: C.zeta_function()
(2/5*T^2 + 2/5*T + 1/5)/(2*T^2 - 3*T + 1)
sage: C.zeta_polynomial()
2/5*T^2 + 2/5*T + 1/5
```

An example of Komichi (from his master's thesis, unpublished, but see [10]) is illustrated using Sage:

──────────────── Sage ────────────────
```
sage: C = HammingCode(3,GF(2))
sage: C1 = HammingCode(3,GF(2))
sage: C2 = C1.extended_code()
sage: C3 = (C2.direct_sum(C2)).direct_sum(C2)
sage: R.<T> = PolynomialRing(CC, "T")
sage: f = C3.zeta_polynomial(); f = R(f); rts = f.roots()
```

```
sage: [abs(z[0]*sqrt(2.0)) for z in rts]
[0.733550688875582, 1.36323230986647, 1.00000000000000,
 1.00000000000000, 1.00000000000000, 1.00000000000000,
 1.00000000000000, 1.00000000000000, 1.00000000000000,
 1.00000000000000, 1.00000000000000, 1.00000000000000,
 1.00000000000000, 1.00000000000000, 1.00000000000000,
 1.00000000000000, 1.00000000000000, 1.00000000000000]
```

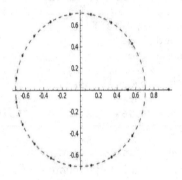

Fig. 0.1. Zeros of the Duursma zeta function of Komichi's code.

Please see Figure 0.1 for the plot of these zeros. More on these Duursma zeta functions later.

0.3.4. *More code constructions*

code constructions	dual_code, extended_code, direct_sum, punctured, shortened, permuted_code, galois_closure

extended_code simply adds a check-bit at the end.

―――――――――――――― Sage ――――――――――――――
```
sage: C = HammingCode(3,GF(2))
sage: Cx = C.extended_code()
sage: Cx.is_self_orthogonal()
True
sage: Cx.is_self_dual()
```

```
True
sage: Cx.divisor()
4
sage: Cx.spectrum()
[1, 0, 0, 0, 14, 0, 0, 0, 1]
```

More on self-dual codes later.

`galois_closure` of a code C defined over $GF(p^k)$ returns the smallest code defined over $GF(p^k)$ closed under the Galois action of $\text{Gal}(GF(p^k)/GF(p))$.

──────── Sage ────────
```
sage: C = HammingCode(3,GF(4,'a'))
sage: Cc = C.galois_closure(GF(2))
sage: C; Cc
Linear code of length 21, dim 18 over Finite Field in a of size 2^2
Linear code of length 21, dim 20 over Finite Field in a of size 2^2
sage: C.is_subcode(Cc)
True
sage: Cc.is_galois_closed()
True
```

0.3.5. Automorphism group of a code

What is an automorphism of a code?

Let S_n denote the symmetric group on n letters. The (permutation) automorphism group of a code C of length n is simply the group

$$\text{Aut}(C) = \{\sigma \in S_n \mid (c_1, ..., c_n) \in C \implies (c_{\sigma(1)}, ..., c_{\sigma(n)}) \in C\}.$$

There are no known methods for computing these groups which are polynomial time in the length n of C.

If

(a) $C_1, C_2 \subset \mathbb{F}^n$ are codes, and
(b) there exists $\sigma \in S_n$ for which $(c_1, ..., c_n) \in C_1 \iff (c_{\sigma(1)}, ..., c_{\sigma(n)}) \in C_2$,

then $C_1 \cong C_2$ (i.e., C_1 and C_2 are permutation equivalent).

```
sage: C = HammingCode(3,GF(2))
sage: g = SymmetricGroup(7).random_element(); g
(1,2)(3,7,4)
sage: Cg = C.permuted_code(g)
sage: Cg.is_permutation_equivalent(C)
True
sage: G = C.automorphism_group_binary_code(); G
Permutation Group with generators [(3,4)(5,6), (3,5)(4,6),
 (2,3)(5,7), (1,2)(5,6)]
sage: g = G("(2,3)(5,7)")
sage: Cg = C.permuted_code(g)
sage: C == Cg
True
sage: g = SymmetricGroup(7).random_element(); g
(1,6,4,7,3)(2,5)
sage: C.is_permutation_automorphism(g)
False
sage: Cg = C.permuted_code(g)
sage: Cg.is_permutation_equivalent(C)
True
```

The permutation automorphism group of the extended ternary Golay code is the Mathieu group M_{11}.

```
sage: C = ExtendedTernaryGolayCode(); C; C.minimum_distance()
Linear code of length 12, dimension 6 over Finite Field of size 3
6
sage: G = C.permutation_automorphism_group(); G
Permutation Group with generators [(5,7)(6,11)(8,9)(10,12),
 (4,6)(5,10)(7,8)(9,12), (3,4)(6,8)(9,11)(10,12),
 (2,3)(5,7)(8,10)(9,12), (1,2)(5,12)(6,11)(7,10)]
sage: G.order(); G.is_simple()
7920
True
sage: M11 = MathieuGroup(11); G.is_isomorphic(M11)
True
```

(The full "monomial" automorphism group is larger, but Sage lacks the functionality to compute that at this point.)

All this is expected behavior.

coding theory functions (group theoretical)	module_composition_factors, automorphism_group_binary_code

module_composition_factors prints the GAP record of the Meataxe composition factors module in Meataxe notation.

Example: C.module_composition_factors(G) computes the MeatAxe format of a code C as a G-module, where G is a subgroup of the automorphism group of C.

──────── Sage ────────
```
sage: C = HammingCode(3,GF(2))
sage: Cx = C.extended_code()
sage: G = Cx.automorphism_group_binary_code()
sage: G.order()
1344
sage: Cx.module_composition_factors(G)
[ rec(
      field := GF(2),
      isMTXModule := true,
      dimension := 1,
      generators := [ [ [ Z(2)^0 ] ], [ [ Z(2)^0 ] ], [ [ Z(2)^0 ] ],
          [ [ Z(2)^0 ] ], [ [ Z(2)^0 ] ], [ [ Z(2)^0 ] ] ],
<snip>
      IsIrreducible := true ) ]
```

0.3.6. Even more code constructions

Examples: punctured, shortened

The code C^L obtained from C by puncturing at the positions in L is the code of length $n - |L|$ consisting of codewords of C which have their i-th coordinate deleted if $i \in L$ and left alone if $i \notin L$.

──────── Sage ────────
```
sage: C = HammingCode(3,GF(2))
sage: C.punctured([1,2])
Linear code of length 5, dimension 4 over Finite Field of size 2
sage: C.shortened([1,2])
Linear code of length 5, dimension 2 over Finite Field of size 2
```

The subcode $C(L)$ is all codewords $c \in C$ which satisfy $c_i = 0$ for all $i \in L$. The punctured code $C(L)^L$ is called the shortened code on L and is denoted C_L.

coding theory functions (boolean)	`is_self_dual`, `==` `is_self_orthogonal`, `is_subcode`, `is_permutation_automorphism`, `is_permutation_equivalent`, `is_galois_closed`

Examples of most of these have already been seen.

0.3.7. Block designs and codes

coding theory functions (combinatorial)	`assmus_mattson_designs`

- A block design: a pair (X, B), where X is a non-empty finite set of $v > 0$ elements called points, and B is a non-empty finite multiset of size b whose elements are called blocks, such that each block is a non-empty finite multiset of k points.
- If every subset of points of size t is contained in exactly λ blocks the block design is called a $t - (v, k, \lambda)$ design.
- When $\lambda = 1$ then the block design is called a $S(t, k, v)$ Steiner system.

Theorem 0.1. *(Assmus and Mattson)* Let $A_0, A_1, ..., A_n$ be the weights of the codewords in a binary linear $[n, k, d]$ code C, and let $A_0^*, A_1^*, ..., A_n^*$ be the weights of the codewords in its dual $[n, n-k, d^*]$ code C^*. Fix a t, $0 < t < d$, and let $s = |\{i \mid A_i^* \text{ not} = 0, 0 < i \leq n - t\}|$. Assume $s \leq d - t$.

- If $A_i \neq 0$ and $d \leq i \leq n$ then $C_i = \{c \in C \mid wt(c) = i\}$ holds a simple t-design.
- If $A_i^* \neq 0$ and $d^* \leq i \leq n - t$ then $C_i^* = \{c \in C^* \mid wt(c) = i\}$ holds a simple t-design.

―――――――――― Sage ――――――――――
```
sage: C = HammingCode(3,GF(2))
sage: Cx = C.extended_code()
sage: Cx.assmus_mattson_designs(3)
['weights from C: ', [4, 8],
```

```
'designs from C: ',  [[3, (8, 4, 1)], [3, (8, 8, 1)]],
'weights from C*: ',  [4],
'designs from C*: ',  [[3, (8, 4, 1)]]]
```

0.3.8. Special constructions

A number of well-known families of codes are implemented in Sage. A few examples are discussed below.

Special constructions	BinaryGolayCode, ExtendedBinaryGolayCode, TernaryGolayCode, ExtendedTernaryGolayCode, CyclicCode, BCHCode, CyclicCodeFromCheckPolynomial, DuadicCodeEvenPair, DuadicCodeOddPair, HammingCode, QuadraticResidueCodeEvenPair, QuadraticResidueCodeOddPair, QuadraticResidueCode, ExtendedQuadraticResidueCode, ReedSolomonCode, self_dual_codes_binary, ToricCode, WalshCode

Example: `ReedSolomonCode` - Also called a generalized Reed-Solomon code: Let

- $\mathbb{F} = GF(q)$,
- n and k be such that $1 \leq k \leq n \leq q$, and
- pick n distinct elements of $\mathbb{F}$, $\{x_1, x_2, ..., x_n\}$.

Define the generalized Reed-Solomon code by

$$C = \{(f(x_1), f(x_2), ..., f(x_n)) \mid f \in \mathbb{F}[x], \deg(f) < k\}.$$

This is an $[n, k, n - k + 1]$ code.

```
sage: C = ReedSolomonCode(6,4,GF(7)); C
Linear code of length 6, dimension 4 over Finite Field of size 7
sage: C.minimum_distance()
3
sage: F.<a> = GF(3^2,"a")
sage: pts = [0,1,a,a^2,2*a,2*a+1]
sage: len(Set(pts)) == 6 # to make sure there are no duplicates
True
sage: C = ReedSolomonCode(6,4,F,pts); C
Linear code of length 6, dimension 4 over Finite Field in a of size 3^2
sage: C.minimum_distance()
3
```

Example: `ToricCode`. `ToricCodes` can be bad or very good.

```
sage: C = ToricCode([[-2,-2],[-1,-2],[-1,-1],[-1,0],[0,-1],
    [0,0],[0,1],[1,-1],[1,0]],GF(5))
sage: C
Linear code of length 16, dimension 9 over Finite Field of size 5
sage: C.minimum_distance()
6
```

There are (much larger) toric codes which have best possible parameters. See, for example, papers of Diego Ruano if you are interested in learning more about this family of codes arising from algebraic geometry (see for example [15]).

0.3.9. Coding theory bounds

Sage has implemented a number of the most common bounds for linear codes.

code bounds	best_known_linear_code_www, bounds_minimum_distance codesize_upper_bound(n,d,q), dimension_upper_bound(n,d,q) gilbert_lower_bound(n,q,d), plotkin_upper_bound(n,q,d) griesmer_upper_bound(n,q,d), elias_upper_bound(n,q,d) hamming_upper_bound(n,q,d), singleton_upper_bound(n,q,d) gv_info_rate(n,delta,q), gv_bound_asymp(delta,q) plotkin_bound_asymp(delta,q), elias_bound_asymp(delta,q) hamming_bound_asymp(delta,q), singleton_bound_asymp(delta,q) mrrw1_bound_asymp(delta,q)

Example: best_known_linear_code_www interface with http://www.codetables.de since A. Brouwer's online tables have been disabled. It explains the construction of the best known linear code over $GF(q)$ with length n and dimension k, courtesy of the www page http://www.codetables.de/.

INPUT:

- n – integer, the length of the code
- k – integer, the dimension of the code
- F – finite field, whose field order must be in [2, 3, 4, 5, 7, 8, 9]
- verbose – bool (default=False), print verbose message

─────────────────── Sage ───────────────────
```
sage: L = best_known_linear_code_www(72, 36, GF(2)) # uses internet
sage: print L
Construction of a linear code [72,36,15] over GF(2):
[1]:    [73, 36, 16] Cyclic Linear Code over GF(2)
        CyclicCode of length 73 with generating polynomial x^37 + x^36
        + x^34 + x^33 + x^32 + x^27 + x^25 + x^24 + x^22 + x^21 + x^19
        + x^18 + x^15 + x^11 + x^10 + x^8 + x^7 + x^5 + x^3 + 1
[2]:    [72, 36, 15] Linear Code over GF(2)
        Puncturing of [1] at 1
```

```
last modified: 2002-03-20
```

0.3.10. *Asymptotic bounds*

Theorem 0.2. *(Manin) There exists a continuous decreasing function*

$$\alpha_q : [0,1] \to [0,1],$$

such that

- α_q *is strictly decreasing on* $[0, \frac{q-1}{q}]$,
- $\alpha_q(0) = 1$,
- *if* $\frac{q-1}{q} \leq x \leq 1$ *then* $\alpha_q(x) = 0$,
- $\Sigma_q = \{(\delta, R) \in [0,1]^2 \mid 0 \leq R \leq \alpha_q(\delta)\}$.

Not a single value of $\alpha_q(x)$ is known for $0 < x < \frac{q-1}{q}$! It is not known whether the maximum value of the bound, $R = \alpha_q(\delta)$ is attained by a sequence of linear codes. It is not known whether $\alpha_q(x)$ is differentiable for $0 < x < \frac{q-1}{q}$, nor is it known if $\alpha_q(x)$ is convex on $0 < x < \frac{q-1}{q}$.

Theorem 0.3. *(Gilbert-Varshamov) We have*

$$\alpha_q(x) \geq 1 - x\log_q(q-1) - x\log_q(x) - (1-x)\log_q(1-x).$$

In other words, for each fixed $\epsilon > 0$, there exists an (n,k,d)-code C (which may depend on ϵ) for which $R(C) + \delta(C)$ is at least

$$1 - \delta(C)\log_q(\frac{q-1}{q}) - \delta(C)\log_q(\delta(C)) - (1-\delta(C))\log_q(1-\delta(C)) - \epsilon.$$

The curve $(\delta, 1 - \delta\log_q(\frac{q-1}{q}) - \delta\log_q(\delta) - (1-\delta)\log_q(1-\delta))$ is called the Gilbert-Varshamov curve.

Sage has excellent plotting functionality. The output of the following commands can be found in Figure 0.2.

──────── Sage ────────
```
sage: f = lambda x: gv_bound_asymp(x,2)
sage: P1 = plot(f,0,1/2)
sage: P2 = list_plot([[(3/7,4/7)])
sage: P3 = text('$Hamm(7,4,3)$', (0.4,0.62), rgbcolor=(0,1,0))
sage: P4 = text('$*$', (4/8,4/8), rgbcolor=(1,0,0))
sage: P5 = text('$Hamm^+(8,4,4)$', (0.45,0.4), rgbcolor=(0,1,0))
```

```
sage: show(P1+P2+P3+P4+P5)
```

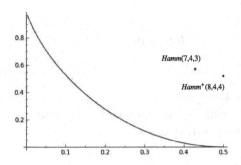

Fig. 0.2. Gilbert-Varshamov curve plotted with the $[7,4,3]_2$ and extended $[8,4,4]_2$ Hamming codes.

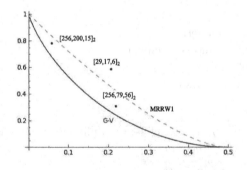

Fig. 0.3. Gilbert-Varshamov curve and MRRW1 curve plotted with some "good" codes.

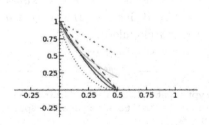

Fig. 0.4. (Color online) Plot of the Gilbert-Varshamov (dotted), Elias (red), Plotkin (dashed), Singleton (dash-dotted), Hamming (green), and MRRW (blue) curves using Sage.

0.4. Cryptography in Sage

This will be a very brief selection of topics.

A cryptosystem is an injection

$$E : KS \to \text{Hom}_{\text{Set}}(MS, CS),$$

where

- KS is the key space,
- MS is the plaintext (or message) space, and
- CS is the ciphertext space.

0.4.1. *Classical cryptography*

Sage's modules on "classical" ciphers was created by David Kohel and Minh van Nyugen.

- Hill cipher,
- substitution cipher,
- transposition and shift ciphers,
- affine cipher, and
- Vigenere cryptosystem,

are implemented.

Let $A = \{a_0, a_1, a_2, \ldots, a_{n-1}\}$ be an alphabet. Define an injection $f : A \longrightarrow \mathbb{Z}/n\mathbb{Z}$ given by $f(a_i) = i$.

Set $MS = CS = \mathbb{Z}/n\mathbb{Z} \cong A$

key space: $KS = \{(a, b) \in \mathbb{Z}/n\mathbb{Z} \times \mathbb{Z}/n\mathbb{Z} \mid \gcd(a, n) = 1\}$. Let $(a, b) \in KS$.

Encryption: For $p \in MS$, define $c \in CS$ by $c \equiv ap + b \pmod{n}$.

Decryption: For $c \in CS$, define $p \in MS$ by $p \equiv a^{-1}(c - b) \pmod{n}$ where a^{-1} is the inverse of a modulo n.

Here is an example of an affine cipher:

```
sage: A = AffineCryptosystem(AlphabeticStrings())
sage: P = A.encoding("`Hello` to everyone in Spain!!"); P
HELLOTOEVERYONEINSPAIN
sage: a, b = (3, 7)
sage: C = A.enciphering(a, b, P); C
CTOOXMXTSTGBXUTFUJAHFU
```

```
sage: L = A.brute_force(C)
sage: sorted(L.items())[30:35]
[((3, 4), IFMMPUPFWFSZPOFJOTQBJO), ((3, 5), ZWDDGLGWNWJQGFWAFKHSAF),
((3, 6), QNUUXCXNENAHXWNRWBYJRW), ((3, 7), HELLOTOEVERYONEINSPAIN),
((3, 8), YVCCFKFVMVIPFEVZEJGRZE)]
sage: L = A.brute_force(C, ranking="chisquare")
sage: L[0]
((3, 7), HELLOTOEVERYONEINSPAIN)
```

Here is an example of a substitution cipher:

──────── Sage ────────

```
sage: S = ShiftCryptosystem(AlphabeticStrings())
sage: P = S.encoding("Shift from Mathematica to Sage!."); P
SHIFTFROMMATHEMATICATOSAGE
sage: K = 3
sage: C = S.enciphering(K, P); C
VKLIWIURPPDWKHPDWLFDWRVDJH
sage: S.enciphering(26-K, C)
SHIFTFROMMATHEMATICATOSAGE
sage: S.deciphering(K, C)
SHIFTFROMMATHEMATICATOSAGE
```

0.4.2. Algebraic cryptosystems

Sage's modules on algebraic cryptosystems was created by Martin Albrecht and Minh van Nyugen. Sage has significant functionality in the following areas.

- mini-DES, based on: Schaefer [16].
- mini-AES, based on: Phan [14].
- Small Scale Variants of the AES Polynomial System Generator; see [4].
- Multivariate Polynomial Systems; see [3].

See also the excellent paper Albrecht [1].

0.4.3. RSA

RSA is a deterministic public key encryption algorithm which relies on

- the extended Euclidean algorithm, and

- Euler's theorem in the special case of a modulus which is a product of two primes.

Public key cryptosystem generalities:

- Two keys - a public key and a private key.
- public key - known to everyone, used for encryption.
- private key - Known only to the receiver, ciphertext can only be decrypted using the private key.
- The security of the RSA cryptosystem relies on that belief that it is computationally infeasible to compute the private key from the public key.

It is important to know who generates the keys. Suppose Alice wants to send a message to Bob using RSA. She says, "Bob, I need to tell you something." Bob says, "Hang on a second while I generate the keys." Clear? Bob then

- chooses two distinct prime numbers p and q (only Bob knows these),
- computes $n = pq$ (n is used for both the public and private keys),
- computes $\phi(pq) = (p-1)(q-1)$ (ϕ = Euler's function),
- chooses an integer e such that $1 < e < \phi(pq)$ and $gcd(e, \phi(pq))$ (e is the public key exponent),
- determines d which satisfies $de \equiv 1 \pmod{\phi(pq)}$ (d is the private key exponent).

The public key consists of (n, e). The private key consists of (n, d).

Example 0.1. Alice wants to send a message to Bob.

Bob selects $p = 1009$ and $q = 1013$, so $n = pq = 1022117$. Bob computes $\phi(n) = 1020096$. If he selects $e = 123451$, then he can compute $d = 300019$.

Alice wants to send Bob the message $m = 46577$. She encrypts it using 46577^{123451} (mod 1022117), which is the ciphertext $c = 622474$.

──────── Sage ────────
```
sage: p = next_prime(1000); q = next_prime(1010); n = p*q; n
1022117
sage: k = euler_phi(n); e = 123451
sage: k; xgcd(k, e)
1020096
(1, -36308, 300019)
sage: x = xgcd(k, e)[1]; y = xgcd(k, e)[2]
sage: d = y%k
sage: y*e%k; d*e%k
```

```
1
1
sage: m = randint(100, k); m
46577
sage: c = power_mod(m,e,n) # faster than m^e%n
622474
sage: power_mod(c,d,n)
46577
```

As the above Sage computation shows, m was correctly decrypted.

0.4.4. Discrete logs

The discrete logarithm problem is the following: Let G be a multiplicative abelian group and let $a, b \in G$. Find $x \in \mathbb{Z}$ such that

$$b^x = a,$$

if it exists. Sage has some good functionality for computing discrete logs, which allows you to implement various discrete log-based cryptosystems/key exchanges in Sage.

───── Sage ─────
```
sage: p = next_prime(10^30)
sage: F = GF(p)
sage: b = F(2); b.multiplicative_order()
500000000000000000000000000028
sage: b = F(3); b.multiplicative_order()
416666666666666666666666666669
sage: b = F(5); b.multiplicative_order()
100000000000000000000000000056
sage: a = F.random_element(); a
837776537981704766224734890062
sage: time a.log(b)
CPU times: user 0.04 s, sys: 0.01 s, total: 0.06 s
Wall time: 0.22 s
972953394163188347701109599170
```

0.4.5. Diffie-Hellman

Alice and Bob want to share a secret key (which they generate *together*).

- Alice and Bob agree on a finite cyclic group G and a generating element $g \in G$. (g is assumed to be known by all attackers.) Assume G has order n.
- Alice picks a random a, $1 < a < n$, and sends g^a to Bob.

- Bob picks a random b, $1 < b < n$, and sends g^b to Alice.
- Alice computes $(g^b)^a$.
- Bob computes $(g^a)^b$.
- Both Alice and Bob possess a shared secret key, g^{ab}.

──────── Sage ────────
```
sage: G = IntegerModRing(101)
sage: g = G.random_element(); g; g.multiplicative_order()
3
100
sage: a = randint(1,50); b = randint(1,50)
sage: a; b
35
36
sage: ga = g^a; gb = g^b
sage: ga^b; ga^b == gb^a
36
True
```

0.4.6. *Linear feedback shift registers*

Let q be a prime power, $\ell > 1$ be an integer, and let $c_1, ..., c_\ell$ are given elements of $GF(q)$. See Brock [2] for details.

A linear feedback shift register sequence (LFSR) modulo p of *length* ℓ is a sequence $s_0, s_1, s_2, ... \in GF(q)$ such that

- $s_0, s_1, ..., s_{\ell-1}$ are given, and
- $s_n + c_1 s_{n-1} + c_2 s_{n-2} + ... + c_\ell s_{n-\ell} = 0$, $n \geq \ell$.

These are used to create very fast pseudorandom number generators (which can then be used as a streamcipher).

Terminology:

- key - the list of coefficients $[c_1, c_2, ..., c_\ell]$
- fill - the list of initial values $s_0, s_1, ..., s_{\ell-1}$.
- connection polynomial - $c(x) = 1 + c_1 x + ... c_\ell x^\ell$.

──────── Sage ────────
```
sage: F = GF(2); l = F(1); o = F(0)
sage: fill = [o,l]; key = [1,1]; n = 20
sage: c = lfsr_sequence(key, fill, n); c
[0, 1, 1, 0, 1, 1, 0, 1, 1, 0, 1, 1, 0, 1, 1, 0, 1, 1, 0, 1]
sage: f = lfsr_connection_polynomial(c); f
x^2 + x + 1
```

```
sage: f.is_primitive()
True
```

Notice that this Fibonacci sequence mod 2 seems to be periodic with period 3 ($= q^{\deg(c(x))} - 1$).

───── Sage ─────
```
sage: F = GF(3); l = F(1); o = F(0)
sage: fill = [o,l]; key = [1,1]; n = 20
sage: c = lfsr_sequence(key, fill, n); c
[0, 1, 1, 2, 0, 2, 2, 1, 0, 1, 1, 2, 0, 2, 2, 1, 0, 1, 1, 2]
sage: f = lfsr_connection_polynomial(c); f
2*x^2 + 2*x + 1
sage: f.is_primitive()
True
```

Notice that this Fibonacci sequence mod 3 seems to be periodic with period 8 ($= q^{\deg(c(x))} - 1$).

Theorem 0.4. *Let $S = \{s_i\}$ be a LFSR over $GF(p)$. The period of S is at most $p^k - 1$. Its period is exactly $P = p^k - 1$ if and only if the characteristic polynomial of*

$$A = \begin{pmatrix} 0 & 1 & 0 & \ldots & 0 \\ 0 & 0 & 1 & \ldots & 1 \\ \vdots & & \ldots & & \\ 0 & 0 & \ldots & 0 & 1 \\ -c_\ell & -c_{\ell-1} & \ldots & & -c_1 \end{pmatrix},$$

is irreducible and primitive over $GF(p)$.

0.4.7. BBS streamcipher

The Blum-Blum-Shub (BBS) steam cipher is a pseudorandom number generator which outputs a sequences of 0's and 1's. It has a very long period (for certain reasonable choices of the parameters). See, for example, Hogan [8] for details.

Definition 0.1. *Let p, q be two distinct prime numbers such that $p \equiv 3 \pmod{4}$ and $q \equiv 3 \pmod{4}$. Let $n = pq$ and let $0 < r < n$ be a random number. We define x_0, the first "seed" of the Blum-Blum-Shub pseudorandom number generator as*

$$x_0 = r^2 \pmod{n}.$$

Each proceeding seed can be defined as

$$x_{i+1} = x_i^2 \pmod{n}.$$

The streamcipher, $b = b_1 b_2 \ldots b_t$, is created by setting $b_i = x_i \mod 2$.

```
─────────────────────────── Sage ───────────────────────────
sage: from sage.crypto.stream import blum_blum_shub
sage: p0 = next_prime(1015); q0 = next_prime(1100)
sage: blum_blum_shub(length=50, seed=999, p=p0, q=q0)
11111000110010101001001100100001010100101001011010
sage: from sage.crypto.util import carmichael_lambda as carmichael
sage: carmichael(carmichael(p0*q0))
32004
```

The last output tells us the maximum possible value of period of the BBS sequence.

0.4.8. *Blum-Goldwasser cryptosystem*

This is a public key cryptosystem which uses the BBS streamcipher.

- Alice wants to send a message m to Bob.
- Bob generates two distinct prime numbers p and q such that $p \equiv 3 \pmod{4}$, $q \equiv 3 \pmod{4}$.
- Bob computes $n = pq$.
- Using the extended Euclidean algorithm, Bob computes a, b such that $ap + bq = 1$.

The public key is n. The private key is (p, q, a, b).

Let x_0 be a randomly selected quadratic residue $\pmod{n}$.

- Plaintext: $m = m_1 m_2 \ldots m_t$ - a binary string of length t.
- Let $b = b_1 b_2 \ldots b_t$ be the BBS streamcipher of length t associated to x_0, n.
- Ciphertext: $c = b \oplus m$, where $\oplus$ indicates the XOR operation.

Alice sends the ciphertext c along with a number $y = x_0^{2^{t+1}} \pmod{n}$.

```
─────────────────────────── Sage ───────────────────────────
sage: from sage.crypto.public_key.blum_goldwasser import BlumGoldwasser
sage: bg = BlumGoldwasser(); bg
The Blum-Goldwasser public-key encryption scheme.
sage: p = 499; q = 547
```

```
sage: pubkey = bg.public_key(p, q); pubkey
272953
sage: prikey = bg.private_key(p, q); prikey
(499, 547, -57, 52)
sage: p*q; p*prikey[2]+q*prikey[3]
272953
1
sage: M = "10011100000100001100"
sage: C = bg.encrypt(M, pubkey, seed=159201); C
([[0, 0, 1, 0], [0, 0, 0, 0], [1, 1, 0, 0], [1, 1, 1, 0],
 [0, 1, 0, 0]], 139680)
sage: M0 = bg.decrypt(C, prikey); M0
[[1, 0, 0, 1], [1, 1, 0, 0], [0, 0, 0, 1], [0, 0, 0, 0], [1, 1, 0, 0]]
sage: M = "".join(map(lambda x: str(x), flatten(M))); M
'10011100000100001100'
```

See for example Hogan [8] for more details on BG cryptosystems.

0.5. Miscellaneous topics

This section discussed Duursma zeta functions, self-dual codes, and ends with some cool examples.

0.5.1. *Duursma zeta functions*

In Sage, computing Duursma zeta functions of codes is implemented.

C is an $[n, k, d]_q$ code

$C^\perp$ is an $[n, k^\perp, d^\perp]_q$ code

Motivated by local class field theory, Iwan Duursma [5] introduced the zeta function $Z = Z_C$ associated to C:

$$Z(T) = \frac{P(T)}{(1-T)(1-qT)}, \qquad (0.2)$$

where $P(T)$ is a polynomial of degree $n + 2 - d - d^\perp$, called the zeta polynomial.

The *genus*[b] of an $[n, k, d]_q$-code C is defined by

$$\gamma(C) = n + 1 - k - d$$
$$= \text{"distance code is from being MDS"}.$$

---
[b] For AG codes, it is often equal to the genus of the associated curve.

Note that if C is a self-dual code then its genus satisfies

$$\gamma = n/2 + 1 - d.$$

Recall the weight enumerator polynomial:

$$A_C(x,y) = \sum_{i=0}^{n} A_i x^{n-i} y^i = x^n + A_d x^{n-d} y^d + \cdots + A_n y^n,$$

where

$$A_i = |\{c \in C \mid \text{wt}(c) = i\}| = \# \text{ of codewds wt } i.$$

denotes the number of codewords of weight i. If $b > 1$ is an integer, then we say C is *b-divisible* provided A_i is only non-zero for multiples of b. Recall C is a *formally self-dual code* if and only if $A_C(x,y) = A_{C^\perp}(x,y)$.

A polynomial $P(T)$ for which

$$\frac{(xT + (1-T)y)^n}{(1-T)(1-qT)} P(T) = \cdots + \frac{A_C(x,y) - x^n}{q-1} T^{n-d} + \cdots$$

is called a *Duursma zeta polynomial of C*. (The Duursma zeta polynomial $P = P_C$ exists and is unique, provided the minimum distance of both C and its dual code are > 1.)

The functional equation holds:

$$P^\perp(T) = P(\frac{1}{qT}) q^g T^{g+g^\perp}, \qquad (0.3)$$

where $g = n/2 + 1 - d$ and $g^\perp = n/2 + 1 - d^\perp$.

The Riemann hypothesis (RH) is the statement that all zeros of $P(T)$ lie on the circle $|T| = 1/\sqrt{q}$.

Let C be a formally self-dual b-divisible $[n,k,d]_q$-code. We say C is *Type I* if $q = b = 2$, and n is even. We say C is *Type II* if $q = 2$, $b = 4$, and $8|n$. We say C is *Type III* if $q = b = 3$, and $4|n$. If $q = 4$, $b = 2$, and n is even then C is said to be *Type IV*.

Theorem 0.5. *(Mallows-Sloane bounds) If C is self-dual then*

$$d \leq \begin{cases} 2[n/8] + 2, & \text{if } C \text{ is Type } I, \\ 4[n/24] + 4, & \text{if } C \text{ is Type } II, \\ 3[n/12] + 3, & \text{if } C \text{ is Type } III, \\ 2[n/6] + 2, & \text{if } C \text{ is Type } IV. \end{cases}$$

More definitions: *Virtual weight enumerator* - a homogeneous polynomial $F(x,y) = x^n + \sum_{i=1}^{n} f_i x^{n-i} y^i$ of degree n with complex coefficients. If $F(x,y) = x^n + \sum_{i=d}^{n} f_i x^{n-i} y^i$ with $f_d \neq 0$ then we say that the *length* of F is n and the *minimum distance* of F is d. A *formally self-dual weight enumerator* is an $F(x,y)$ as above of even degree and invariant under $\sigma = \frac{1}{\sqrt{q}} \begin{pmatrix} 1 & 1 \\ q-1 & -1 \end{pmatrix}$. The *genus* of a formally self-dual weight enumerator: $\gamma(F) = n/2 + 1 - d$.

A virtual weight enumerator F is formally identified with an object we call a *virtual code* C subject only to the following condition: we formally extend the definition of $C \longmapsto A_C$ to all virtual codes by $A_C = F$. The above theorem of Mallows and Sloane can be extended to virtual weight enumerators (see Joyner-Kim [11] for details).

Theorem 0.6. *If F is a formally self-dual weight enumerator with length n and minimum distance d, then*

$$d \leq \begin{cases} 2[n/8] + 2, & \text{if } F \text{ is Type } I, \\ 4[n/24] + 4, & \text{if } F \text{ is Type } II, \\ 3[n/12] + 3, & \text{if } F \text{ is Type } III, \\ 2[n/6] + 2, & \text{if } F \text{ is Type } IV. \end{cases}$$

A formally self-dual weight enumerator F (i.e., a virtual self-dual code) is called *extremal* if the bound in the theorem holds with equality.

A code is called *optimal* if its minimum distance is maximal among all linear codes of that length and dimension.

It is known that any two extremal codes (if they exist) have the same weight enumerator polynomial.

Conjecture 0.1. *(Duursma)* The RH holds for $Z(T)$ for all extremal virtual codes.

```
sage: C = HammingCode(3,GF(2))
sage: C.zeta_function()
(1/5 + 2/5*T + 2/5*T^2)/(1 - 3*T + 2*T^2)
sage: C = ExtendedTernaryGolayCode()
sage: C.zeta_function()
(1/7 + 3/7*T + 3/7*T^2)/(1 - 4*T + 3*T^2)
```

These satisfy the RH.

Consider the $[26, 13, 6]_{13}$ code with weight distribution

$$[1, 0, 0, 0, 0, 0, 39, 0, 455, 0, 1196, 0, 2405,$$

$$0, 2405, 0, 1196, 0, 455, 0, 39, 0, 0, 0, 0, 0, 1].$$

This is an optimal formally self-dual code C with zeta polynomial

$$P(T) = \tfrac{3}{17710} + \tfrac{6}{8855}T + \tfrac{611}{336490}T^2 + \tfrac{9}{2185}T^3 + \tfrac{3441}{408595}T^4$$
$$+ \tfrac{6448}{408595}T^5 + \tfrac{44499}{1634380}T^6 + \tfrac{22539}{520030}T^7 + \tfrac{66303}{1040060}T^8$$
$$+ \tfrac{22539}{260015}T^9 + \tfrac{44499}{408595}T^{10} + \tfrac{51584}{408595}T^{11} + \tfrac{55056}{408595}T^{12}$$
$$+ \tfrac{288}{2185}T^{13} + \tfrac{19552}{168245}T^{14} + \tfrac{768}{8855}T^{15} + \tfrac{384}{8855}T^{16}.$$

Using Sage, it can be checked that only 8 of the 12 zeros of this function have absolute value $\sqrt{2}$.

Regarding the RH,

- Duursma has "explicitly" computed all zeta functions of extremal virtual self-dual codes.

- Duursma verified the RH for Type IV codes.

- For all low values of the parameters, computations using Sage have shown that the RH holds for all "small" virtual extremal self-dual codes.

0.5.2. *Self-dual codes*

Sage has good functionality for working with "small" self-dual codes.

Sage includes a database of all self-dual binary codes of length ≤ 20 (and some of length 22). The main function is self_dual_codes_binary, which is a list of Python dictionaries.

Format of each entry: dictionary with keys order autgp, spectrum, code, Comment, Type, where

- **code** - a self-dual code C of length n, dimension $n/2$, over $GF(2)$,
- **order autgp** - order of the permutation autom. group of C,
- **Type** - the type of C (which can be "I" or "II", in the binary case),
- **spectrum** - the spectrum $[A_0, A_1, ..., A_n]$,
- **Comment** - possibly an empty string.

```
─────────── Sage ───────────
sage: C = self_dual_codes_binary(10)
sage: C.keys()
['10']
sage: C['10'].keys()
['1', '0']
sage: C['10']['0']
{'Comment': 'No Type II of this length.', 'Type': 'I',
 'code': Linear code of length 10, dimension 5 over Finite Field of size 2,
 'order autgp': 3840, 'spectrum': [1, 0, 5, 0, 10, 0, 10, 0, 5, 0, 1]}
sage: C = self_dual_codes_binary(10)
sage: C = C['10']['0']['code']
sage: C
Linear code of length 10, dimension 5 over Finite Field of size 2
sage: C.divisor()
2
sage: C = self_dual_codes_binary(10)["10"]
sage: C["0"]["code"] == C["0"]["code"].dual_code()
True
sage: C["1"]["code"] == C["1"]["code"].dual_code()
True
sage: len(C.keys()) # number of inequiv sd codes of length 10
2
sage: C = self_dual_codes_binary(12)["12"]
sage: C["0"]["code"] == C["0"]["code"].dual_code()
True
sage: C["1"]["code"] == C["1"]["code"].dual_code()
True
sage: C["2"]["code"] == C["2"]["code"].dual_code()
True
```

These commands check that some of the database entries (of length 10 and of length 12), are indeed self dual.

For classification of doubly even self-orthogonal codes using Sage, see http://www.rlmiller.org/de_codes/, [13].

The number of permutation equivalence classes of all doubly even $[n, k]$-codes is shown in the table at http://www.rlmiller.org/de_codes/, and the list of codes so far discovered is linked from the list entries. See Figure 0.5.

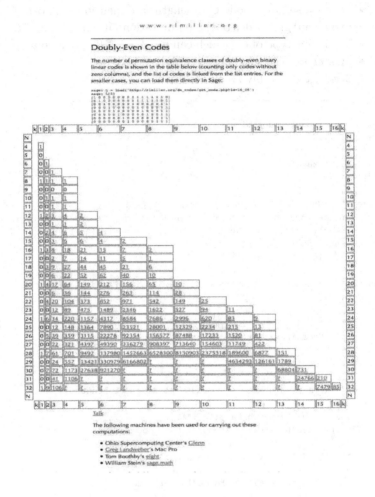

Fig. 0.5. http://www.rlmiller.org/de_codes/.

Each link on that webpage points to a Sage object file when loaded. For example

```
sage:    L = load('24_12_de_codes.sobj')
```
is a list of matrices in standard form.

0.5.3. Cool example (on self-dual codes)

Recall:

- $W_5(x, y) = x^8 + 14x^4y^4 + y^8$ is the weight enumerator of Type II $[8, 4, 4]$ code C constructed by extending the binary $[7, 4, 3]$ Hamming code by a check bit. This is the smallest Type II code.
- $W_6(x, y) = x^{24} + 759x^{16}y^8 + 2576x^{12}y^{12} + 759x^8y^{16} + y^{24}$ is the weight enumerator of the extended binary Golay code with parameters $[24, 12, 8]$.

Theorem 0.7. *Assume C is a formally self-dual divisible code of Type II. Then $A_C(x, y)$ is invariant under the group*

$$G_{II} = \langle \frac{1}{\sqrt{2}} \begin{pmatrix} 1 & 1 \\ 1 & -1 \end{pmatrix}, \begin{pmatrix} 1 & 0 \\ 0 & i \end{pmatrix} \rangle$$

of order 192. *Moreover*, $\mathbb{C}[x, y]^{G_{II}} = \mathbb{C}[W_5, W_6]$.

Group theory arises naturally in the study of self-dual codes.

Consider the group G generated by

$$g_1 = \begin{pmatrix} 1/\sqrt{q} & 1/\sqrt{q} \\ (q-1)/\sqrt{q} & -1/\sqrt{q} \end{pmatrix}, g_2 = \begin{pmatrix} i & 0 \\ 0 & 1 \end{pmatrix}, g_3 = \begin{pmatrix} 1 & 0 \\ 0 & i \end{pmatrix},$$

with $q = 2$. This group leaves invariant the weight enumerator of any self-dual doubly even binary code, e.g., `ExtendedBinaryGolayCode`.

Sage code below implicitly calls GAP to construct the matrix group.

```
─────────────────────────── Sage ───────────────────────────
sage: F = CyclotomicField(8)
sage: z = F.gen()
sage: a = z+1/z
sage: b = z^2
sage: MS = MatrixSpace(F,2,2)
sage: g1 = MS([[1/a,1/a],[1/a,-1/a]])
sage: g2 = MS([[1,0],[0,b]])
sage: g3 = MS([[b,0],[0,1]])
sage: G = MatrixGroup([g1,g2,g3])
```

```
sage: G.order()
192
```

Sage code below implicitly calls **Singular** for computing the invariants of G. We see that the invariants are indeed as predicted.

```
───────────────── Sage ─────────────────
sage: G.invariant_generators()
[x1^8 + 14*x1^4*x2^4 + x2^8,
 x1^24 + 10626/1025*x1^20*x2^4 + 735471/1025*x1^16*x2^8\
 + 2704156/1025*x1^12*x2^12 + 735471/1025*x1^8*x2^16\
 + 10626/1025*x1^4*x2^20 + x2^24]
```

The above result implies that any such weight enumerator must be a polynomial in

$$x^8 + 14x^4y^4 + y^8$$

and

$$1025x^{24} + 10626x^{20}y^4 + 735471x^{16}y^8 + 2704156x^{12}y^{12} + \\ 735471x^8y^{16} + 10626x^4y^{20} + 1025y^{24}.$$

0.6. Coding theory not implemented in Sage

Of course, a lot of coding theory functionality is lacking in Sage. This is a very selective example of this author's wish-list.

- *Minimum distance for non-binary codes*
  - The only fast implementation of `minimum_distance` is in the binary case (and due to Robert Miller).
  - Guava has a fast implementation of `MinimumDistance` in the ternary case.
  - Sage needs a much faster implementation of `minimum_distance` in the non-binary case.
- *Automorphism groups for non-binary codes*

- The only fast implementation of automorphism groups is in the binary case (and also due to Robert Miller).
  Sage needs a much faster implementation of automorphism groups in the non-binary case.
- *algebraic-geometric (AG) codes*
  - AG codes are implemented in Singular, but not yet completely implemented in Sage.
  - There is a module ag_code in Sage's coding directory but it does not work at present and is not imported.
  - This needs to be fixed! (See also Sage trac ticket # 8997.)
- *Decoding*
  - Sage has no special decoding algorithms. (Not even for Hamming codes!)
  - Guava has some but still is very limited.
  - Sage needs a lot of work in this area!
- *Gray codes*
  - Sage has nothing on Gray codes.
  - Lack of developers in this area is the main problem.
- *Cycle and cocycle codes*
  - Sage has nothing on graph-theoretic cycle or cocycle codes.
- *LDPC codes*
  - Sage has nothing on LDPC codes!
- Finish porting or "wrapping" everything in Guava to Sage. Guava is not part of Sage, though it can be loaded easily.

```
sage: install_package("gap_packages")
sage: gap.eval('LoadPackage("guava")')
'true'
sage: C = gap("HammingCode(3,GF(2))")
sage: C.MinimumDistance()
3
```

- *Codes over finite rings*
  - Sage has nothing on ring codes!

- There is Cython code written (mostly) by Cesar A. Garcia-Vazquez.
- Cesar's code can go in with some extra effort (see Sage trac #6452).

──────── Sage ────────
```
sage: M = Matrix(IntegerModRing(12), [[0, 1, 6, -1],
[1, 6, 1, 2], [6, 1, 1, 0]])
sage: C = RingCode(M) ; C
(4, 1728, 2)-code over the Ring of integers modulo 12
sage: c = C.minimum_weight_codeword(); c
(0, 1, 0, 5)
sage: c in C
True
```

- *Circuit and cocircuit codes from matroids*
  - Sage has nothing on matroids (at the time of this writing), much less circuit or cocircuit codes.

References

[1] M. Albrecht, *Algebraic Attacks against the Courtois Toy Cipher*, in Cryptologia (2008) 220-276.
http://www.informatik.uni-bremen.de/~malb/papers/algebraic-attacks-on-the-courtois-toy-cipher--diplomarbeit.pdf

[2] T. Brock, *Linear Feedback Shift Registers and Cyclic Codes in* Sage, USNA Math Honors paper, 2006, http://www.usna.edu/Users/math/wdj/brock/.

[3] J. Buchmann, A. Pychkine, R.-P. Weinmann, *Block ciphers sensitive to Groebner basis attacks*, in **Topics in Cryptology RSA'06**, http://eprint.iacr.org/2005/200.

[4] C. Cid, S. Murphy, M. Robshaw, *Small scale variants of the AES*, in **Proceedings of Fast Software Encryption**, 2005,
http://www.isg.rhul.ac.uk/~sean/smallAES-fse05.pdf.

[5] I. Duursma, *Weight distributions of geometric Goppa codes*, Transactions of the AMS, vol. 351, pp. 3609-3639, September 1999.
http://www.math.uiuc.edu/~duursma/pub/amst-duursma.ps.

[6] The GAP Group, **GAP** – *Groups, Algorithms, and Programming*, Version 4.4.10; 2007. http://www.gap-system.org.

[7] GUAVA, a coding theory package for GAP, http://sage.math.washington.edu/home/wdj/guava/.

[8] M. Hogan, "The Blum-Goldwasser cryptosystem," USNA Math Honors paper, 2010, http://www.usna.edu/Users/math/wdj/hogan/.

[9] W. C. Huffman, V. Pless, **Fundamentals of error-correcting codes**, Cambridge Univ. Press, 2003.

[10] T. Harada, M. Tagami, *A Riemann hypothesis analogue for invariant rings*, Discrete Mathematics, vol. 307, pp. 2552-2568, 2007.

[11] D. Joyner, J.-L. Kim, **Selected unsolved problems in coding theory**, Birkhaüser Press, to appear.

[12] R. Miller, *Graph automorphism computation*, http://www.rlmiller.org/talks/nauty.pdf, March 2007.

[13] R. Miller, *Doubly even codes*, http://www.rlmiller.org/talks/June_Meeting.pdf, June 2007.

[14] R. C.-W. Phan, *Mini advanced encryption standard*, Cryptologia (2002) 283-306.

[15] D. Ruano, *On the structure of generalized toric codes*, Journal of Symbolic Computation, 44 (2009)499-506.
http://arxiv.org/abs/cs/0611010.

[16] E. Schaefer, *A simplified data encryption algorithm*, Cryptologia(1996)77-84.

[17] G.-M. Greuel, G. Pfister, H. Schönemann. SINGULAR 3.0. A Computer Algebra System for Polynomial Computations. Centre for Computer Algebra, University of Kaiserslautern (2005).
http://www.singular.uni-kl.de.

[18] W. Stein, Sage: *Mathematical software*, version 4.5.
http://www.sagemath.org/.

[19] S. Shokranian, M.A. Shokrollahi, **Coding theory and bilinear complexity**, Scientific Series of the International Bureau, KFA Jülich Vol. 21, 1994.

[20] J. van Lint, **Introduction to coding theory, 3rd ed.**, Springer-Verlag GTM, 86, 1999.

[21] M. A. Tsfasman, S. G. Vladut, D. Nogin, **Algebraic geometric codes: basic notions**, AMS, Math. Surveys, 2007.

Chapter 1

Aspects of Random Network Coding

Olav Geil and Casper Thomsen

Department of Mathematical Sciences
Aalborg University
olav@math.aau.dk, caspert@math.aau.dk

This chapter gives an introduction to basic network coding and treats the estimation of success probability when linear encoding functions are chosen by random in an error-free acyclic network. With regard to the latter we give a coherent description.

Contents

1.1 Introduction 47
1.2 The network coding problem 50
    1.2.1 Linear network coding for multicast 52
    1.2.2 A polynomial time algorithm for solving the multicast problem 59
1.3 Random network coding 62
    1.3.1 The algebraic approach 62
    1.3.2 The combinatorial approach 70
1.4 Bibliographic notes 79
References 80

1.1. Introduction

Consider the directed graph $G = (V, E)$ in Fig. 1.1 where edges are enumerated $1, \ldots, |E| = 9$. The sender s wants to send a set of messages $\{a, b\}$ to receiver r_1 as well as to receiver r_2. Concentrating first on r_1, we find a set of two edge disjoint paths from s to r_1, namely

$$\{(1,5), (2,4,6,8)\}. \tag{1.1}$$

Such a set is called a flow of size two and can be used to get information through to r_1. Sender s forwards message a to edge 1 and forwards message b to edge 2. Message a then propagates along the path $(1, 5)$ to r_1 and b

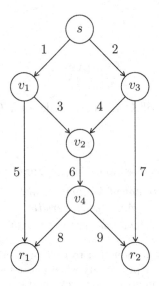

Fig. 1.1. The butterfly network.

propagates along path $(2, 4, 6, 8)$ to r_1. A similar situation holds regarding receiver r_2, the flow now being

$$\{(1, 3, 6, 9), (2, 7)\}. \tag{1.2}$$

So we have two partial solutions to the full communication problem of getting messages a and b through to both receivers r_1 and r_2 simultaneously. Trying to combine the two partial solutions into a solution of the full communication problem we see that if sender s forwards a to edge 1 and forwards b to edge 2, then using the flow (1.1) edge 6 is supposed to forward message b, and using the flow (1.2) edge 6 should forward message a. The situation is illustrated in Fig. 1.2. So the two partial solutions do not play together. We have actually inspected all combinations of possible flows of size two from s to r_1 and r_2 and can conclude that the full communication problem cannot be solved by the above naive approach.

The problem is, however, solvable as we shall now see. Assume for simplicity that the messages are binary symbols. As before a is forwarded to edge 1 and b is forwarded to edge 2. The vertex v_1 forwards a to edge 5 as well as to edge 3. Symmetrically, vertex v_3 forwards message b to edge 4 as well as to edge 7. Vertex v_2 receives a from edge 3 and b from edge 4. It now forwards the sum $a + b$ to edge 6 which then propagates along edge 8 and 9 respectively to receiver r_1 and r_2 respectively. The situation

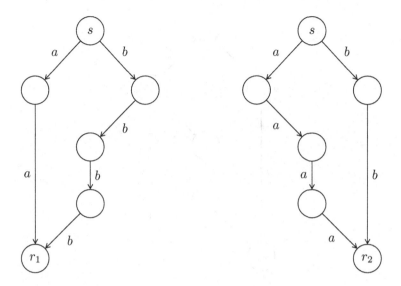

Fig. 1.2. The two partial solutions.

is illustrated in Fig. 1.3. Besides receiving a from edge 5 r_1 can recover b by calculating the sum of what is received from edges 5 and 8. We will say that r_1 can decode the message set $\{a, b\}$. A similar situation holds for r_2.

The above solution is an example of what is known as network coding. Here coding refers to the encoding functions that given incoming information to a vertex decide what should be forwarded along the various outgoing edges and it refers to the decoding functions at the receiving ends. Network coding was invented only a decade ago by Ahlswede, Cai, Li, and Yeung in the seminal paper [1], see also [23]. By now network coding is a huge research area with many different aspects. This for instance includes random network coding where encoding functions are chosen in a distributed manner. Other examples are error/erasure correction in networks and cryptographic issues.

In Section 1.2 we give an introduction to the basics of network coding. Our focus will be on the situation of linear network coding in multicast. In Section 1.3 we then explain the concept of random network coding and describe the main results from the literature. The study of random network coding involves simple probability theory, some linear algebra, and obviously also a portion of graph theory. A great deal of papers on the topic use algebraic methods in addition, for instance the seminal paper [13]. Later papers [2, 3] analyze the problem using combinatorial arguments instead.

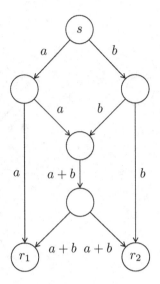

Fig. 1.3. Applying network coding.

In Section 1.3 we do not only describe known material but we also present some new results obtained by using the combinatorial approach. These new results will help us clarify a connection between the results in the literature.

1.2. The network coding problem

Let $G = (V, E)$ be a directed graph and let $S = \{s_1, \ldots, s_{|S|}\} \subseteq V$ and $R = \{r_1, \ldots, r_{|R|}\} \subseteq V$ be given. The vertices in S are called senders and the vertices in R are called receivers. A message vector is a variable $\vec{X} = (X_1, \ldots, X_h)$ that can take on all possible values in A^h, where A is some finite abelian group. The entries of $\vec{X}$ are called messages. We have a surjective map $K : \{X_1, \ldots, X_h\} \to S$ called the source map. If $K(X_i) = s_j$ then X_i is said to be generated at s_j. Let out(v) be the out edges of the vertex v and for an edge $e = (u, v)$ let out$(e) = $ out(v). We define in(v) and in(e) in a similar way. Throughout this paper we always make the crucial assumption that G is cycle free. This means that we can always order the edges in such a way that whenever a directed path exists with the edge e_1 visited before the edge e_2, then $e_1 \prec e_2$. Such an ordering is called an ancestral ordering. For every edge j in the network we have a variable called $Y(j)$ that takes on values in A. The relation among the variables in the network is that if we visit the edges in an ancestral order then for every edge $j = (u,v)$ we have a function f_j such that

$$Y(j) = f_j\bigg(\big(Y(i) \mid i \in \text{in}(j)\big), \big(X_k \mid X_k \text{ is generated at } u\big)\bigg).$$

If the argument of f_j is empty, $Y(j)$ will always take on the value 0. The function f_j is called an encoding function. Finally, we have the so-called demand function D which maps R to the set of ordered non-empty subsets of $\{X_1, \ldots, X_h\}$. If $D(r) = (X_{i_1}, \ldots, X_{i_t})$ for a receiver $r \in R$ then we will say that r demands the messages $X_{i_1}, \ldots, X_{i_t}$. For every receiver r we have $|D(r)|$ variables $Z_1^{(r)}, \ldots, Z_{|D(r)|}^{(r)}$ that take on values in A. These variables are functions of the variables emerging at r. More formally, for $j = 1, \ldots, |D(r)|$ we have a function $d_j^{(r)}$ such that

$$Z_j^{(r)} = d_j^{(r)}\bigg(\big(Y(i) \mid i \in \text{in}(r)\big), \big(X_k \mid K(X_k) = r\big)\bigg). \qquad (1.3)$$

A network $G = (V, E)$ with sources S, receivers R, message vector $\vec{X} = (X_1, \ldots, X_h)$, source function K, and demand function D is called a network coding problem. It is said to be solvable if there exists a non-trivial alphabet A, and a set of choices of encoding functions f_j and decoding functions $d_j^{(r)}$ such that

$$(Z_1^{(r)}, \ldots, Z_{|D(r)|}^{(r)}) = D(r)$$

holds for all $r \in R$. If $D(r) = (X_1, \ldots, X_h)$ for all $r \in R$ (all receivers demand all information) then we speak about multicast. If the alphabet is a finite field $\mathbf{F}_q$ (merely as an abelian group) and if all the encoding functions f_j as well as all the decoding functions $d_j^{(r)}$ are linear over $\mathbf{F}_q$ then we talk about linear network coding. For linear network coding we have the scheme

$$Y(j) = \sum_{i \in \text{in}(j)} f_{i,j} Y(i) + \sum_{\substack{X_i \in S \\ K(X_i) = u}} a_{i,j} X_i$$

where $j = (u, v)$, and

$$Z_j^{(r)} = \sum_{i \in \text{in}(r)} b_{i,j}^{(r)} Y(i) + \sum_{\substack{X_i \in S \\ K(X_i) = r}} \tilde{b}_{i,j}^{(r)} X_i. \qquad (1.4)$$

In the following we shall for simplicity always assume that S and R are disjoint sets. Hence, the last part of (1.3) and (1.4) is absent.

Given a network coding problem let $r \in R$, and $D(r) = (X_{i_1}, \ldots, X_{i_t})$. A flow for r is a set of t mutually edge disjoint paths from S to r, where for every $s \in S$ exactly $v^{(r)}(s)$ of them originate from s. Here, $v^{(r)}(s)$ is the number of elements in $D(r)$ that are generated at s. We will think of a flow as a collection of edges. Hence, given a flow F by $j \in F$ we mean that j is an edge that appears in F. It is clear [5] that for the network coding problem to be solvable, a flow must exist for all receivers $r \in R$. If not we would have a bottleneck too small for enough information to pass. The amazing result which we already experienced in an example in Section 1.1 is that for the case of multicast the existence of a flow for each receiver is actually a sufficient condition for solvability. Even more, linear network coding will do.

1.2.1. *Linear network coding for multicast*

In the remaining part of the paper we concentrate on multicast network coding problems. We start by taking a closer look at linear network coding. Let A be the $h \times |E|$ matrix with $a_{i,j}$ in the (i,j)th entry if $j \in \text{out}(K(X_i))$ and 0 otherwise. Let F be the $|E| \times |E|$ matrix with $f_{i,j}$ in the (i,j)th entry if $j \in \text{out}(i)$ and 0 otherwise. For $r \in R$ let $B^{(r)}$ be the $|E| \times h$ matrix with $b_{i,j}^{(r)}$ in the (i,j)th entry if $i \in \text{in}(r)$ and 0 otherwise[a].

For any non-negative integer n the (i,j)th entry of F^n is

$$\sum_{\substack{(i = j_0, j_1, \ldots, j_n = j) \\ \text{is a path} \\ \text{in } G}} f_{i=j_0, j_1} f_{j_1, j_2} \cdots f_{j_{n-1}, j_n = j}. \tag{1.5}$$

So F^n holds information about all paths of length n (we define the length of a path to be the set of edges herein). The directed graph being acyclic $F^N = 0$ for some large enough $N \in \mathbf{N}$. Therefore, the matrix $(I + F + \cdots + F^{N-1})$ holds information about all paths in the network. We conclude that if values are given to the $a_{i,j}$'s and the $f_{i,j}$'s then

$$(Y(1), \ldots, Y(|E|)) = (X_1, \ldots, X_h) A (I + F + \cdots + F^{N-1})$$

holds. If values are given to the $b_{i,j}^{(r)}$'s then the output of the decoding function at receiver r is

$$(Y(1), \ldots, Y(|E|)) B^{(r)} = (X_1, \ldots, X_h) A (I + F + \cdots + F^{N-1}) B^{(r)}$$

---
[a]Our notation deviates slightly from the notation in the literature. Here, often the transpose of $B^{(r)}$ is considered rather than $B^{(r)}$.

or alternatively
$$(X_1, \ldots, X_h) M^{(r)}$$
where
$$\begin{aligned} M^{(r)} &= A(I + F + \cdots + F^{N-1})B^{(r)} \\ &= A(I - F)^{-1}B^{(r)} \end{aligned}$$
is called the transfer matrix for r. Here, we used the fact that
$$(I - F)(I + F + \cdots + F^{N-1}) = I$$
which is seen by applying the information that $F^N = 0$.

Relaxing the condition for successful coding and decoding slightly we will - rather than demanding $M^{(r)}$ to equal I for all $r \in R$ - simply require $M^{(r)}$ to be of full rank for all $r \in R$. That is, we require
$$\prod_{r \in R} \det M^{(r)} \neq 0. \tag{1.6}$$

Given coefficients $a_{i,j}, f_{i,j}, b_{i,j}^{(r)}$ such that (1.6) holds, then it is clear that one can change the values of $b_{i,j}^{(r)}$ such that $M^{(r_1)} = \cdots = M^{(r_{|R|})} = I$ holds. It is shown in [13] that $\det M^{(r)}$ either equals $\det E^{(r)}$ or equals $-\det E^{(r)}$ where
$$E^{(r)} = \begin{bmatrix} A & 0 \\ I - F & B^{(r)} \end{bmatrix}$$
is called the Edmonds matrix for r. The advantage of the Edmonds matrix over the transfer matrix is that to establish the first we do not need to detect a number N such that $F^N = 0$. For the purpose of deciding if $\prod_{r \in R} \det M^{(r)} \neq 0$ it is clearly enough to decide if $\prod_{r \in R} \det E^{(r)} \neq 0$.

It is not surprising that the determinant of the transfer matrix has a particular topological meaning. The topological meaning is easiest explained if given a multicast network coding problem we start by extending the graph $G = (V, E)$ as follows (see Figs. 1.4 and 1.5). First we add h new vertices $\chi_1, \ldots, \chi_h$ to V. Then we add to E for every $i = 1, \ldots, h$ an edge from χ_i to the source where X_i is generated. For every receiver $r \in R$ we add h new vertices $r^{(1)}, \ldots, r^{(h)}$. Finally, from r to every vertex $r^{(i)}$, $i = 1, \ldots, h$ we add an edge. Define now for each receiver r an extended flow to $r^{(1)}, \ldots, r^{(h)}$ to be a set of h mutually edge disjoint paths originating

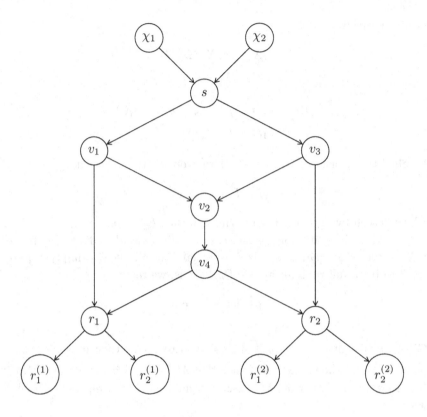

Fig. 1.4. The extended butterfly network.

in $\{\chi_1, \ldots, \chi_h\}$ and ending in $\{r^{(1)}, \ldots, r^{(h)}\}$. A path in the extended flow corresponds in a natural way to a product of one $a_{i,j}$, some $f_{i,j}$'s, and one $b_{i,j}$. Using (1.5) and the discussion following it we see that the (i,j)th entry of $M^{(r)}$ holds information about all the paths from χ_i to $r^{(j)}$. An extended flow in a similar way corresponds to a product of $a_{i,j}$'s, $f_{i,j}$'s and $b_{i,j}$'s. It is clear that there is a one-to-one correspondence between the flows to $r^{(1)}, \ldots, r^{(h)}$ in the extended graph and the monomial expressions occurring in $\det M^{(r)}$. Every monomial expression appears with the coefficient 1 or -1. Turning to the product $\prod_{r \in R} \det M^{(r)}$ the coefficients are no longer only 1's and -1's, but are members of $\mathbf{F}_p$ where p is the characteristic of the field $\mathbf{F}_q$ under consideration. Actually, terms may even cancel out each other depending on the characteristic of the field. Note, however, that of course $\det M^{(r)} \neq 0$ holds for all $r \in R$ if and only if $\prod_{r \in R} \det M^{(r)} \neq 0$ holds.

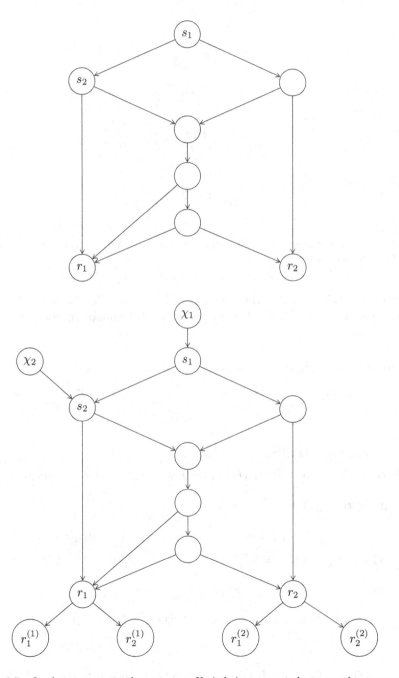

Fig. 1.5. In the upper network a message X_1 is being generated at s_1 and a message X_2 is being generated at s_2. The lower graph is the corresponding extended network.

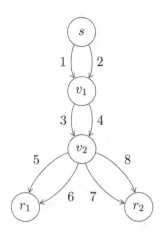

Fig. 1.6. Network where two equal monomials occur in the product of the determinants of the Edmonds matrices.

Example 1.1. Consider the network in Fig. 1.6. The edges are conveniently simply numbered. The determinant of Edmonds matrices for the two receivers are

$$(a_{1,1}a_{2,2} - a_{1,2}a_{2,1})(b_{5,1}^{(r_1)}b_{6,2}^{(r_1)} - b_{6,1}^{(r_1)}b_{5,2}^{(r_1)})$$
$$(f_{1,3}f_{2,4}f_{3,5}f_{4,6} - f_{2,3}f_{1,4}f_{3,5}f_{4,6} - f_{1,3}f_{2,4}f_{4,5}f_{3,6} + f_{2,3}f_{1,4}f_{4,5}f_{3,6})$$

and

$$(a_{1,1}a_{2,2} - a_{1,2}a_{2,1})(b_{7,1}^{(r_2)}b_{8,2}^{(r_2)} - b_{8,1}^{(r_2)}b_{7,2}^{(r_2)})$$
$$(f_{1,3}f_{2,4}f_{3,7}f_{4,8} - f_{2,3}f_{1,4}f_{3,7}f_{4,8} - f_{1,3}f_{2,4}f_{4,7}f_{3,8} + f_{2,3}f_{1,4}f_{4,7}f_{3,8}).$$

By fixing the $a_{i,j}$'s and $b_{i,j}^{(r)}$'s as

$$a_{1,1} = a_{2,2} = 1 \qquad a_{1,2} = a_{2,1} = 0$$
$$b_{5,1}^{(r_1)} = b_{6,2}^{(r_1)} = b_{7,1}^{(r_2)} = b_{8,2}^{(r_2)} = 1 \qquad b_{6,1}^{(r_1)} = b_{5,2}^{(r_1)} = b_{8,1}^{(r_2)} = b_{7,2}^{(r_2)} = 0$$

we get

$$(f_{1,3}f_{2,4}f_{3,5}f_{4,6} - f_{2,3}f_{1,4}f_{3,5}f_{4,6} - f_{1,3}f_{2,4}f_{4,5}f_{3,6} + f_{2,3}f_{1,4}f_{4,5}f_{3,6})$$

and

$$(f_{1,3}f_{2,4}f_{3,7}f_{4,8} - f_{2,3}f_{1,4}f_{3,7}f_{4,8} - f_{1,3}f_{2,4}f_{4,7}f_{3,8} + f_{2,3}f_{1,4}f_{4,7}f_{3,8}).$$

Notice that in the product of these, we have

$$2f_{1,3}f_{2,4}f_{3,5}f_{4,6}f_{2,3}f_{1,4}f_{4,7}f_{3,8} - 2f_{1,3}f_{2,4}f_{3,5}f_{4,6}f_{2,3}f_{1,4}f_{3,7}f_{4,8}$$
$$- 2f_{1,3}f_{2,4}f_{4,5}f_{3,6}f_{2,3}f_{1,4}f_{4,7}f_{3,8} + 2f_{1,3}f_{2,4}f_{4,5}f_{3,6}f_{2,3}f_{1,4}f_{3,7}f_{4,8}$$

which cancel out if and only if the characteristic is 2.

When the $a_{i,j}$'s, $f_{i,j}$'s and $b_{i,j}^{(r)}$'s are unknown to us we can think of them as variables. The polynomial $\prod_{r \in R} \det M^{(r)}$ is then called the transfer polynomial. Above we explained what is an extended flow. Considering now the original (the non extended) network by a flow to r of size h we mean a set of h edge disjoint paths from S to r such that exactly $v(s)$ of them originate in s. Here, $v(s)$ is[b] the number of messages being generated at s. From our discussion we have

Proposition 1.1. *The transfer polynomial is non-zero if and only if to every receiver there exists a flow of size h.*

We call a collection of $|R|$ flows of size h, one to each receiver, a flow system of size h. Hence, Proposition 1.1 tells us that the transfer polynomial is non-zero if and only if a flow system of size h can be found in the network. As for a flow system we will use the notation $\mathcal{F} = \{F_1, \ldots, F_{|R|}\}$ where F_i is the flow to receiver r_i for $i = 1, \ldots, |R|$. In the transfer polynomial the $a_{i,j}$'s and the $f_{i,j}$'s appear at most in power $|R|$ whereas the $b_{i,j}^{(r)}$'s appear in power at most 1. This observation will allow us to determine precisely when a solution can be found to the network coding problem. We shall need the footprint bound from Gröbner basis theory.

Definition 1.1. Let $\prec$ be a monomial ordering and consider an ideal $I \subseteq \mathbf{F}[X_1, \ldots, X_m]$. Here, $\mathbf{F}$ is any field. The footprint of I with respect to $\prec$ is the set of monomials that cannot be found as leading monomial of any polynomial in the ideal. It is denoted by $\Delta_\prec(I)$.

By $\bar{\mathbf{F}}$ we denote the algebraic closure of the field $\mathbf{F}$. The following result is known as the footprint bound [10].

Proposition 1.2. *If $I \subseteq \mathbf{F}[X_1, \ldots, X_m]$ is finite dimensional (meaning that $\Delta_\prec(I)$ is finite) then the variety $\mathcal{V}_{\bar{\mathbf{F}}}(I)$ (that is the set of common zeros from $\bar{\mathbf{F}}^m$ of polynomials in I) is of size at most $|\Delta_\prec(I)|$.*

---

[b]The definition of $v(s)$ is in agreement with the definition of $v^{(r)}(s)$ on page 52.

Corollary 1.1. *Let $F(X_1, \ldots, X_m) \in \mathbf{F}[X_1, \ldots, X_m]$ be a non-zero polynomial and consider some fixed monomial ordering $\prec$. Here, $\mathbf{F}$ is some field containing $\mathbf{F}_q$. Assume that the leading monomial of F is $X_1^{i_1} \cdots X_m^{i_m}$, with $0 \leq i_1, \ldots, i_m < q$. The number of non-zeros of $F(X_1, \ldots, X_m)$ in $\mathbf{F}_q^m$ is at least $(q - i_1) \cdots (q - i_m)$.*

Proof. We apply the footprint bound and note that

$$\Delta_\prec(\langle X_1^q - X_1, \ldots, X_m^q - X_m, F(X_1, \ldots, X_m)\rangle)$$
$$\subseteq \{X_1^{s_1} \cdots X_m^{s_m} \mid 0 \leq s_t < q \text{ for } t = 1, \ldots m \text{ and}$$
$$i_t \leq s_t \text{ does not hold for all } t = 1, \ldots, m\}. \qquad \square$$

We are now ready for an important conclussion regarding solvability.

Theorem 1.1. *A multicast network coding problem is solvable if and only if the corresponding transfer polynomial is non-zero. When the network coding problem is solvable then linear network coding is guarenteed to work for all fields $\mathbf{F}_q$ with $q \geq |R|$.*

Proof. For multicast to be possible a flow system of size h must exist. By Proposition 1.1 the "only if" part is proven. To prove the "if" part assume that a flow system of size h exists. Then by Proposition 1.1 the transfer polynomial is non-zero and as no variable can occur in powers exceeding $|R|$, by Corollary 1.1 there exists a non-zero solution to the linear network coding problem if the field size exceeds $|R|$. In the next section we shall see that field size equal to $|R|$ is enough. $\qquad \square$

Clearly, the first part of Theorem 1.1 could also have been formulated:

Theorem 1.2. *A multicast network coding problem is solvable if and only if the graph contains a flow system of size h.*

Recall that all the coefficients in the transfer polynomial belongs to $\mathbf{F}_p$ where p is the characteristic under consideration. If we seek the smallest power m of p such that encoding and decoding coefficients can be successfully chosen from $\mathbf{F}_{p^m}$ then we can proceed as follows. For increasing powers of p we reduce the transfer polynomial modulo $X^{p^m} - X$ where X runs through all variables $a_{i,j}$ and $f_{i,j}$ (we need not consider the $b_{i,j}^{(r)}$'s as they do not appear in powers higher than 1). The reduction is indeed fast as we only need to replace all occurrences of X^{p^m} with X. If and only if the remainder is a non-zero polynomial we can declare success. It is an NP-complete problem to find the minimum feasible field [8]. The only possible conclussion (unless NP=P) is that in the worst case the transfer polynomial can have exponentially many monomials which is indeed the case.

1.2.2. *A polynomial time algorithm for solving the multicast problem*

There exist simple and fast algorithms for finding flows in a network [5]. Hence, given a solvable multicast network coding problem it is not hard to find a flow system. Now it is a fortunate fact that given a flow system of size h then a simple low complexity algorithm can find a solution over $\mathbf{F}_q$, for any $q \geq |R|$. This algorithm described by Jaggi *et al.* in [14] uses global coding vectors.

Definition 1.2. Assume coefficients $a_{i,j}$, $f_{i,j}$ are chosen for a network coding problem. For every edge j we define the corresponding global coding vector to be $c_g(j) = (c_1, \ldots, c_h)$ if $Y(j) = c_1 X_1 + \cdots + c_h X_h$.

It is clear that if for each receiver r the global coding vectors of $in(r)$ span the whole of $\mathbf{F}_q^h$ then the encoding coefficients are part of a solution to the network coding problem. The only thing missing to have a full solution is to choose the $b_{i,j}^{(r)}$'s which can be done by using standard linear algebra techniques. The algorithm starts by modifying the network coding problem slightly. First a new vertex s' is added to the network. Then for each $s_i \in S$, $v(s_i)$ edges are added from s' to s_i. Here $v(s_i)$ is the number of messages generated at s_i. Altogether h edges are being added. Next, we assume all processes to be generated at s'. Denote by $\mathcal{F} = (F_1, \ldots, F_{|R|})$ a flow system. Here, F_i is the flow to receiver r_i. The algorithm works on the modified network coding problem. In the initialization part the following is done:

- If (i, j) is not part of any path in the flow system we set $f_{i,j}$ to 0
- Encoding functions at the newly added point s' are chosen in such a way that the global coding vectors of the h added edges are

$$(1, 0, \ldots, 0), (0, 1, 0, \ldots, 0), \ldots, (0, \ldots, 0, 1) \qquad (1.7)$$

(information about the encoding functions at s' are not remembered by the algorithm as they are not relevant for the original network).

Note that the h edges constitute a cut in the flow to each receiver. We initialize $C_1 = \cdots = C_{|R|}$ to be the ordered set of the h added edges and $B_1, \ldots, B_{|R|}$ to be the corresponding ordered set of global coding vectors (which then corresponds to (1.7)). The main part of the algorithm is a loop in which the ordered sets $C_1, \ldots, C_{|R|}$ are updated and where coding coefficients related to the update are chosen in such a way that for $l = 1, \ldots, |R|$ the following loop invariant holds:

- C_l is a cut in F_l
- The ordered set B_l of global coding vectors corresponding to C_l span the whole of $\mathbf{F}_q^h$.

The updates are done such that at the end (after finitely many steps) $C_l = \text{in}(r_l) \cap F_l$ holds for $l = 1, \ldots, |R|$. This means that by the second loop invariant each receiver has enough information to calculate by use of simple linear algebra methods appropriate $b_{i,j}^{(r)}$'s.

Updating is done as follows. We visit the edges $j \in \mathcal{F}$ according to an ancestral ordering. Considering an edge j we are concerned with the receivers r_l such that $j \in F_l$. We add j to C_l and remove from C_l its predecessor in F_l. That is, we remove the unique edge $i \in C_l \cap \text{in}(j)$ for which (i, j) is part of a path in F_l. Let $l_1, \ldots, l_k$ be values of l for which C_l is updated and let $i_1, \ldots, i_k$ (possibly $i_s = i_t$ for some $s \neq t$) be the edges removed from $C_{l_1}, \ldots, C_{l_k}$. The task now is to choose coding coefficients $f_{i_1, j}, \ldots, f_{i_k, j}$ in such a way that all sets of global coding vectors $B_{l_1}, \ldots, B_{l_k}$ for $C_{l_1}, \ldots, C_{l_k}$ span the whole of $\mathbf{F}_q^h$. Choosing the coding coefficients by random the probability of success is positive as guaranteed by the following lemma.

Lemma 1.1. *Given a basis $\{\vec{b}_1, \ldots, \vec{b}_h\}$ for $\mathbf{F}_q^h$ and $\vec{c} \in \mathbf{F}_q^h$, there is exactly one choice of $a \in \mathbf{F}_q$ such that*

$$\vec{c} + a\vec{b}_h \in \text{Span}_{\mathbf{F}_q}\{\vec{b}_1, \ldots, \vec{b}_{h-1}\}. \tag{1.8}$$

Proof. Write $\vec{c}$ as a linear combination $\vec{c} = c_1 \vec{b}_1 + \cdots + c_h \vec{b}_h$. The only a for which (1.8) holds is $a = -c_h$. □

For each i_t we now apply the lemma with the basis being B_{l_t} (as it is before the update), with $\vec{b}_h = c_g(i_t)$, and with

$$\vec{c} = \sum_{l \in \{i_1, \ldots, i_k\} \setminus \{i_t\}} f_{l,j} c_g(l).$$

Algorithm 1.1.

Input: A solvable multicast (acyclic) network coding problem with $G = (V, E)$. A field $\mathbf{F}_q$ with $q \geq |R|$. An ancestral ordering on E.
Output: Encoding coefficients $a_{i,j}, f_{i,j}$.
Initialization
Let $V' := V \cup \{s'\}$ and $E' := E \cup \{e_1, \ldots, e_h\}$ where the new edges are as described in the text;
Find a flow system $\mathcal{F} = \{F_1, \ldots, F_{|R|}\}$ in $G' := (V', E')$ of size h from s' to the receivers;
for $j \in E$ and $i = 1, \ldots, h$
  if $j \in \text{out}(e_i)$ and $(e_i, j) \notin \mathcal{F}$ **then**
    $a_{i,j} = 0$;
  end
endfor
for $(i, j) \in E \times E$ where $i \in \text{in}(j)$ and $(i, j) \notin \mathcal{F}$
  $f_{i,j} = 0$;
endfor
for $l = 1, \ldots, |R|$
  $C_l := (e_1, \ldots, e_h)$;
  $B_l := ((1, 0, \ldots, 0), (0, 1, 0, \ldots, 0), \ldots, (0, \ldots, 0, 1))$;
endfor
Update
for $j \in \mathcal{F}$ by the ancestral ordering
  Let $(r_{l_1}, \ldots, r_{l_k})$ be the receivers that use j in $\mathcal{F}$;
  Let $(i_1, \ldots, i_k)$ be the corresponding predecessors of j in $\mathcal{F}$;
  for $v = 1, \ldots, k$
    $C_{l_v} := (C_{l_v} \setminus \{i_v\}) \cup \{j\}$;
  endfor
  Choose encoding coefficients (notation with respect to extended network) $(f_{i_1,j}, \ldots, f_{i_t,j})$ by random until B_{l_v} spans $\mathbf{F}_q^h$ for all $v = 1, \ldots, k$;
  for $v = 1, \ldots, k$
    if i_v equals e_k for some $k \in \{1, \ldots, h\}$ **then**
      rename $f_{i_v,j}$ to $a_{k,j}$;
    end
  endfor
endfor
return all values $a_{i,j}, f_{i,j}$ for which values have been allocated

Let $k' = |\{i_1, \ldots, i_k\}| = |(\text{in}(j) \cap \mathcal{F})|$. We deduce that the probability of success in a step of the algorithm is at least

$$\frac{q^{k'} - kq^{k'-1} + (k-1)}{q^{k'}} = 1 - \frac{k}{q} + \frac{k-1}{q^{k'}}. \tag{1.9}$$

This is found by assuming as a worst case that if a choice of coding coefficients are not successful for r_{l_t} then it is successful for $r_{l_1}, \ldots, r_{l_{t-1}}, r_{l_{t+1}}, \ldots, r_k$. Doing this we have counted the choice $f_{i_1,j} = \cdots = f_{i_k,j} = 0$ k times.

The expression in (1.9) is positive for $q \geq k$ and as clearly $k \leq |R|$ we have proved that for a solvable network coding problem $\mathbf{F}_q$ is big enough when $q \geq |R|$.

In Algorithm 1.1 we give a pseudo code description of the algorithm.

1.3. Random network coding

Consider a solvable network coding problem. From Corollary 1.1 it is clear that if for q large enough we choose the coding coefficients and the decoding coefficients by random then with high probability the network coding problem is solved. In random network coding we start by fixing *a priori* a (possible empty) subset of the encoding coefficients. For instance if in$(j) = \{i\}$ then we might choose *a priori* $f_{i,j} = 1$. The remaining coding coefficients are now chosen by random (independently and uniformly). We can view this as being done in a distributed manner. Having fixed the encoding coefficients the receivers must learn how to decode (if possible). This is done by injecting into the system the message vectors

$$\vec{X}_1 = (1, 0, \ldots, 0), \vec{X}_2 = (0, 1, 0, \ldots, 0), \ldots, \vec{X}_h = (0, \ldots, 0, 1).$$

A receiver now simply reads what is transmitted on each incoming edge. The ordered list of information arriving on a particular edge by construction is the global coding vector for this edge. If (and only if) every receiver r receives on its incoming edges a collection of global coding vectors that spans the whole of $\mathbf{F}_q^h$ then it is possible to choose the $b_{i,j}^{(r)}$'s in such a way that the network coding problem is solved. Calculating the $b_{i,j}^{(r)}$'s is only a matter of performing simple linear algebra operations.

The remaining part of this paper deals with the question what can be said about the probability that randomly chosen coding coefficients do actually work. This probability is denoted P_{succ}. As mentioned in the introduction we take an algebraic approach as well as a pure combinatorial one.

1.3.1. *The algebraic approach*

Consider a solvable network coding problem where some encoding functions are chosen *a priori* in such a way that they are part of some solution. Let

η be the number of encoding coefficients chosen randomly from $\mathbf{F}_q$. As we explain below Corollary 1.1 implies the bound

$$P_{\text{succ}} \geq P_{\text{Ho0}} := \left(\frac{q-|R|}{q}\right)^\eta \tag{1.10}$$

whenever $q \geq |R|$ holds. The notation P_{Ho0} and the notations $P_{\text{Ho1}}, P_{\text{Ho2}}$ to follow refer to the authors of [13]. To prove (1.10) consider the transfer polynomial $\prod_{r \in R} \det M^{(r)}$ having variables $a_{i,j}, f_{i,j}, b_{i,j}^{(r)}$. Plug in the *a priori* chosen encoding functions and view the $b_{i,j}^{(r)}$'s as constants, i.e. we will consider the polynomial as a polynomial over the field $\mathbf{F}_q(Y_1, \ldots, Y_w)$ where $Y_1, \ldots, Y_w$ are the $b_{i,j}^{(r)}$'s. The polynomial viewed in this way is called the *a priori* transfer polynomial. It has η variables each occurring in powers of at most $|R|$. Hence from Corollary 1.1 we find that for at least $(q-|R|)^\eta$ of the q^η possible choices of randomly chosen coefficients the transfer polynomial will evaluate to a non-zero value. This concludes the proof of (1.10).

Consider some fixed monomial ordering. If we know the leading monomial of the *a priori* transfer polynomial to be $X_1^{i_1} \cdots X_m^{i_m}$ then we can deduce the improved bound

$$P_{\text{succ}} \geq P_{\text{FP2}} := \prod_{j=1}^m \frac{q-i_j}{q}. \tag{1.11}$$

If on the other hand we know the *a priori* transfer polynomial but do not want to detect which of the monomials is leading then we can apply the weaker bound:

$$P_{\text{succ}} \geq P_{\text{FP1}} := \min\{\prod_{j=1}^m \left(\frac{q-i_j}{q}\right) \mid X_1^{i_1} \cdots X_m^{i_m} \text{ is a monomial}$$

$$\text{in the transfer polynomial}\}. \tag{1.12}$$

The FP in the notation P_{FP1} and P_{FP2} refers to the footprint bound (Proposition 1.2). The estimate P_{Ho0} is not very good but relatively simple to calculate. The estimates P_{FP1} and P_{FP2} are much better but more complicated. We shall now see how to improve upon P_{Ho0} without having to pay a price regarding ease of calculation. We will need the following lemma

Lemma 1.2. *Consider a field* $\mathbf{F}$ *that contains* $\mathbf{F}_q$. *Let* $F(X_1, \ldots, X_m) \in \mathbf{F}[X_1, \ldots, X_m]$ *be a non-zero polynomial such that all monomials* $X_1^{j_1} \cdots X_m^{j_m}$ *in the support of* F *satisfy*

(1) $j_1, \ldots, j_m \leq d$, *where* d *is some fixed integer* $d \leq q$.

(2) $j_1 + \cdots + j_m \leq dN$ for some fixed integer N with $N \leq m$.

Assume $(x_1, \ldots, x_m) \in \mathbf{F}_q^m$ is chosen by random then $F(x_1, \ldots, x_m) \neq 0$ happens with probability at least

$$\left(\frac{q-d}{q}\right)^N.$$

Proof. With Eq. (1.12) in mind we want to establish the minimal value of

$$\prod_{t=1}^{m} \left(\frac{q - j_t}{q}\right) \tag{1.13}$$

taken over all possible choices of $j_1, \ldots, j_m$ such that (1) and (2) hold. We start by observing that if $x \geq y$ then

$$(q - x)(q - y) > (q - (x+1))(q - (y-1))$$

holds. Now assume $j_1, \ldots, j_m$ are chosen such that (1.13) attains its minimal value. Clearly, $j_1 + \cdots + j_m = dN$ must hold. Assuming without loss of generality that $j_1 \geq \cdots \geq j_m$ holds we get by applying the observation from the beginning of the proof that $j_1 = \cdots = j_N = d$ and that $j_{N+1} = \cdots = j_m = 0$. □

Define η' to be the maximal number of edges j in a flow such that some coefficients $a_{i,j}$ or some coefficients $f_{i,j}$ are chosen by random, the maximum is taken over all possible receivers $r \in R$ and all possible flows to r of size h. If no encoding coefficients are chosen a priori then η' is the maximal number of edges in a flow to some receiver in the network. The number of encoding coefficients in the monomial corresponding to the flow equals η'. As a flow system consists of $|R|$ flows we get that the a priori transfer polynomial satisfies conditions (1) and (2) in Lemma 1.2 with $d = |R|$ and $N = \eta'$. We therefore get

$$P_{\text{succ}} \geq P_{\text{Ho2}} := \left(\frac{q - |R|}{q}\right)^{\eta'}. \tag{1.14}$$

The number of edges in a flow is obviously at most equal to the number of edges in the network. We therefore get the weaker bound

$$P_{\text{succ}} \geq P_{\text{Ho1}} := \left(\frac{q - |R|}{q}\right)^{\eta''}$$

where η'' is the number of edges j in E such that some $a_{i,j}$ or $f_{i,j}$ are not chosen *a priori*. If no encoding coefficients are chosen *a priori* then we have $\eta'' = |E|$. We collect the results obtained so far in a theorem.

Theorem 1.3. *The success probability P_{succ} satisfies*

$$P_{\text{Ho0}} \leq P_{\text{Ho1}} \leq P_{\text{Ho2}} \leq P_{\text{FP1}} \leq P_{\text{FP2}} \leq P_{\text{succ}}.$$

Example 1.2. Consider the network in Fig. 1.7. There is one sender s

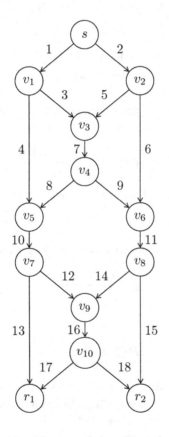

Fig. 1.7. The network from Example 1.2.

and two receivers r_1 and r_2. Two messages X_1 and X_2 are assumed to be generated at s. To each receiver there exist flows of size 2 hence the network coding problem is indeed solvable. In this example we only consider fields of characteristic 2. *A priori* we choose $a_{1,1} = a_{2,2} = 1$ and $a_{1,2} = a_{2,1} = 0$.

Also we choose *a priori* $f_{i,j} = 1$ except

$$f_{3,7}, f_{5,7}, f_{4,10}, f_{8,10}, f_{9,11}, f_{6,11}, f_{12,16}, f_{14,16}$$

which we choose by random. The *a priori* transfer polynomial becomes

$$F = (b^2 c^2 e^2 gh + c^2 f^2 gh + a^2 d^2 f^2 gh)Q ,$$

where

$$a = f_{3,7} \quad b = f_{5,7} \quad c = f_{4,10} \quad d = f_{8,10}$$
$$e = f_{9,11} \quad f = f_{6,11} \quad g = f_{12,16} \quad h = f_{14,16}$$

and $Q = \det \tilde{B}^{(r_1)} \det \tilde{B}^{(r_2)}$. Here, $\tilde{B}^{(r_1)}$ and $\tilde{B}^{(r_2)}$, respectively, is the matrix consisting of the non-zero rows of $B^{(r_1)}$ and $B^{(r_2)}$, respectively. Restricting to fields $\mathbf{F}_q$ of size at least 4 we can immediately apply the bounds established prior to the example. Considering all monomials in F we get

$$P_{\mathrm{FP1}} = \frac{(q-2)^3(q-1)^2}{q^5} .$$

Choosing then as monomial ordering the lexicographic ordering $\prec$ with

$$a \prec b \prec d \prec e \prec g \prec h \prec f \prec c$$

the leading monomial of F becomes $c^2 f^2 gh$ and therefore we have

$$P_{\mathrm{FP2}} = \frac{(q-2)^2(q-1)^2}{q^4} .$$

For comparison

$$P_{\mathrm{Ho2}} = \frac{(q-2)^4}{q^4} .$$

We see that P_{FP2} exceeds P_{Ho2} with a factor $(q-1)^2/(q-2)^2$, which is larger than 1. Also P_{FP1} exceeds P_{Ho2}. In Table 1.1 we list values of P_{FP1}, $P_{\mathrm{FP2}}(q)$ and P_{Ho2} for various choices of q.

Table 1.1. Estimates on the success probability

q	4	8	16	32	64
P_{FP2}	0.140	0.430	0.672	0.893	0.909
P_{FP1}	0.703×10^{-1}	0.322	0.588	0.773	0.880
P_{Ho2}	0.625×10^{-1}	0.316	0.586	0.772	0.880

We next consider the field $\mathbf{F}_2$. We reduce the *a priori* transfer polynomial modulo $(a^2 - a, \ldots, h^2 - h)$ to get

$$\widehat{F} = (bcegh + cfgh + adfgh)Q .$$

Choosing as monomial ordering the lexicographic ordering described above and applying Corollary 1.1 we get that the success probability is at least 2^{-4}. For comparison the bounds $P_{\text{Ho}0}, P_{\text{Ho}1}, P_{\text{Ho}2}$ do not apply as we do not have $q > |R|$.

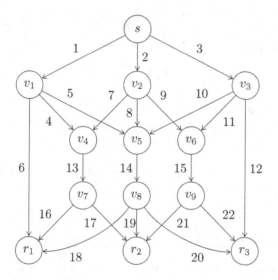

Fig. 1.8. The network from Example 1.3

Example 1.3. Consider the network in Fig. 1.8. Three messages X_1, X_2, X_3 are generated at the sender s. The vertices r_1, r_2 and r_3 are the receivers. We will apply network coding over various fields of characteristic two. We start by considering random linear network coding over fields of size at least 4. As $4 > |R| = 3$ we know that this can be done successfully.

A Priori we set $a_{1,1} = a_{2,2} = a_{3,3} = 1$ and $a_{i,j} = 0$ in all other cases. Furthermore we let $f_{i,j} = 1$ except $f_{4,13}, f_{7,13}, f_{5,14}, f_{8,14}, f_{10,14}, f_{9,15}, f_{11,15}$, which we choose by random. Therefore the *a priori* transfer polynomial F is a polynomial in the seven variables $f_{4,13}, f_{7,13}, f_{5,14}, f_{8,14}, f_{10,14}, f_{9,15}, f_{11,15}$. We have

$$F = (abcdefg + abce^2 f^2 + b^2 c^2 efg)Q ,$$

where

$$a = f_{4,13} \qquad b = f_{5,14} \qquad c = f_{7,13} \qquad d = f_{8,14}$$
$$e = f_{9,15} \qquad f = f_{10,14} \qquad g = f_{11,15}$$

and $Q = \det \tilde{B}^{(r_1)} \det \tilde{B}^{(r_2)} \det \tilde{B}^{(r_3)}$. Here, $\tilde{B}^{(r_1)}$, $\tilde{B}^{(r_2)}$, and $\tilde{B}^{(r_3)}$, respectively, is the matrix consisting of the non-zero rows of $B^{(r_1)}$, $B^{(r_2)}$, and $B^{(r_3)}$, respectively. Choosing a lexicographic ordering with d being larger than the other variables we get that the success probability is at least

$$P_{\mathrm{FP2}} = \frac{(q-1)^7}{q^7}$$

which is larger than

$$P_{\mathrm{FP1}} = \frac{(q-1)^3 (q-2)^2}{q^5}.$$

For comparison we have

$$P_{\mathrm{Ho2}} = \frac{(q-3)^3}{q^3}.$$

Both P_{FP1} and P_{FP2} exceed P_{Ho2} for all values of $q \geq 4$. In Table 1.2 we list P_{FP1}, P_{FP2} and P_{Ho2} for various values of q.

Table 1.2. Estimates on the success probability

q	4	8	16	32	64
P_{FP2}	0.133	0.392	0.636	0.800	0.895
P_{FP1}	0.105	0.376	0.630	0.799	0.895
P_{Ho2}	0.156×10^{-1}	0.244	0.536	0.744	0.865

We next consider the field $\mathbf{F}_2$. We reduce F modulo $(a^2 - a, \ldots, g^2 - g)$ to get

$$\widehat{F} = (abcdefg + abcef + bcefg)Q.$$

Choosing a proper monomial ordering we see from Corollary 1.1 that the success probability is at least 2^{-5}. For comparison neither of the bounds P_{Ho0}, P_{Ho1}, P_{Ho2} apply.

We conclude this section with a discussion of the two bounds P_{FP1} and P_{FP2}. For both bounds it holds that in order to use them we need detailed information about the transfer polynomial. In the next section we describe how to derive information on the success probability from an arbitrary flow system in the network. Therefore it is a natural question to ask whether the situation is similar regarding the bound P_{FP2}. More precisely, we ask if given any flow system of size h the corresponding monomial can be made leading in the transfer polynomial by choosing an appropriate monomial ordering. Unfortunately this is not the case at all. Firstly, already in Example 1.1 we saw that networks exist having flow systems for which the corresponding monomials cancel out each other when calculating $\prod_{r \in R} \det M^{(r)}$. Even if the monomial corresponding to a flow system is present in $\prod_{r \in R} \det M^{(r)}$ there may not exist a monomial ordering with respect to which it is leading. To illustrate this we revisit Example 1.1.

Example 1.4. This is a continuation of Example 1.1 where we considered the network in Fig. 1.6. We will assume that the characteristic of $\mathbf{F}_q$ is different from 2. This implies that the *a priori* transfer polynomial in its support contains the following three monomials

$$f_{1,3}f_{3,5}f_{2,4}f_{4,6} \cdot f_{1,3}f_{3,7}f_{2,4}f_{4,8} \tag{1.15}$$

$$f_{1,3}f_{3,5}f_{2,4}f_{4,6} \cdot f_{2,3}f_{3,7}f_{1,4}f_{4,8} \tag{1.16}$$

$$f_{2,3}f_{3,5}f_{1,4}f_{4,6} \cdot f_{2,3}f_{3,7}f_{1,4}f_{4,8}. \tag{1.17}$$

Recall that by definition a monomial ordering $\prec$ is a well-order such that for all monomials M, N, K it holds that if $M \prec N$ then $MK \prec NK$. Let $\prec$ be any monomial ordering. Now either

$$f_{2,3}f_{1,4}f_{3,5}f_{4,6}f_{3,7}f_{4,8} \prec f_{1,3}f_{2,4}f_{3,5}f_{4,6}f_{3,7}f_{4,8} \tag{1.18}$$

or

$$f_{2,3}f_{1,4}f_{3,5}f_{4,6}f_{3,7}f_{4,8} \succ f_{1,3}f_{2,4}f_{3,5}f_{4,6}f_{3,7}f_{4,8}. \tag{1.19}$$

Assume first that (1.18) holds. Multiplying both sides of (1.18) by $f_{1,3}f_{2,4}$ gives us (1.16) on the left-hand side and (1.15) on the right-hand side. Hence, if (1.18) holds then (1.16) cannot be the leading monomial of the transfer polynomial. So assume instead that (1.19) holds. Multiplying both sides of (1.19) by $f_{2,3}f_{1,4}$ yields (1.17) on the left-hand side and (1.16) on the right-hand side. In conclusion (1.16) cannot be leading monomial of the *a priori* transfer polynomial.

1.3.2. The combinatorial approach

In the previous section we estimated the success probability by algebraic means. We shall now see how to do it by using a pure combinatorial approach. We will present two methods of which the first is related to Jaggi et al.'s Algorithm 1.1. This method was used in [14] to deal with the success probability in case of an erroneous network. Here, we apply it for the first time to the case of an error-free network. The resulting bounds are called flow bounds. The second method is due to Balli, Yan and Zhang [3]. It should be mentioned that the drawback of both combinatorial methods are that only to a very limited extend they allow for *a priori* chosen encoding coefficients.

1.3.2.1. Flow bounds

Recall the probabilistic nature of Algorithm 1.1, the correctness of the algorithm relying on Lemma 1.1. We shall use our insight regarding the algorithm by Jaggi et al. to produce bounds on the success probability. We shall assume that no encoding coefficients are chosen *a priori*. Although it is actually possible to give conditions under which *a priori* chosen encoding coefficients allow for flow bounds we will not describe them here as they are rather technical. The only result of this nature is given at the end of the section in an example where the flow bound technique is applied to a network with *a priori* chosen encoding coefficients.

Given a solvable network coding problem consider an arbitrary flow system of size h. Having chosen all encoding coefficients by random we apply a modified version of Algorithm 1.1 corresponding to Fig. 5 in [14]. Recall, that Algorithm 1.1 works on the extended graph defined by adding a new vertex s', by adding h new edges and by moving all sources to s'. The algorithm initialize $C_1, \ldots, C_{|R|}$, $B_1, \ldots, B_{|R|}$ as usual. The encoding functions at s' leading to the initial value of $B_1, \ldots, B_{|R|}$ have got nothing to do with the original network and therefore the initialization part is not in conflict with the assumption that all encoding coefficients of the original network are chosen by random. The only job for the algorithm now is to check if for every update of the sets $B_1, \ldots, B_{|R|}$ the randomly chosen encoding coefficients work. That is, the algorithm checks if all sets $B_1, \ldots, B_{|R|}$ span $\mathbf{F}_q^h$. If a set B_i is detected that does not span the whole of $\mathbf{F}_q^h$ then the algorithm returns "failure" and aborts. If this does not happen the algorithm returns "success" after having visited all edges in the flow system.

There is a slight difference in the analysis of the algorithm as now an edge j in the flow system will also receive information from outside the flow system. This is indeed not a problem. Actually, we already formulated Lemma 1.1 so that it can also handle this situation. Obviously, we can lower bound the success probability by the probability that the algorithm returns success. Denote for every edge j in the flow system $\mathcal{F} = \{F_1, \ldots, F_{|R|}\}$ by $R_{\mathcal{F}}(j)$ the number of receivers that in the flow system uses edge j. Assume the algorithm has run successfully up to the step where edge j is visited, i.e. assume that for every receiver at this stage of the algorithm we have a full basis of global coding vectors. The ratio of choices of coefficients $f_{i,j}$, $a_{i,j}$ (notation with respect to the original network) that would now work is

$$\frac{q - R_{\mathcal{F}}(j)}{q} + \frac{R_{\mathcal{F}}(j) - 1}{q^{|\text{in}'(j)|}} \quad (1.20)$$

where $\text{in}'(j)$ is the in-edges to j in the extended network. That is, $|\text{in}'(j)|$ corresponds in the original network to the number of in-edges to j plus the number of messages X_i generated in u where $j = (u, v)$. The last part of (1.20) ensures that we do not count the case of all encoding coefficients being 0 more than once. In conclusion we have the following theorem:

Theorem 1.4.

$$P_{\text{succ}} \geq P_{\text{FB2}} := \prod_{j \in \mathcal{F}} \left(\frac{q - R_{\mathcal{F}}(j)}{q} + \frac{R_{\mathcal{F}}(j) - 1}{q^{|\text{in}'(j)|}} \right) \quad (1.21)$$

$$\geq P_{\text{FB1}} := \prod_{j \in \mathcal{F}} \frac{q - R_{\mathcal{F}}(j)}{q}. \quad (1.22)$$

It is possible to improve on the bounds P_{FB1} and P_{FB2} by analyzing the flow system more thoroughly. For two receivers r_1 and r_2 which in the flow system use edge j, consider the set of global coding vectors B_{r_1}, B_{r_2} as they are before the update. Let i_1, i_2 be the corresponding (not necessarily different) edges that are going to be replaced with j in C_{r_1} and C_{r_2}. If

$$C_{r_1} \backslash \{i_1\} = C_{r_2} \backslash \{i_2\}$$

holds then indeed

$$B_{r_1} \backslash \{c_g(i_1)\} = B_{r_2} \backslash \{c_g(i_2)\}.$$

Hence, we can replace $R_{\mathcal{F}}(j)$ in (1.21) and (1.22) with

$$r_{\mathcal{F}}(j) := |\{C_v \backslash \{\text{in}'(j) \cap F_v\} \mid \text{where } v \text{ is such that } j \in F_v\}|.$$

Note that actually $r_\mathcal{F}(j)$ depends not only on the flow system and the edge but also on the ancestral ordering with respect to which we visit the edges. We shall comment on this in an example below. Replacing $R_\mathcal{F}(j)$ with $r_\mathcal{F}(j)$ in (1.21) and (1.22), respectively, we get expressions that we call P_{FB4} and P_{FB3}, respectively.

Next we compare the flow bounds to the bounds from the previous section. Consider an edge e in a flow system $\mathcal{F}$. Assume that there are m' indeterminates of the form $f_{e',e}$ or $a_{e',e}$ in the flow system. We label the indeterminates $W_1, \ldots, W_{m'}$ and denote for $t = 1, \ldots, m'$ by k_t the number of times the variable W_t appears in $\mathcal{F}$. By definition

$$\sum_{t=1,\ldots,m'} k_t = R_\mathcal{F}(e).$$

Applying Lemma 1.2 with $\eta = 1$ we get

$$\prod_{t=1,\ldots,m'} \left(\frac{q-k_t}{q}\right) \geq \frac{q - R_\mathcal{F}(e)}{q}. \tag{1.23}$$

Assuming the monomial corresponding to $\mathcal{F}$ can be chosen as leading monomial of the transfer polynomial we see from (1.23) and (1.11) that the Gröbner basis based bound P_{FP2} is at least as sharp as P_{FB1}.

Now we want to compare the first flow bound P_{FB1} with P_{Ho2}. To this end consider variables $Y_e, e \in \mathcal{F}$. For every fixed e we identify $W_1^{k_1} \cdots W_{m'}^{k_{m'}}$ with $Y_e^{R_\mathcal{F}(e)}$. The monomial corresponding to the flow system in this way defines a monomial in the variables Y_e, $e \in \mathcal{F}$. If we consider a polynomial with leading monomial $\prod_{e \in \mathcal{F}} Y_e^{R_\mathcal{F}(e)}$ then from Corollary 1.1 we get a number which is equal to P_{FB1}. The number η' defined on page 64 satisfies $\sum_{e \in \mathcal{F}} R_\mathcal{F}(e) \leq d\eta'$ and therefore by Lemma 1.2 $P_{\text{FB1}} \geq P_{\text{Ho2}}$ holds.

Example 1.5. In this example we consider a network with *a priori* chosen encoding coefficients. The topology is such that the flow bounds can still be applied directly. Consider the network in Fig. 1.9. We assume that two messages X_1 and X_2 are generated at s. Receivers are r_1 and r_2. Let $f_{1,3}$ and $f_{2,3}$ be the only randomly chosen coding coefficients. All other encoding coefficients $f_{i,j}$ are chosen *a priori* to 1 and the $a_{i,j}$'s are chosen such that the global coding vectors on $\text{out}(s)$ are pairwise linearly independent. We see that the leading monomial of the *a priori* transfer polynomial is either $f_{1,3}^2$ or $f_{2,3}^2$ depending on the monomial ordering. This gives the bound $P_{\text{FP2}} = \frac{q-2}{q}$. For comparison no matter which flow system we consider we get $P_{\text{FB1}} = \frac{q-2}{q}$, $P_{\text{FB2}} = \frac{q-2}{q} + \frac{1}{q^2}$ and $P_{\text{FB3}} = \frac{q-1}{q}$. We see that already P_{FB2} exceeds P_{FP2}. The true success probability is $\frac{q-1}{q}$.

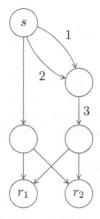

Fig. 1.9. Network where flow bounds are better than Gröbner basis based bounds.

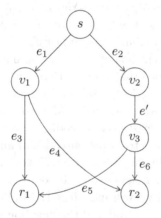

Fig. 1.10. A network where the ancestral ordering has implication for P_{FB3} and P_{FB4}

The next example illustrates that choosing different ancestral orderings may result in different values of P_{FB3} and P_{FB4}.

Example 1.6. Consider the network in Fig. 1.10. Here two messages X_1 and X_2 are generated at s. The receivers are r_1 and r_2. Of the ancestral orderings $e_1 \prec e_2 \prec e' \prec e_3 \prec e_4 \prec e_5 \prec e_6$ and $e_1 \prec e_2 \prec e_3 \prec e_4 \prec e' \prec e_5 \prec e_6$ the first produces the result

$$P_{\text{FB3}} = P_{\text{FB4}} = (1 - 1/q)^7$$

and the latter

$$P_{\text{FB3}} = (1-1/q)^6(1-2/q), \quad P_{\text{FB4}} = (1-1/q)^7.$$

1.3.2.2. *The bounds by Balli, Yan, and Zhang*

As explained in Section 1.3.1 the seminal paper on random network coding [13] used algebraic methods to establish bounds on P_{succ}. In the above we have shown that thinking instead in terms of Jaggi *et al.*'s algorithm we get the flow bounds from pure combinatorial arguments. The expression P_{Ho2} (corresponding to Eq. (1.14)) and the expression P_{FB1} (corresponding to Eq. (1.22)) has got very much the same flavour. This establishes a connection between the algebraic and the combinatorial approach. The next bounds that we shall be concerned with are the Balli-Yan-Zhang bounds. The idea behind these bounds is a clever modification of Jaggi *et al.*'s algorithm, and it is therefore clear that there is a strong connection between the recent flow bounds and the Balli-Yan-Zhang bounds. Altogether this chapter establishes via the flow bounds a connection between the algebraic approach and the pure combinatorial approach.

Balli *et al.* both present bounds that take into account the redundancy of the network and bounds that do not. Here, we shall concentrate on the latter. We will not as in [3] restrict ourselves to consider only the case of single source and receivers not being allowed to forward messages, but consider the general multicast network coding problem. As was the case for the flow bounds the bounds to be presented below allow for certain patterns of *a priori* chosen encoding functions. We will, however, for simplicity again assume that all encoding functions are chosen by random.

As already indicated we will introduce the Balli-Yan-Zhang bounds in much the same way as we introduced the flow bounds. We start by modifying the algorithm by Jaggi *et al.* to a new algorithm which still solves the problem of finding feasible encoding coefficients. Consider the extended network as in the algorithm by Jaggi *et al.* To the original network $G = (V, E)$ we added the vertex s', we added h edges from s' to S, and we changed the set-up such that all h messages are now generated in s'. After some initialization the algorithm then did an updating procedure for every edge in $\mathcal{F}$ visited in an ancestral ordering. Recall that the existence of ancestral orderings on E is guaranteed by the assumption that the network is cycle free. The first observation that we make is that this assumption likewise guarantees the existence of ancestral orderings on the set of vertices V. Given a flow system $\mathcal{F} = \{F_1, \ldots, F_{|R|}\}$ let $V(F_i)$ and $V(\mathcal{F})$, respectively, be the vertices that appear as start/end points of edges in F_i and $\mathcal{F}$, respectively. We shall now design a Jaggi et al. like algorithm that in the initialization part do exactly what the original algorithm does. Rather than running the updating procedure for every edge $j \in \mathcal{F}$ we shall run a similar updating procedure for every $w \in V(\mathcal{F})$. The vertices will be visited according to the ancestral ordering. The sets C_l and B_l play the same role as before. However, rather than replacing in C_l the edge i with the unique edge $j \in \text{out}(i) \cap \mathcal{F}$ we shall for the w under consideration replace all edges $\text{in}(w) \cap \mathcal{F}$ in C_l with the edges $\text{out}(w) \cap \mathcal{F}$ unless of course $w = r_l$. Again encoding coefficients are to be chosen in such a way that the resulting set of global coding vectors $B_l = c_g(C_l)$ span the whole of $\mathbf{F}_q^h$. The alternative version of the algorithm by Jaggi et al. is presented as Algorithm 1.2 below.

The correctness of the algorithm is obvious of the following reason. Given an ancestral ordering on the vertices in V then we can define a related ancestral ordering on the edge set E by choosing first the edges that start in the first vertex of V. Then we choose the edges that start in the second vertex of V and so on. Given such ancestral orderings the condition for successful choice of encoding coefficients in Algorithm 1.2 is less restrictive than the corresponding condition in the algorithm by Jaggi et al. Therefore fields of size $|R|$ or more are large enough.

We next consider the random network coding problem. Hence, assume we are given a solvable network coding problem and that all encoding coefficients have been chosen by random. We shall apply a modified version of Algorithm 1.2 in an attempt to check the feasibility of the encoding coefficients. As long as every set B_l span the whole of $\mathbf{F}_q^h$ the algorithm continues. In case of an early abort it returns "failure". Otherwise it returns "success" when having visited all points in $V(\mathcal{F})$. The initialization part of the new algorithm is exactly the same as for the original one. The only change is that in the updating part we rather than choosing encoding coefficients related to the vertex $w \in V(\mathcal{F})$ now only check whether the already chosen coefficients related to w work. Of course the probability of success for random coding is at least equal to the probability that the modified algorithm runs successfully. We shall need the following generalization of Lemma 1.1 to estimate the latter probability.

Algorithm 1.2.

Input: A solvable multicast (acyclic) network coding problem with $G = (V, E)$. A field $\mathbf{F}_q$ with $q \geq |R|$. An ancestral ordering on V.
Output: Encoding coefficients $a_{i,j}, f_{i,j}$.
Initialization
Let $V' := V \cup \{s'\}$ and $E' := E \cup \{e_1, \ldots, e_h\}$ where the new edges are as described in the text;
Find a flow system $\mathcal{F} = \{F_1, \ldots, F_{|R|}\}$ in $G' := (V', E')$ of size h from s' to the receivers;
for $j \in E$ and $i = 1, \ldots, h$
   if $j \in out(e_i)$ and $(e_i, j) \notin \mathcal{F}$ **then**
      $a_{i,j} = 0$;
   end
endfor
for $(i, j) \in E \times E$ where $i \in in(j)$ and $(i, j) \notin \mathcal{F}$
   $f_{i,j} = 0$;
endfor
for $l = 1, \ldots, |R|$
   $C_l := (e_1, \ldots, e_h)$;
   $B_l := ((1, 0, \ldots, 0), (0, 1, 0, \ldots, 0), \ldots, (0, \ldots, 0, 1))$;
endfor
Update
for $w \in V(\mathcal{F}) \setminus \{s'\}$ by the ancestral ordering
   for $l = 1, \ldots, |R|$
      if $w \neq r_l$ **then**
         $I_l := in(w) \cap F_l$;
         $J_l := out(w) \cap F_l$;
      else
         $I_l := J_l := \emptyset$;
      end
      $C_l := (C_l \setminus I_l) \cup J_l$;
   endfor
   Choose encoding functions (notation with respect to extended network) $\{f_{i,j} \mid (i, j) \in I_l \times J_l \text{ for some } l\}$ by random until $B_l := c_g(C_l)$ spans all of $\mathbf{F}_q^h$ for $l = 1, \ldots, h$;
   for $f_{i,j}$ that was updated
      if $i = e_k$ for some $k \in \{1, \ldots, h\}$ **then**
         rename $f_{i,j}$ to $a_{k,j}$;
      end
   endfor
endfor
return all values $a_{i,j}, f_{i,j}$ for which values have been allocated

Lemma 1.3. *Consider integers k, h, μ with $1 \leq k < h$ and $k + \mu < h$. Let $\{\vec{b}_1, \ldots, \vec{b}_h\}$ be a basis for $\mathbf{F}_q^h$ and let $\vec{b}'_{k+1}, \ldots, \vec{b}'_{k+\mu}$ be such that*

$$V = \mathrm{Span}_{\mathbf{F}_q}\{\vec{b}_1, \ldots, \vec{b}_k, \vec{b}'_{k+1}, \ldots, \vec{b}'_{k+\mu}\}$$

is of dimension $h + \mu$. Given $\vec{c} \in \mathbf{F}_q^h$ the number of choices of $(a_{k+1}, \ldots, a_h)$, $a_i \in \mathbf{F}_q$ such that

$$\vec{c} + a_{k+1}\vec{b}_{k+1} + \cdots + a_h\vec{b}_h \in V$$

equals q^μ.

Assume the algorithm has run successfully up to the step where vertex w is visited. That is, assume we for every receiver r_l at this stage of the algorithm have that B_l spans the whole of $\mathbf{F}_q^h$. Consider for a particular receiver r_l the sets I_l and J_l. Recall from the algorithm that if $w \neq r_l$ then $I_l := \mathrm{in}(w) \cap F_l$ where the set $\mathrm{in}(w)$ is the edges (u, w) with $u \in V'$. Similarly J_l is the set $\mathrm{out}(w) \cap F_l$. If $w = r_l$ there is no updating to be done regarding r_l and we therefore have $I_l = J_l = \emptyset$. Assuming $J_l \neq \emptyset$ the ratio of encoding functions related to w that makes the new set B_l span the whole of $\mathbf{F}_q^h$ by Lemma 1.3 is

$$\prod_{i=1}^{|J_l|} \frac{q^{|J_l|} - q^{i-1}}{q^{|J_l|}} = \prod_{i=1}^{|J_l|} \left(1 - \frac{1}{q^{|J_l|-i+1}}\right) \geq 1 - \frac{1}{q-1}.$$

Assuming the worst case, the probability that the algorithm returns "success" when visiting w is at least

$$1 - \sum_{\substack{l \in \{1, \ldots, h\} \\ J_l \neq \emptyset}} \left(1 - \prod_{i=1}^{|J_l|}\left(1 - \frac{1}{q^{|J_l|-i+1}}\right)\right) \qquad (1.24)$$

$$\geq 1 - \frac{\rho(w)}{q-1} \qquad (1.25)$$

$$\geq 1 - \frac{|R|}{q-1}. \qquad (1.26)$$

Here, $\rho(w)$ is the number of receivers r_l such that $J_l \neq \emptyset$. Taking the product over all points in

$$\mathcal{I} = \{w \in V \mid \mathrm{out}(w) \cap \mathcal{F} \neq \emptyset\}$$

we get from (1.24), (1.25) and (1.26) bounds on the overall success probability. We have the following theorem

Theorem 1.5.

$$P_{\text{succ}} \geq P_{\text{Balli2}} := \prod_{w \in \mathcal{I}} \left(1 - \frac{\rho(w)}{q-1}\right)$$

$$\geq P_{\text{Balli1}} := \left(1 - \frac{|R|}{q-1}\right)^{|\mathcal{I}|}$$

$$\geq P_{\text{Balli0}} := \left(1 - \frac{|R|}{q-1}\right)^{|V|}.$$

Observe that the expression P_{Balli0} relies on the number of receivers and the number of vertices whereas the expression P_{Ho1} in the case of no *a priori* chosen encoding coefficients relies on the number of receivers and the number of edges.

Example 1.7. Let the graph in Fig. 1.11 be the flow system of a network coding problem where s is the sender, where two messages are being generated at s, and r_1 and r_2 are the two receivers. Assume that $f_{i,j}$ where $i = 1, 2$ and $j = 3, 4, 5, 6$ are chosen independently and uniformly from $\mathbf{F}_q$ and that $a_{1,1} = a_{2,2} = 1$ and $a_{1,2} = a_{2,1} = 0$. The Balli-Yan-Zhang bounds also apply to this case. For $q = 3$ $P_{\text{FB1}} = P_{\text{FB2}} = P_{\text{FP2}} = P_{\text{FP1}} \simeq 0.1975$, $P_{\text{Ho2}} \simeq 0.1111$, $P_{\text{Balli2}} = P_{\text{Balli1}} = 0$. From (1.24) we can produce the bound 0.1850. For $q = 4$ we have $P_{\text{Balli2}} = P_{\text{Balli1}} \simeq 0.3333$, $P_{\text{FB1}} = P_{\text{FB2}} = P_{\text{FP2}} = P_{\text{FP1}} \simeq 0.3164$, $P_{\text{Ho2}} = 0.25$. From (1.24) we can produce the bound 0.4062. For q larger than 4 the Balli-Yan-Zhang bounds are superior.

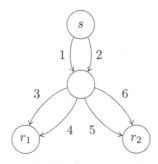

Fig. 1.11. The network from Example 1.7.

In the above example we see that indeed the Balli-Yan-Zhang bounds can be very good. However, we cannot hope to prove that they will always be the best as actually we have examples where they are not. Although the conditions in Algorithm 1.2 are less restrictive than the conditions in Algorithm 1.1 indeed there is a price to be paid when we consider as a worst case that the incident B_{l_1} not spanning the whole of $\mathbf{F}_q^h$ and the incident B_{l_2} not spanning the whole of $\mathbf{F}_q^h$ never happens simultaneously for $l_1 \neq l_2$.

1.4. Bibliographic notes

Kötter and Médard were the first to apply algebraic methods in network coding. This was done in [16] where they showed how to translate the general linear network coding problem (that is, not necessarily multicast) into a problem of finding a common zero of a set of multivariate polynomials. The other way around Dougherty, Freiling and Zeger showed in [6] that given any set S of polynomials with integer coefficients then there exists a linear network coding problem that is solvable if and only if S has got a common zero in the field under consideration.

Based on the algebraic point of view Ho, Médard, Kötter, Karger, Effros, Shi and Leon in [13] coined the concept of random network coding. In our presentation the footprint bound plays a crucial role. It should be mentioned that Ho et al. did not use this bound. They used a weaker bound namely an adaption of the Schwartz-Zippel bound. The idea of using the footprint bound and thereby revealing more information is from [11].

The pure combinatorial approach towards random network coding taken by Balli, Yan and Zhang is very suitable and thereby also fruitful.

We should mention that the papers [13, 16] show how to deal with networks with cycles and delays. Furthermore, the paper [13] contains some results in the direction of taking redundancy into account. Here, by redundancy we mean the difference between the min cut number and the number of generated messages. The papers [2, 3] treat redundancy in detail. This chapter is concerned with linear network coding or more precisely scalar linear network coding. It should be mentioned that vector linear network coding is also considered in the literature [13]. In Section 1.1 we introduced the concept of network coding in very general terms and then concentrated on (scalar) linear network coding in the remaining part of the chapter. The general set-up is in place as demonstrated by the fact that there exists network coding problems which has only non-linear solutions [7].

Error-correction in networks was introduced by Cai and Yeung in [4, 22]. In their model to design an error-correcting code the entire network needs to be known. Random network coding enables a couple of different point of views on error-correction. In robust network coding one considers patterns of errors occurring in the network. The idea is to choose encoding coefficients in such a way that they work for all error patterns in a given family [15, 16]. Estimations on the success probability of robust network coding are presented in [2, 14]. Unfortunately, to guarantee positive success probability one needs quite large fields. This is not the case for the channel oblivious model presented by Kötter and Kschischang in [17]. Here the network is viewed as a black box where errors and erasure occur. Kötter and Kschischang coined the name operator channel to describe this phenomenon and modified Gabidulin's construction of rank-metric codes [9] to work in this setting. It seems as if the paper [17] nearly solves all problems regarding error/erasure correction for multicast on the operator channel. It has been followed up by a number of papers that each contributes in making the theory even more complete. We only mention a few here [18–20].

Finally, we would like to mention that there are excellent tutorials on network coding such as [8, 12, 21].

References

[1] R. Ahlswede, N. Cai, S.-Y. R. Li, and R. W.-H. Yeung, "Network Information Flow," *IEEE Trans. on Inf. Theory*, vol. **46**, 2000, pp. 1204-1216.
[2] H. Balli, X. Yan, and Z. Zhang, "On Randomized Linear Network Codes and Their Error Correction Capabilities," *IEEE Trans. Inform. Theory*, vol. 55, 2009, pp. 3148-3160.
[3] H. Balli and Z. Zhang, "On The Limiting Behavior Of Random Linear Network Codes," in *Proceedings of 2009 Workshop on Network Coding, Theory and Applications, IEEE*, Lausanne, June 15-15, 2009, 5 pages.
[4] N. Cai and R. Yeung, "Network Error Correction, Part II: Lower Bounds," *Communications in Information and Systems*, vol. 6, 2006, pp. 37-54.
[5] G. Chartrand and O. Oellermann, *Applied and Algorithmic Graph Theory*, McGraw-Hill, New York, 1992.
[6] R. Dougherty, C. Freiling, and K. Zeger, "Linear Network Codes and Systems of Polynomial Equations," *IEEE Trans. Inform. Theory*, vol. 54, 2008, pp. 2303-2316.
[7] R. Dougherty, C. Freiling, and K. Zeger, "Insufficiency of Linear Coding in Network Information Flow," *IEEE Trans. Inform. Theory*, vol. 51, 2005, pp. 2745-2759.
[8] C. Fragouli and E. Soljanin, "Network Coding Fundamentals," in *Foundations and Trends in Networking*, vol. 2(1), now Publishers Inc., Hanover, 2007
[9] E. M. Gabidulin, "Theory of codes with maximal rank distance," *Problems of Information Transmission*, vol. 21, 1985, pp. 1-12.

[10] O. Geil and T. Høholdt, "Footprints or Generalized Bezout's Theorem," *IEEE Trans. Inform. Theory*, vol. 46, Mar. 2000, pp. 635-641.
[11] O. Geil, R. Matsumoto, and C. Thomsen, "On Field Size and Success Probability in Network Coding," *Lecture Notes in Computer Science*, (J. von zur Gathen, J. L. Imaa, C. K. Koc Eds.), LNCS 5130, Proceedings of WAIFI 2008, pp. 157-173.
[12] T. Ho and D. S. Lun, *Network Coding - An Introduction*, Cambridge University Press, 2008, 170 pages.
[13] T. Ho, M. Médard, R. Kötter, D. R. Karger, M. Effros, J. Shi, and B. Leong, "A random linear network coding approach to multicast," *IEEE Trans. Inform. Theory*, vol. 52, 2006, pp. 4413-4430.
[14] S. Jaggi, P. Sanders, P. A. Chou, M. Effros, S. Egner, K. Jain, and L. M. G. M. Tolhuizen, "Polynomial time algorithms for multicast network code construction," *IEEE Trans. Inform. Theory*, vol. 51, 2005, pp. 1973-1982.
[15] R. Kötter, M. Médard, "Beyond routing: An algebraic approach to network coding," in *Proc. 21st Annu. Joint Conf. IEEE Computer and Communations Societies (INFOCOMM)*, vol. 1, New York, June 2002, pp. 122-130.
[16] R. Kötter, M. Médard, "An algebraic approach to network coding," *IEEE/ACM Trans. Networking*, vol. 11, 2003, pp. 782-795.
[17] R. Kötter and F. R. Kschischang, "Coding for errors and erasures in random network coding," *IEEE Trans. Inform. Theory*, vol. 54, 2008, pp. 3579-3591.
[18] H. Mahdavifar and A. Vardy, "Algebraic list-decoding on the operator channel," in Proc. of *Information Theory Proceedings (ISIT), 2010 IEEE International Symposium on*, Austin, Texas, June 13-18, 2010, pp. 1193-1197.
[19] D. Silva, F. R. Kschischang, and R. Kötter, "A Rank-Metric Approach to Error Control in Random Network Coding," *IEEE Trans. on Inf. Theory*, vol. 54, 2008, pp. 3951-3967.
[20] D. Silva and F. R. Kschischang, "On Metrics for Error Correction in Network Coding," *IEEE Trans. on Inf. Theory*, vol. 55, 2009, pp. 5479-5490.
[21] R. W. Yeung, *Information Theory and Network Coding*, Springer, 2008, 604 pages.
[22] R. W. Yeung and N. Cai, "Network Error Correction, Part I: Basic Concepts and Upper Bounds," *Communications in Information and Systems*, vol. 6, 2006, pp. 19-36.
[23] R. W.-H. Yeung, and Z. Zhen, "Distributed Source Coding for Satellite Communications," *IEEE Trans. on Inf. Theory*, vol. 45, 1999, pp. 1111-1120.

Chapter 2

Steganography from a Coding Theory Point of View

Carlos Munuera

Department of Applied Mathematics, University of Valladolid
Avda Salamanca SN, 47014 Valladolid, Castilla, Spain

Steganography is the science of invisible communications. It is a very old method that, thanks to recent progress, has begun to play an important role as an alternative and as a complement of Cryptography. This Chapter presents an introduction to the objectives, the methods and some open problems of Steganography, conducted from a Coding Theory point of view.

Contents

2.1 Introduction 84
    2.1.1 What is steganography? 84
    2.1.2 Digital steganography 85
    2.1.3 Steganography, cryptography and watermarking 86
    2.1.4 About this chapter 87
2.2 Steganographic systems 87
    2.2.1 The cover 88
    2.2.2 Steganographic schemes 88
    2.2.3 Selection rules 89
    2.2.4 Parameters 91
    2.2.5 Proper stegoschemes 92
2.3 Error-Correcting codes 93
    2.3.1 Correcting errors 93
    2.3.2 Linear codes over fields 95
    2.3.3 An example: binary Hamming codes 97
    2.3.4 Generalized Hamming weights for linear codes 98
2.4 Linking the problems 99
    2.4.1 Stegoschemes and error-correcting codes 99
    2.4.2 Group codes and stegoschemes 101
    2.4.3 Linear stegoschemes over rings $\mathbb{Z}_q$ 103
    2.4.4 Linear stegoschemes over fields 105
2.5 Bounds 106
    2.5.1 The domain of stegoschemes 106

    2.5.2 Balls and entropy 107
    2.5.3 A Hamming-like bound 109
    2.5.4 Asymptotic bounds 110
    2.5.5 Perfect stegoschemes 114
    2.5.6 Another new problem for coding theory 116
2.6 Nonshared selection rules 117
    2.6.1 Wet paper codes 118
    2.6.2 Solvability and the weight hierarchy of codes 119
    2.6.3 The rank of random matrices 121
2.7 The ZZW embedding construction 123
    2.7.1 Description of the method 123
    2.7.2 Asymptotic behavior 124
2.8 Bibliographical notes and further reading 125
References 126

2.1. Introduction

The field of Information Security has grown and evolved significantly in recent times. Consequently the techniques used for data protection have taken a great interest as a subject of scientific research. Within these techniques Cryptography has the undeniable leading role. However new ideas have emerged with great force in the last years. Steganography is a very old method that, thanks to recent progress, has begun to play an important role as an alternative –and more generally as a complement– of Cryptography. The advantage of Steganography over Cryptography alone is that messages do not attract attention of third parties to themselves. Therefore, whereas Cryptography only protects the information, Steganography protects both the messages and the communicating parties. For instance, Steganography became widely known in 2001, when FBI and CIA reported that Ben Laden used it to covertly distribute information to his supporters.

The purpose of this Chapter is to provide an introduction to the objectives, the methods and the open problems of Steganography.

2.1.1. What is steganography?

Steganography is the science of communicating secret messages in such a way that no one, apart from the sender and receiver, can detect the existence of a message. In other words, Steganography is the science of invisible communications. In the framework of information security it is used when even the fact of communicating has to be kept secret.

The word 'steganography' is derived from the Greek words *steganos* and *graphein*, and means *covered writing*. This name was proposed in 1500 by Johannes Trithemius in his treatise *Steganographia*, Ref. 39, a book on magic and secrecy methods. Its name describes exactly how Steganography works: the secret message we want to protect is embedded into an apparently innocuous object, the *cover*. Covers may be images, texts, private letters, computer files, etc. The typical example of Steganography is invisible ink. Here the cover is a handwritten document and the secret message is written in the margins or between the lines. However the history of Steganography is much older. Ancient Greeks used the trick of shaving the head of slaves and tattooed a message on it. After the hair had grown the message was hidden. Another example of Steganography used for many centuries is the *acrostic*. This is a form of writing in such a way that the first letter or word of each line, paragraph or other recurring feature in the text spells out the message. Often it has been used as a challenge to the reader, but it can also be a way of communicating secret messages.

2.1.2. Digital steganography

Today's typical cover is a computer file: a document, image, a program or a protocol. Media files are ideal for steganographic purposes because of their large size and redundancy. For this reason modern Steganography is sometimes referred as *Digital Steganography*. Let us show two examples.

Example 2.1 (Mimic functions). A *mimic function* f changes a file F to acquire the statistical properties of another file F'. This means that each string s that occurs in F' with probability p', also occurs in $f(F)$ with approximately equal probability $p \approx p'$. Mimic functions play an important role in Steganography, in order to avoid attacks of statistical steganalysis. See Refs. 43 and 44.

Grammar-based mimicking uses grammar models to transform a text (the secret message) imitating a default grammatical structure. A good example is the software *SpamMimic* (on line available at http://www.spammimic.com/) which transforms a text to a spam-like e-mail message that can be transmitted without arousing suspicion.

Example 2.2 (Hiding information in images). It is fairly simple to hide a secret message in a graphic file without obviously altering the visible appearance of the image it contains. The sender uses an innocuous digital image which is stored in the computer as a sequence of bits corresponding to its pixels (see Paragraph 2.2.3 below). Then the sender adjusts the color of some pixels in such a way to correspond to a letter in the alphabet. Usually these changes of color are made by replacing the least significant bits of selected pixels by message bits. For this reason this technique is referred to as LSB Steganography. Due to the limited capacity of the human eye to distinguish colors (or gray levels) the changes made are visually imperceptible. Furthermore, since the original image is deleted after changing it, there is no possibility of comparison. The receivers collect these pixels and recover the information. Software programs for LSB Steganography can be easily (and freely) obtained from the internet as they are the most popular steganographic tools today. These techniques will be studied in greater detail throughout the Chapter. See also Refs. 42 and 48.

For simplicity, from now on we shall use the example of an image file to explain the methods of Digital Steganography.

2.1.3. *Steganography, cryptography and watermarking*

For a better understanding it is worth to compare Steganography with two related disciplines: Cryptography and Watermarking. Although all three share a common purpose –in a broad sense, the information security– there are important differences between them.

As discussed before, Steganography is a form of information hiding so that even the fact of communicating is secret. By contrast, the purpose of Cryptography is to hide the contents of the transmitted messages (but not the messages themselves) so that third parties do not have access to their contents. In particular such third parties aware of the existence of the messages and also know the sender and the receiver, so that they can evaluate the traffic of information. For this reason, more and more both –Cryptography and Steganography– are used together to ensure a comprehensive protection of information and participants.

The difference between Steganography and Watermarking is more subtle. Both deal with information hiding although both have different purposes. A watermark is a hidden message that stores information about the cover in which it is contained and which is the main value to be protected: ownership, copyrights or others. Besides the presence of a watermark into a cover work is, in general, not secret. Thus the main requirement of Watermarking is imperceptibility. On the contrary, for Steganography purposes the cover has no intrinsic value, it is just a container to hide the message. Here the requirement is undetectability: third parties should not be able to determine whether the cover contains, or not, a hidden information.

2.1.4. About this chapter

This Chapter presents an introduction to Steganography conducted from a Coding Theory point of view. As a result of this orientation the focus is on the so-called steganographic schemes. Other major parts of the theory are referenced briefly or omitted at all. Also as a result of this orientation we assume a certain familiarity of the reader with Error Correcting Codes. However in order to make the Chapter self-contained we describe some basic facts about codes in Section 2.3.

Steganography is now a rapidly developing discipline, to the point that almost 90 % of publications on this field have been written in the last five years. In order to guide the reader, along the text we include numerous bibliographic citations in which s/he can extend the information provided herein.

2.2. Steganographic systems

In order to hide a message into a cover, we need two main ingredients, namely

- the cover
- the embedding and recovering methods.

The security of a steganographic system depends on these two factors: (i) the type of the cover and the selection of the places within it might be modified; and (ii) the embedding process and the distortion caused by the embedding changes. Consequently, the design of steganographic systems has (at least) two facets corresponding to these two ingredients. Firstly, the choice of accurate covers and the search for strategies to modify them in an imperceptible way; studying these strategies relies on a variety of methods, including psycho visual and statistical criteria, most of which are quite far from our interests in this report. Secondly, the design of efficient algorithms for embedding and recovering the information. These algorithms form the so-called *steganographic scheme* (or simply *stegoscheme*).

The study of efficient embedding and extracting functions is the main purpose of this Chapter. Before entering this topic, we dedicate a few words to the cover.

2.2.1. *The cover*

In principle the cover is freely chosen by the sender at his/her convenience. The only limitations are given by the type of communication (handwritten, digital, etc.) and the context (to avoid suspicion from third parties). Anyway, the cover may be a preexisting document or it can be created *ex-profeso* from the message. In the first case the embedding process (slightly) changes the cover and we speak about *cover modification*. In the second one, embedding is achieved through the cover making and we speak about *cover synthesis*. Note that these two strategies have already been shown in Examples 2.2 and 2.1 respectively.

Cover modification is the most common method of Steganography. The rest of this Chapter is devoted to it. Cover synthesis is less common but also is used in practice. For instance, in the internet one can find programs that make music, fractal images, spam, etc. from a text message.

2.2.2. *Steganographic schemes*

A steganographic scheme is formed by the embedding and recovering functions. These functions may depend on a key used to increase the system security. Formally we define a steganographic scheme as a five-tuple $\mathcal{S} = (\mathcal{C}, \mathcal{M}, \mathcal{K}, \text{emb}, \text{rec})$, where

(1) $\mathcal{C}$ is a set of possible covers;
(2) $\mathcal{M}$ is a set of possible messages;
(3) $\mathcal{K}$ is a set of possible keys;
(4) $\text{emb} : \mathcal{C} \times \mathcal{M} \times \mathcal{K} \to \mathcal{C}$ is an embedding function;
(5) $\text{rec} : \mathcal{C} \times \mathcal{K} \to \mathcal{M}$ is a recovering function

and it holds that $\text{rec}(\text{emb}(\mathbf{c}, \mathbf{m}, \mathbf{k}), \mathbf{k}) = \mathbf{m}$ for all $\mathbf{m} \in \mathcal{M}, \mathbf{c} \in \mathcal{C}$ and $\mathbf{k} \in \mathcal{K}$. The original cover $\mathbf{c}$ is called the *plain cover* and $\text{emb}(\mathbf{c}, \mathbf{m}, \mathbf{k})$ is the *stegocover*.

In Cryptography a distinction between the set of all possible plain texts and the set of possible cipher texts is made. In Steganography the main requirement is that the stegocover looks like an innocuous message, hence a similar distinction does not make sense.

In practice, there are many (even too many!) possible choices for $\mathcal{C}, \mathcal{M}$ and $\mathcal{K}$. If we restrict to the most common form of Steganography today, that is Digital Steganography, then all involved objects –namely the message, the cover and the key– are digital objects. This means that they are finite sequences of symbols taken from a finite alphabet. From now on, we shall assume that both the cover **c** and **m** the secret we want to hide are sequences of symbols from a finite alphabet $\mathcal{A}$ with q elements, $\mathbf{c} = (c_1, \ldots, c_n)$ and $\mathbf{m} = (m_1, \ldots, m_k)$. They can be seen as 'vectors' in $\mathcal{A}^n$ and $\mathcal{A}^k$ respectively. In most cases the alphabet $\mathcal{A}$ will be the set (field) with two elements $\mathcal{A} = \mathbb{F}_2$, but other alphabets are also possible, as finite rings $\mathbb{Z}_q = \mathbb{Z}/q\mathbb{Z}$, finite fields $\mathbb{F}_q$ (if q is a prime power), etc. This restriction leads to the following simplified definition. A *digital steganographic scheme* $\mathcal{S}$ of *type* $[n,k]$ over the alphabet $\mathcal{A}$ is a pair of functions emb : $\mathcal{A}^n \times \mathcal{A}^k \to \mathcal{A}^n$ and rec : $\mathcal{A}^n \to \mathcal{A}^k$ such that $\mathrm{rec}(\mathrm{emb}(\mathbf{c}, \mathbf{m})) = \mathbf{m}$ for all $\mathbf{c} \in \mathcal{A}^n$ and $\mathbf{m} \in \mathcal{A}^k$. Note that for simplicity we have not included keys as a part of our scheme. The set of possible messages is $\mathcal{A}^k$ and the elements of $\mathcal{A}^n$ are called *cover sequences* or *cover vectors*. The secret message **m** is hidden in the cover vector **c** as $\mathbf{c}' = \mathrm{emb}(\mathbf{c}, \mathbf{m})$ and subsequently recovered by the receiver as $\mathrm{rec}(\mathbf{c}')$.

Proposition 2.1. *Let $\mathcal{S} = (\mathrm{emb}, \mathrm{rec})$ be a steganographic scheme of type $[n, k]$ defined over the alphabet $\mathcal{A}$. Then*

(1) the map rec *is surjective;*
(2) for fixed $\mathbf{c} \in \mathcal{A}^n$, the map $\mathrm{emb}(\mathbf{c}, -) : \mathcal{A}^k \to \mathcal{A}^n$ *is injective.*

In particular, $k \leq n$.

Proof. All statements follow easily from the condition $\mathrm{rec}(\mathrm{emb}(\mathbf{c}, \mathbf{m})) = \mathbf{m}$. □

Let us denote by d the Hamming distance in $\mathcal{A}^n$: for $\mathbf{x}, \mathbf{y} \in \mathcal{A}^n$, $d(\mathbf{x}, \mathbf{y})$ is the number of coordinates in which **x** and **y** differ. More generally, given a set $\mathcal{Y} \subseteq \mathcal{A}^n$, we define $d(\mathbf{x}, \mathcal{Y}) = \min\{d(\mathbf{x}, \mathbf{y}) : \mathbf{y} \in \mathcal{Y}\}$. When embedding a message **m** into a cover **c**, the Hamming distance measures the number of embedding changes, as this number is $d(\mathbf{c}, \mathrm{emb}(\mathbf{c}, \mathbf{m}))$.

2.2.3. *Selection rules*

Imagine we want to embed a secret message **m** in a digital image $\mathcal{I}$. To take a concrete example we can consider the case in which $\mathcal{I}$ is a gray-scale bitmap image. Recall that it is a sequence of gray-scale pixels represented by integer numbers in some range $\{0, \ldots, M-1\}$. Usually M is a power of 2, $M = 2^D$, and each integer in the sequence is written as an array of D bits. D is the *depth* of that image. Bitmap images for computers use $D = 8$ but other depths are also possible (e.g. $D = 1$ for black and white images or $D = 12$ for medical and deep-space images).

During the embedding process, usually not all pixels of the image will store secret information. As the embedding changes should be restricted to the parts of the image where they are more difficult to detect, the sender first selects the placement and intensity of embedding changes (what pixels can be changed and how many bits per pixel). Thus the cover has two different forms: the original image and the sequence of bits (or symbols from a given alphabet $\mathcal{A}$) selected and adapted for use in the embedding process. The first one, $\mathcal{I}$ is called *cover object* or *cover work*, and the second, **c**, is the *cover sequence* or *cover vector*. The process by which the sender gets **c** from $\mathcal{I}$ is the *selection rule* used. As a result of applying a selection rule to a certain cover object, we obtain a cover sequence **c** which contains the information we allow to be distorted in that object and which can be used as a cover vector for a given steganographic scheme.

In general selection rules have two steps. First we select those image pixels that will store the secret message (e.g. pixels in the corners or other suitable parts of the image). After that selection we obtain an n-tuple $\mathbf{s} = (s_1, \ldots, s_n)$ with $s_i \in \{0, \ldots, M-1\}$ (or several such n-tuples). The cover sequence **c** is obtained from **s** in a way that depends on the type of stegoscheme in which we are concerned. Let us show some examples.

Example 2.3 (LSB modulation). Represent each integer s_i in **s** as a binary vector $\mathbf{s}_i = (b_{i,0}, \ldots, b_{i,D-1})$ with $s_i = \sum_j b_{ij} 2^j$. Then **c** is the sequence obtained concatenating the t least significant bits of each s_i, that is $\mathbf{c} = (b_{1,0}, \ldots, b_{1,t-1}; \ldots; b_{n,0}, \ldots, b_{n,t-1}) \in \mathbb{F}_2^{nt}$. See Refs. 40 and 42.

Example 2.4 (± 1-Steganography). In Least Significant Bit modulation ($t = 1$ in the previous Example), the gray level of each selected pixel s_i in the cover image can be changed in only one way in the embedding process, either to $s_i - 1$ or to $s_i + 1$ according to the parity of s_i. Now we want to allow both possibilities for each pixel (plus the possibility of remaining unchanged). This demand leads us to consider the alphabet with three elements $\mathcal{A} = \mathbb{Z}_3$. For $s \in \mathbb{Z}$ let $\bar{s} = s \pmod 3$. Our cover sequence **c** is then $\mathbf{c} = (\bar{s}_1, \ldots, \bar{s}_n)$. Take a stegoscheme $\mathcal{S} = (\text{emb}, \text{rec})$ of type $[n, k]$ over $\mathbb{Z}_3$. The secret we can hide is a k-tuple $\mathbf{m} = (m_1, \ldots, m_k) \in \mathbb{Z}_3^k$. Once $\text{emb}(\mathbf{c}, \mathbf{m}) = (x_1, \ldots, x_n)$ is computed, the gray-scale sequence **s** is changed to $\mathbf{s}' = (s_1', \ldots, s_n')$, where the s_i' are chosen in a way that $\bar{s}_i' = x_i$ and $|s_i' - s_i| \leq 1$ (note that $\mathbb{Z}_3 = \{\overline{s_i - 1}, \bar{s}_i, \overline{s_i + 1}\}$). The number of pixels changed in the cover image is just $d(\mathbf{c}, \text{emb}(\mathbf{c}, \mathbf{m}))$. More details and other methods can be found in Refs. 18, 35 and 46. The same idea can be applied to $\pm t$–Steganography by using the alphabet $\mathbb{Z}_{2t+1}$. See also Ref. 51.

Example 2.5 (JPEG images). Using a JPEG ('Joint Photographic Experts Group') format, the image is split into blocks of 8×8 pixels. For each block, gray-scale data undergoes a discrete cosine transform (DCT) that gives the spatial frequency spectrum. After quantization, the resulting data for all 8×8 blocks is compressed with the Huffmann loss-less algorithm, Ref. 32. When applying LSB Steganography to JPEG images, the cover **c** is the sequence of least significant bits of some selected DCT coefficients. See Ref. 45 for more details.

Thus selection rules are used to determine the placement and intensity of embedding changes in the cover object. Of course, the sender should always be able to obtain the cover sequence **c** from the cover image $\mathcal{I}$. The receiver however does not have $\mathcal{I}$ but an altered image $\mathcal{I}'$. In some situations the receiver will be able to recover the placement of embedding changes from $\mathcal{I}'$. We then speak of *shared* selection rules. In other cases the receiver cannot deduce such placement from $\mathcal{I}'$. It may even be that s/he does not know the selection rule used. Then we say that the selection rule is *nonshared*. These two possibilities, shared and nonshared, correspond –to some extent– with private and public key Cryptography respectively.

We shall study first stegoschemes for shared selection rules. The case of nonshared selection rules will be treated later in Section 2.6.

2.2.4. *Parameters*

The quality of a $[n, k]$ steganographic scheme $\mathcal{S}$ defined over an alphabet $\mathcal{A}$ with q symbols, can be measured in terms of its parameters. We can distinguish absolute and relative parameters. Absolute parameters are

- The *cover length* n.
- The *embedding capacity* k.
- The *embedding radius* ρ, that is maximum possible number of embedding changes
$$\rho = \max\{d(\mathbf{c}, \mathrm{emb}(\mathbf{c}, \mathbf{m})) : \mathbf{c} \in \mathcal{A}^n, \; \mathbf{m} \in \mathcal{A}^k\}.$$

- The *average number of embedding changes* R_a. Assuming all messages and covers equally probably and since $\#\mathcal{A} = q$, we have
$$R_a = \frac{1}{q^{kn}} \sum_{\mathbf{c} \in \mathcal{A}^n, \mathbf{m} \in \mathcal{A}^k} d(\mathbf{c}, \mathrm{emb}(\mathbf{c}, \mathbf{m})).$$

A stegoscheme with cover length n, embedding capacity k and average number of embedding changes R_a, is referred to as an $[n, k, R_a]$ stegoscheme. Relative parameters are

- The *relative payload* $\alpha = k/n$, which measures the amount of message symbols embedded per cover symbol. Sometimes it is convenient to express this measure in bits per cover symbol. Then we consider the *binary relative payload* or *embedding rate*

$$E = \frac{k}{n} \log_2(q).$$

- The *change rate* $c = R_a/n$, that measures the probability that a given symbol in the cover is changed during the embedding process. The change rate is also called *average distortion* by some authors. Note that $0 \leq \alpha, c \leq 1$.
- The *embedding efficiency* e, which is the ratio between relative payload and change rate. Then e is the amount of information embedded in the cover by one single change. We consider also the *lower embedding efficiency* $\underline{e}$.

$$e = \frac{k}{R_a}, \ \underline{e} = \frac{k}{\rho}.$$

Clearly $e \geq \underline{e}$. If we want to measure in bits then we consider the *binary embedding efficiency* and *binary lower embedding efficiency*, which are the ratios embedding rate/change rate and embedding rate/embedding radius respectively.

In general, the performance of a steganographic scheme is measured in terms of the ratios relative payload/change rate or embedding rate/change rate, that is in terms of the embedding efficiency. We shall see some bounds on these parameters in Section 2.5.

2.2.5. *Proper stegoschemes*

The purpose of a good scheme is, of course, to embed as much information in the cover as possible with as few changes as possible. This leads to the following definition. The scheme $\mathcal{S} = (\text{emb}, \text{rec})$ is said to be *proper* if the number of changes produced in the cover by the embedding process is the minimum possible allowed by the recovering map, that is if $d(\mathbf{c}, \text{emb}(\mathbf{c}, \mathbf{m})) = d(\mathbf{c}, \text{rec}^{-1}(\mathbf{m}))$ for all $\mathbf{c} \in \mathcal{A}^n$ and $\mathbf{m} \in \mathcal{A}^k$. Clearly the embedding map of any steganographic scheme can be slightly modified to make it proper.

Proposition 2.2. *Let $\mathcal{S} = (\text{emb}, \text{rec})$ be a steganographic scheme of type $[n, k]$ over $\mathcal{A}$. There exists a proper stegoscheme $\mathcal{S}^* = (\text{emb}^*, \text{rec})$ of the same type $[n, k]$ and such that $R_a(\mathcal{S}^*) \leq R_a(\mathcal{S})$.*

Proof. If there exist $\mathbf{c}, \mathbf{x} \in \mathcal{A}^n$ and $\mathbf{m} \in \mathcal{A}^k$ such that $\text{rec}(\mathbf{x}) = \mathbf{m}$ and $d(\mathbf{c}, \mathbf{x}) < d(\mathbf{c}, \text{emb}(\mathbf{c}, \mathbf{m}))$, define the embedding map emb' by $\text{emb}'(\mathbf{c}, \mathbf{m}) = \mathbf{x}$ and $\text{emb}'(\mathbf{c}', \mathbf{m}') = \text{emb}(\mathbf{c}', \mathbf{m}')$ for all $\mathbf{c}' \in \mathcal{A}^n, \mathbf{m}' \in \mathcal{A}^k$ with $(\mathbf{c}', \mathbf{m}') \neq (\mathbf{c}, \mathbf{m})$. Then the stegoscheme $\mathcal{S}' = (\text{emb}', \text{rec})$ verifies $R_a(\mathcal{S}') < R_a(\mathcal{S})$. After a finite number of steps of this type we arrive at a proper scheme. □

Since proper schemes have better parameters, in the following we restrict to this kind of schemes. Now the question is how to construct them. We shall show that Error-Correcting Codes are a good tool to achieve this purpose.

2.3. Error-Correcting codes

The theory of Error-Correcting Codes is a rich and deep branch of Information Theory. Its purpose is to develop methods for detecting and correcting errors produced when transmitting information over noisy channels. The aim of this Section is therefore not to provide a methodical exposition of this theory (the reader may find it in just the 762 pages of Ref. 28 for example) but to remember some essential facts that we shall use in our study of Steganography. For the same reason, much of the theory developed in this Section will be presented through exercises for the reader.

2.3.1. *Correcting errors*

An *Error-Correcting Code* of length n over the alphabet $\mathcal{A}$ is a subset $\mathcal{C} \subseteq \mathcal{A}^n$. Its elements are called *codewords* or simply *words*. If $\mathcal{C}$ has M elements we say that it is an (n, M) code. The third important parameter of $\mathcal{C}$ is its *minimum distance*, defined as

$$d = d(\mathcal{C}) = \min\{d(\mathbf{x}, \mathbf{y}) : \mathbf{x}, \mathbf{y} \in \mathcal{C}, \mathbf{x} \neq \mathbf{y}\}$$

where d is the Hamming distance in $\mathcal{A}^n$.

Exercise 2.1. A *distance* in $\mathcal{A}^n$ is a map $\delta : \mathcal{A}^n \times \mathcal{A}^n \to \mathbb{R}$ verifying the following properties

(1) for all $\mathbf{x}, \mathbf{y} \in \mathcal{A}^n$ we have $\delta(\mathbf{x}, \mathbf{y}) \geq 0$. Equality holds if and only if $\mathbf{x} = \mathbf{y}$;
(2) for all $\mathbf{x}, \mathbf{y} \in \mathcal{A}^n$ we have $\delta(\mathbf{x}, \mathbf{y}) = \delta(\mathbf{y}, \mathbf{x})$; and
(3) for all $\mathbf{x}, \mathbf{y}, \mathbf{z} \in \mathcal{A}^n$ we have $\delta(\mathbf{x}, \mathbf{y}) + \delta(\mathbf{y}, \mathbf{z}) \geq \delta(\mathbf{x}, \mathbf{z})$.

Prove that d is a true distance.

The *support* of a vector $\mathbf{x} \in \mathcal{A}^n$ is the set of nonzero coordinates of $\mathbf{x}$,
$$\mathrm{supp}(\mathbf{x}) = \{i : x_i \neq 0\}$$
and the *weight* of $\mathbf{x}$ is the number of nonzero coordinates, $\mathrm{wt}(\mathbf{x}) = \#\mathrm{supp}(\mathbf{x}) = d(\mathbf{x}, \mathbf{0})$. Recall that wt is the norm associated to the distance d.

A (minimum distance) *decoding map* for the code $\mathcal{C}$ is a map $\mathrm{dec}_\mathcal{C} : \mathcal{A}^n \to \mathcal{C}$ such that for all $\mathbf{x} \in \mathcal{A}^n$, $\mathrm{dec}_\mathcal{C}(\mathbf{x})$ is one of the nearest codewords to $\mathbf{x}$, that is to say $\mathrm{dec}_\mathcal{C}(\mathbf{x}) \in \mathcal{C}$ and $d(\mathbf{x}, \mathrm{dec}_\mathcal{C}(\mathbf{x})) = d(\mathbf{x}, \mathcal{C})$. For most codes it is difficult to determine complete decoding maps, so we consider partial ones defined over some subset $\mathcal{X} \subset \mathcal{A}^n$. In practice, this means that some error-patterns cannot be decoded.

When transmitting a message $\mathbf{c} \in \mathcal{C}$ over a noisy channel, errors occur so that $\mathbf{c}$ is corrupted. Let $\mathbf{x} = \mathbf{c} + \mathbf{e}$ be the received vector, where $\mathbf{e}$ is the error. If $\mathbf{x} \notin \mathcal{C}$, the receiver knows that errors occurred. Then s/he *decodes* $\mathbf{x}$ by $\mathrm{dec}_\mathcal{C}(\mathbf{x})$. If not too many errors have occurred then the decoding succeeds, $\mathrm{dec}_\mathcal{C}(\mathbf{x}) = \mathbf{c}$.

Proposition 2.3. *Let $\mathbf{x} = \mathbf{c} + \mathbf{e} \in \mathcal{A}^n$ with $\mathbf{c} \in \mathcal{C}$. If $2\mathrm{wt}(\mathbf{e}) < d(\mathcal{C})$ then $\mathrm{dec}_\mathcal{C}(\mathbf{x}) = \mathbf{c}$.*

Proof. Let $t = \lfloor (d-1)/2 \rfloor$. Since $2\mathrm{wt}(\mathbf{e}) < d$, then $d(\mathbf{c}, \mathbf{x}) \leq t$, while for any other codeword $\mathbf{c}' \in \mathcal{C}$, $\mathbf{c}' \neq \mathbf{c}$, it holds that $d(\mathbf{c}, \mathbf{x}) > t$. Otherwise $d(\mathbf{c}, \mathbf{c}') \leq d(\mathbf{c}, \mathbf{x}) + d(\mathbf{x}, \mathbf{c}') < d$ contradicting that d is the minimum distance of $\mathcal{C}$. Then $\mathbf{c}$ is the only nearest codeword to $\mathbf{x}$. □

The number $t = \lfloor (d-1)/2 \rfloor$ is the *error-correcting capacity* of $\mathcal{C}$ and $\mathcal{C}$ said to be a *t-error-correcting code*. Sometimes error locations are known in advance. These errors are called *erasures*. A communications channel for which the only possible errors are erasures is called *erasure channel*. In this model, when a transmitter sends a symbol $x \in \mathcal{A}$, the receiver either receives the symbol x or a message that it was not received ("erased"). Thus for erasure channels decoding is simply to find the error values.

Exercise 2.2. A code of minimum distance d corrects up to $d-1$ erasures.

Given $c \in \mathcal{A}^n$ and a nonnegative integer t, the *ball* of center c and radius t is the set of words in $\mathcal{A}^n$ that have Hamming distance at most t to c, $B(c,t) = \{x \in \mathcal{A}^n : d(c,x) \le t\}$. The cardinality of this set is denoted $V_q(n,t)$. If $\#\mathcal{A} = q$, then

$$V_q(n,t) = \#B(c,t) = \sum_{i=0}^{t} \binom{n}{i}(q-1)^i.$$

Proposition 2.4 (The Hamming bound for Error-Correcting Codes).
Let $\mathcal{C}$ be a t-error correcting (n, M) code over an alphabet $\mathcal{A}$ with q elements. Then

$$M \sum_{i=0}^{t} \binom{n}{i}(q-1)^i \le q^n.$$

Proof. The balls of radius t centered at the codewords of $\mathcal{C}$ are disjoint. Then the sum of their cardinalities is smaller than or equal to the number of elements in the whole space $\mathcal{A}^n$. □

Codes reaching equality in the Hamming bound are called *perfect*. Therefore a code is perfect if and only if each element of the space $\mathcal{A}^n$ belongs to one (and only one) ball of radius t centered at a codeword. Related to this bound we can consider the minimum number ρ such that the union of balls of radius ρ centered at the codewords of $\mathcal{C}$ cover the whole space $\mathcal{A}^n$. This number is the *covering radius* of $\mathcal{C}$. Thus

$$\rho = \max\{d(x, \mathcal{C}) : x \in \mathcal{A}^n\}.$$

According to the Hamming bound, the covering radius of an t error correcting code $\mathcal{C}$ verifies $\rho \ge t$. Then $\mathcal{C}$ is perfect if and only if $\rho = t$.

2.3.2. Linear codes over fields

A good code should have good parameters n, M, d, and an efficient decoding system. These purposes may be achieved by considering codes with underlying algebraic structure. If the cardinality of the alphabet $\mathcal{A}$ is a prime power, $q = p^s$, we can assume that $\mathcal{A}$ is a finite field with q elements $\mathcal{A} = \mathbb{F}_q$. Then the methods of linear algebra over fields can be applied.

A *linear code* $\mathcal{C}$ of length n and dimension k over $\mathbb{F}_q$ is a k-dimensional subspace of $\mathbb{F}_q^n$. If that code has minimum distance d we say it is a $[n, k, d]$ code. $\mathcal{C}$ is often described by a so-called *generator matrix* G, that is a matrix whose rows form a basis of $\mathcal{C}$. Then $\mathcal{C} = \{xG : x \in \mathbb{F}_q^k\}$, so encoding is just multiplying by G. Alternatively C can be characterized by a *parity check matrix*, which is a $(n-k) \times n$ matrix H of full rank such that

$$C = \{\mathbf{x} \in \mathbb{F}_q^n : H\mathbf{x}^t = \mathbf{0}\}.$$

In other words, H is the matrix of some implicit equations for C. Obviously $GH^t = 0$.

Exercise 2.3. Let C be a $[n, k, d]$ linear code and let H be a parity check matrix for C. (a) Define the *minimum weight* of C as $\mathrm{wt}(C) = \min\{\mathrm{wt}(\mathbf{x}) : \mathbf{x} \in C, \mathbf{x} \neq \mathbf{0}\}$. Prove that $\mathrm{wt}(C) = d$. (b) Prove that d equals the minimum number of linearly dependent columns in H. (c) Deduce that $k + d \leq n + 1$ (the *Singleton bound*). Codes reaching equality are called *maximum distance separable* (or simply MDS). (d) Consider the code having H as a generator matrix. This code is called the *dual* of C, denoted $C^\perp$. Prove that $C^\perp$ is the $[n, n-k]$ code $C^\perp = \{\mathbf{x} \in \mathbb{F}_q^n : \mathbf{x} \cdot \mathbf{c} = 0 \text{ for all } \mathbf{c} \in C\}$, where $\cdot$ is the usual inner product in $\mathbb{F}_q^n$, $\mathbf{x} \cdot \mathbf{c} = \sum x_i c_i$.

There exists a general decoding method for linear codes known as *syndrome decoding*. Let C be a $[n, k, d]$ linear code and let H be a parity check matrix for C. Consider the *syndrome* map $r : \mathbb{F}_q^n \to \mathbb{F}_q^{n-k}$ given by $r(\mathbf{x}) = H\mathbf{x}^t$. This is a linear map whose kernel is C. Related to this map we can consider the quotient $\mathbb{F}_q^n/C$ which is a vector space of dimension $n - k$, hence it contains q^{n-k} cosets $\mathbf{x} + C$. Since for any two vectors $\mathbf{x}, \mathbf{y} \in \mathbb{F}_q^n$ we have $\mathbf{x} + C = \mathbf{y} + C$ if and only if $r(\mathbf{x}) = r(\mathbf{y})$, we can manage cosets by the corresponding syndromes. Fix a minimum weight element $\mathbf{l}$ in each coset $r(\mathbf{x})$. It is called the *leader* of that coset, denoted $\mathbf{l} = \mathrm{cl}(r(\mathbf{x}))$. In particular $r(\mathrm{cl}(r(\mathbf{x}))) = r(\mathbf{l}) = r(\mathbf{x})$.

Exercise 2.4. Let C be a linear code. (a) Prove that the Hamming distance is translation-invariant, that is for all $\mathbf{x}, \mathbf{y}, \mathbf{z} \in \mathbb{F}_q^n$, it holds that $d(\mathbf{x}, \mathbf{y}) = d(\mathbf{x} + \mathbf{z}, \mathbf{y} + \mathbf{z})$. (b) Deduce that $d(\mathbf{x}, C) = \mathrm{wt}(\mathrm{cl}(r(\mathbf{x})))$. (c) Conclude that the covering radius ρ of C verifies

$$\rho = \max\{\mathrm{wt}(\mathbf{l}) : \mathbf{l} \text{ is a coset leader}\}.$$

A received vector $\mathbf{x}$ with errors in it can be written as $\mathbf{x} = \mathbf{c} + \mathbf{e}$, where $\mathbf{c} \in C$ is the transmitted word and $\mathbf{e}$ the *error-vector*. Again since r is linear we have $r(\mathbf{x}) = r(\mathbf{c} + \mathbf{e}) = r(\mathbf{e})$, hence the syndrome of the error-vector is known.

Proposition 2.5. *The map* $\mathrm{dec}_C : \mathbb{F}_q^n \to C$ *given by* $\mathrm{dec}_C(\mathbf{x}) = \mathbf{x} - \mathrm{cl}(r(\mathbf{x}))$ *is a decoding map for* C.

Proof. Since $r(\mathbf{x}-\mathrm{cl}(r(\mathbf{x}))) = r(\mathbf{x})-r(\mathrm{cl}(r(\mathbf{x}))) = \mathbf{0}$ we have $\mathrm{dec}_\mathcal{C}(\mathbf{x}) \in \mathcal{C}$ and hence $\mathrm{dec}_\mathcal{C}$ is a well-defined map. If there exists $\mathbf{c}' \in \mathcal{C}$ such that $d(\mathbf{x}, \mathbf{c}') < d(\mathbf{x}, \mathrm{dec}_\mathcal{C}(\mathbf{x}))$ then $\mathrm{wt}(\mathbf{x} - \mathbf{c}') < \mathrm{wt}(\mathrm{cl}(r(\mathbf{x})))$. Since $r(\mathbf{x} - \mathbf{c}') = r(\mathbf{x})$, this contradicts the definition of leader. See Exercise 2.4. □

In order to compute coset leaders in practice, often we make a list of all q^{n-k} syndromes and the corresponding coset leaders. The size of that list is the main drawback of this method when q^{n-k} is a big number. There are also general decoding methods for linear codes that do not require the use of tables and where the decoding is obtained through an iterative process (gradient decoding). See Ref. 1.

2.3.3. *An example: binary Hamming codes*

Le us introduce some notation. Let s be a positive integer and let $n = 2^s - 1$. For a given integer a, $0 \leq a \leq n$, let $[a]_2$ be the binary expression of a with s bits. We can consider $[a]_2 \in \mathbb{F}_2^s$. Conversely, for $\mathbf{a} \in \mathbb{F}_2^s$ let $[\mathbf{a}]_{10}$ be the integer, in decimal form, whose binary expression is $\mathbf{a}$ (then $0 \leq [\mathbf{a}]_{10} \leq n$). Finally, let $\mathbf{e}(i)$ the i-th vector of the canonical basis of $\mathbb{F}_2^n$ and $\mathbf{e}(0) = \mathbf{0}$ the all-zero vector.

The *binary Hamming code of redundancy* s, $\mathcal{H}_2(s)$, is the code over $\mathbb{F}_2$ whose parity check matrix H has the vectors $[1]_2, \ldots, [n]_2$ as columns. Clearly this code has length $n = 2^s - 1$ and dimension $k = n - s = 2^s - s - 1$. To compute its minimum distance we can use the result stated in Example 2.3 (b). Since all columns of H are distinct we have $d \geq 3$. It is not difficult to find three linearly dependent columns (e.g. $[1]_2 + [2]_2 = [3]_2$), hence $d = 3$. Thus $\mathcal{H}_2(s)$ is one-error correcting.

A direct computation shows that equality holds in the Hamming bound (Proposition 2.4), hence $\mathcal{H}_2(s)$ is a perfect code with covering radius $\rho = 1$. In particular this implies that all coset leaders have weight at most one. Since there are $n + 1 = 2^s$ such vectors and cosets, we conclude that the coset leaders are exactly the vectors in $\mathbb{F}_2^n$ with weight at most one, $\{\mathbf{e}(0), \ldots, \mathbf{e}(n)\}$. Furthermore, for $i = 0, \ldots, n$, we have $H\mathbf{e}(i)^t = [i]_2$ hence $\mathrm{cl}([i]_2) = \mathbf{e}(i)$. This leads to the following map for decoding $\mathcal{H}_2(s)$

$$\mathrm{dec}_\mathcal{H}(\mathbf{x}) = \mathbf{x} + \mathbf{e}([H\mathbf{x}^t]_{10}).$$

Exercise 2.5. The dual codes to the binary Hamming codes $\mathcal{H}_2(s)$ are called *simplex codes*. Prove that every nonzero codeword in $\mathcal{H}_2(s)^\perp$ has weight 2^{s-1}. Thus $\mathcal{H}_2(s)^\perp$ is a $[2^s - 1, s, 2^{s-1}]$ code.

Exercise 2.6. Alternatively we can define $\mathcal{H}_2(s)$ as the code having parity check matrix whose columns are the coordinates of all points of the projective space $\mathbb{P}^{s-1}(\mathbb{F}_2)$. This interpretation allows us to generalize the Hamming codes to any field $\mathbb{F}_q$. Prove that the q-ary Hamming code of redundancy s, $\mathcal{H}_q(s)$, has parameters

$$[\frac{q^s-1}{q-1}, \frac{q^s-1}{q-1} - s, 3]$$

and hence it is a perfect code.

2.3.4. Generalized Hamming weights for linear codes

Let $\mathcal{C}$ be an $[n,k]$ linear code defined over a finite field $\mathbb{F}_q$. For a subset $\mathcal{Y} \subseteq \mathbb{F}_q^n$, the *support* of $\mathcal{Y}$ is defined as

$$\mathrm{supp}(\mathcal{Y}) = \bigcup_{\mathbf{y} \in \mathcal{Y}} \mathrm{supp}(\mathbf{y}).$$

For $1 \leq r \leq k$, the r-th *generalized Hamming weight* of $\mathcal{C}$ is

$d_r = d_r(\mathcal{C}) = \min\{\#\mathrm{supp}(L) : L \text{ is a linear subcode of } \mathcal{C} \text{ and } \dim(L) = r\}$.

Note that $d_1(\mathcal{C})$ is the usual minimum distance of $\mathcal{C}$. The *weight hierarchy* of $\mathcal{C}$ is the sequence $d_1(\mathcal{C}), \ldots, d_k(\mathcal{C})$. The study of the weight hierarchy is an important topic in Coding Theory and much is known about it. In the next Exercise we state some of its main properties. The reader is addressed to Ref. 47 for the proofs.

Exercise 2.7. Let $\mathcal{C}$ be a linear $[n,k]$ code. Prove the following properties of its weight hierarchy $d_1(\mathcal{C}), \ldots, d_k(\mathcal{C})$: (a) (the monotonicity) $d_1(\mathcal{C}) < \cdots < d_k(\mathcal{C})$; (b) (the generalized Singleton bound) $d_r(\mathcal{C}) \leq n - k + r$; and (c) (the duality) $\{d_1(\mathcal{C}), \ldots, d_k(\mathcal{C})\} \cup \{n+1-d_1(\mathcal{C}^\perp), \ldots, n+1-d_{n-k}(\mathcal{C}^\perp)\} = \{1, \ldots, n\}$, where $\mathcal{C}^\perp$ is the dual of $\mathcal{C}$.

Note that (b) in the previous Exercise gives another proof of the Singleton bound, see Exercise 2.3. We say that $\mathcal{C}$ has *full support* if $d_k(\mathcal{C}) = n$. Otherwise, if $d_k(\mathcal{C}) < n$ then there exists a coordinate i such that $x_i = 0$ for all $\mathbf{x} \in \mathcal{C}$ and $\mathcal{C}$ is called *degenerate*. If $\mathcal{C}$ has full support, then the smallest integer r for which $d_r(\mathcal{C}) = n - k + r$ is called the *MDS rank* of $\mathcal{C}$.

Proposition 2.6. *The MDS rank of a full support $[n,k]$ code $\mathcal{C}$ is $k - d_1(\mathcal{C}^\perp) + 2$.*

Proof. Let r be the MDS rank of $\mathcal{C}$. By the duality property $n-k+r-1 = n+1-d_1(\mathcal{C}^\perp)$. □

Example 2.6. Let us consider the binary Hamming code $\mathcal{H}_2(s)$. Since its dual code has minimum distance $d^\perp = 2^{s-1}$ (Exercise 2.5), we conclude that $\mathcal{H}_2(s)$ has MDS rank $2^{s-1} - s + 1$.

2.4. Linking the problems

In this Section we make clear the relation between steganographic schemes and Error-Correcting Codes. As a consequence we obtain a way to design efficient stegoschemes.

2.4.1. *Stegoschemes and error-correcting codes*

Let us begin by showing an example.

Example 2.7 (F5). In Section 2.2 we remarked that the goal of a good steganographic scheme is to embed as much information as possible in the cover with as few changes as possible. Let us assume now that we want to embed k bits of information into a string $\mathbf{c}$ of n bits allowing one change at most. What is the minimum possible length n of the cover sequence to perform this purpose? The answer is simple. As there are $n+1$ possibilities of changing at most one bit of $\mathbf{c}$ and 2^k messages of length k, necessarily we have $n + 1 \geq 2^k$.

F5 is a steganographic system developed by D. Westfeld in 2001, Ref. 45. F5 allows one to hide sequences of k bits into covers of $n = 2^k - 1$ bits (least significant bits of a JPEG image as described in Example 2.5) by changing at most one of them. Let us briefly explain how the F5 scheme works. Keeping the notation stated in Section 2.3, for a given integer a, $0 \leq a \leq 2^k - 1$, let $[a]_2$ be the binary expression of a with k bits. Conversely, for $\mathbf{a} \in \mathbb{F}_2^k$ let $[\mathbf{a}]_{10}$ be the integer in decimal form whose binary expression is $\mathbf{a}$. Let $\mathbf{e}(i)$ be i-th vector of the canonical basis of $\mathbb{F}_2^n$ and $\mathbf{e}(0) = \mathbf{0}$. The embedding and recovering maps are as follows

$$\text{emb} : \mathbb{F}_2^n \times \mathbb{F}_2^k \to \mathbb{F}_2^n \;;\; \text{emb}(\mathbf{c}, \mathbf{m}) = \mathbf{c} + \mathbf{e}\left(\left[\mathbf{m} + \sum_{i=1}^n c_i[i]_2\right]_{10}\right).$$

$$\text{rec} : \mathbb{F}_2^n \to \mathbb{F}_2^k \;;\; \text{rec}(\mathbf{c}) = \sum_{i=1}^n c_i[i]_2.$$

Exercise 2.8. (a) Check that for the above defined embedding and recovering maps it holds that $\mathrm{rec}(\mathrm{emb}(\mathbf{c}, \mathbf{m})) = \mathbf{m}$ for all $\mathbf{c} \in \mathbb{F}_2^n$ and $\mathbf{m} \in \mathbb{F}_2^k$.
(b) If the reader has not noticed yet the relation of F5 with a well-known error-correcting code, note that the recovering map is linear. Compute its matrix. What is that code? What can be said about the embedding and recovering processes?

Let us develop part (b) of the above exercise for general stegoschemes. Remember that a family of nonempty subsets $\mathcal{Y}_1, \ldots, \mathcal{Y}_t \subseteq \mathcal{A}^n$ gives a *partition* of $\mathcal{A}^n$ if each element of $\mathcal{A}^n$ belongs to one and only one of these subsets, that is to say if they are pairwise disjoint and $\mathcal{Y}_1 \cup \cdots \cup \mathcal{Y}_t = \mathcal{A}^n$.

Proposition 2.7 (From stegoschemes to codes). *Let $\mathcal{S} = (\mathrm{emb}, \mathrm{rec})$ be a proper $[n, k]$ stegoscheme over the alphabet $\mathcal{A}$. For each $\mathbf{m} \in \mathcal{A}^k$ we consider the code $\mathcal{C}_{\mathbf{m}} = \{\mathbf{x} \in \mathcal{A}^n : \mathrm{rec}(\mathbf{x}) = \mathbf{m}\}$. Then the family $\{\mathcal{C}_{\mathbf{m}} : \mathbf{m} \in \mathcal{A}^k\}$ gives a partition on $\mathcal{A}^n$. Furthermore for all $\mathbf{m} \in \mathcal{A}^k$ the map $\mathrm{dec}_{\mathbf{m}} : \mathcal{A}^n \to \mathcal{C}_{\mathbf{m}}$ defined by $\mathrm{dec}_{\mathbf{m}}(\mathbf{x}) = \mathrm{emb}(\mathbf{x}, \mathbf{m})$ is a decoding map for the code $\mathcal{C}_{\mathbf{m}}$.*

Proof. Since rec is surjective, then all sets $\mathcal{C}_{\mathbf{m}}$ are nonempty. It is clear that the family of all of them gives a partition of $\mathcal{A}^n$. The condition $\mathrm{rec}(\mathrm{emb}(\mathbf{x}, \mathbf{m})) = \mathbf{m}$ implies that $\mathrm{dec}_{\mathbf{m}}(\mathbf{x}) \in \mathcal{C}_{\mathbf{m}}$ for all $\mathbf{x} \in \mathcal{A}^n$ and $\mathbf{m} \in \mathcal{A}^k$. Besides, since the stegoscheme is proper we have $d(\mathbf{x}, \mathrm{dec}_{\mathbf{m}}(\mathbf{x})) = d(\mathbf{x}, \mathrm{emb}(\mathbf{x}, \mathbf{m})) = d(\mathbf{x}, \mathcal{C}_{\mathbf{m}})$ and thus $\mathrm{dec}_{\mathbf{m}}$ is a minimum distance decoding map for $\mathcal{C}_{\mathbf{m}}$. □

The converse is also true.

Proposition 2.8 (From codes to stegoschemes). *Let $\{\mathcal{C}_{\mathbf{m}} : \mathbf{m} \in \mathcal{A}^k\}$ be a family of codes indexed by $\mathcal{A}^k$ and giving a partition of $\mathcal{A}^n$. For each $\mathbf{m} \in \mathcal{A}^k$ let $\mathrm{dec}_{\mathbf{m}}$ be a minimum distance decoding map of $\mathcal{C}_{\mathbf{m}}$. Consider the maps $\mathrm{emb} : \mathcal{A}^n \times \mathcal{A}^k \to \mathcal{A}^n$ and $\mathrm{rec} : \mathcal{A}^n \to \mathcal{A}^k$ defined by $\mathrm{emb}(\mathbf{x}, \mathbf{m}) = \mathrm{dec}_{\mathbf{m}}(\mathbf{x})$ and $\mathrm{rec}(\mathbf{x}) = \mathbf{m}$ if $\mathbf{x} \in \mathcal{C}_{\mathbf{m}}$. Then $\mathcal{S} = (\mathrm{emb}, \mathrm{rec})$ is a proper $[n, k]$ stegoscheme over $\mathcal{A}$.*

Proof. Since $\mathrm{dec}_{\mathbf{m}}(\mathbf{x}) \in \mathcal{C}_{\mathbf{m}}$, we have $\mathrm{rec}(\mathrm{emb}(\mathbf{x}, \mathbf{m})) = \mathbf{m}$. Furthermore $d(\mathbf{x}, \mathrm{emb}(\mathbf{x}, \mathbf{m})) = d(\mathbf{x}, \mathrm{dec}_{\mathbf{m}}(\mathbf{x})) = d(\mathbf{x}, \mathcal{C}_{\mathbf{m}})$, hence $\mathcal{S} = (\mathrm{emb}, \mathrm{rec})$ is a proper $[n, k]$ stegoscheme. □

As a consequence of the last two Propositions we conclude that the following objects are equivalent

- a proper $[n, k]$ stegoscheme $\mathcal{S} = (\mathrm{emb}, \mathrm{rec})$ over the alphabet $\mathcal{A}$; and

- a family $\{(\mathcal{C}_\mathbf{m}, \text{dec}_\mathbf{m}) : \mathbf{m} \in \mathcal{A}^k\}$ indexed by $\mathcal{A}^k$, where for every $\mathbf{m}$ $\mathcal{C}_\mathbf{m} \subseteq \mathcal{A}^n$ is a code, $\text{dec}_\mathbf{m}$ is a minimum distance decoding map for $\mathcal{C}_\mathbf{m}$, and the family $\{\mathcal{C}_\mathbf{m} : \mathbf{m} \in \mathcal{A}^k\}$ gives a partition of $\mathcal{A}^n$.

Assuming that we were able to find decoding maps for arbitrary codes (which is far to be true. Remember that decoding a general code is an NP-complete problem, see Refs. 3, 4 and 22), an $[n,k]$ stegoscheme over $\mathcal{A}$ would be equivalent to a partition of $\mathcal{A}^n$ into q^k sets. Such a partition is given by a surjective map rec : $\mathcal{A}^n \to \mathcal{A}^k$ and this map is precisely the recovering map of the corresponding stegoscheme. Thus, in some sense, a stegoscheme can be viewed as a recovering function. In particular this implies that when designing a stegoscheme, we have to pay attention first to the recovering map and subsequently to the embedding map.

Propositions 2.7 and 2.8 provide a method for constructing and studying stegoschemes. Schemes obtained by means of this construction are called *coding-based* stegoschemes. Some authors use the term *matrix encoding*, although we prefer the first name because, as seen before, no matrix is needed.

Remark that the decoding map $\text{dec}_\mathbf{m}$ is used in a completely different way in Coding Theory and Steganography. The purpose in Coding Theory is to correct errors (introduced by the channel). Thus it is important to guarantee that under certain conditions the original sent word and the decoded word coincide. In particular when the nearest codeword to a given received vector is not unique then decoding fails (soft decision decoding). In contrast, for use in Steganography, the purpose is to 'introduce errors' (as few as possible) in the cover. As a consequence uniqueness of the nearest codeword is not important: if there is more than one nearest codeword simply take one of them at random.

2.4.2. Group codes and stegoschemes

As we have seen, a general stegoscheme is quite a complicated object. We have to compute and handle q^k codes and decoding maps simultaneously. These codes and maps may be very different which is impractical. The simplest case arises when the codes are related in a way that all of them can be deduced from only one of them, $\mathcal{C}$, and all decoding maps can be deduced from the decoding map of $\mathcal{C}$, $\text{dec}_\mathcal{C}$. In this Paragraph we shall show a general situation in which these conditions happen. That situation is obtained by adding some algebraic structure to the alphabet and the codes.

Let us assume that the alphabet $\mathcal{A}$ is a finite abelian group. Then the direct product $\mathcal{A}^n$ is also an abelian group. Let us denote by $\star$ the group law in both $\mathcal{A}$ and $\mathcal{A}^n$. Given an element $\mathbf{x} \in \mathcal{A}^n$ and a subset $\mathcal{Y} \subseteq \mathcal{A}^n$ we write $\mathbf{x} \star \mathcal{Y} = \{\mathbf{x} \star \mathbf{y} : \mathbf{y} \in \mathcal{Y}\}$.

Lemma 2.1. *If $\mathcal{A}$ is an abelian group then the Hamming distance in $\mathcal{A}^n$ is translation-invariant, that is $d(\mathbf{x}, \mathbf{y}) = d(\mathbf{x} \star \mathbf{z}, \mathbf{y} \star \mathbf{z})$ for all $\mathbf{x}, \mathbf{y}, \mathbf{z} \in \mathcal{A}^n$. As a consequence, given a subset $\mathcal{C} \subseteq \mathcal{A}^n$ it holds that $d(\mathbf{x}, \mathbf{z} \star \mathcal{C}) = d(\mathbf{x} \star \mathbf{z}^{-1}, \mathcal{C})$.*

Proof. Let $x, y, z \in \mathcal{A}$. Due to the cancellation property of abelian groups it holds that $x \star z = y \star z$ if and only if $x = y$. The second statement is now clear. $\square$

Let $\mathcal{C} \subset \mathcal{A}^n$ be a code with q^{n-k} elements which is also a subgroup of $\mathcal{A}^n$. We say that $\mathcal{C}$ is a *group code*. To derive a family of codes from $\mathcal{C}$ simply consider the quotient group $\mathcal{A}^n/\mathcal{C}$. This quotient has q^k cosets, each of them of cardinality q^{n-k}. Let $\mathcal{R} = \{\mathbf{z}_1, \mathbf{z}_2, \ldots, \mathbf{z}_{q^k}\}$ be a complete set of representatives of these cosets with $\mathbf{z}_1 \in \mathcal{C}$. Then $\mathcal{A}^n/\mathcal{C} = \{\mathcal{C}, \mathbf{z}_2 \star \mathcal{C}, \ldots, \mathbf{z}_{q^k} \star \mathcal{C}\}$ and the family of codes $\{\mathcal{C}, \mathbf{z}_2 \star \mathcal{C}, \ldots, \mathbf{z}_{q^k} \star \mathcal{C}\}$ gives a partition of $\mathcal{A}^n$. Let us see how to derive a decoding map for these codes from a decoding map for $\mathcal{C}$.

Proposition 2.9. *Let $\mathcal{C}$ be a subgroup of $\mathcal{A}^n$ and $\mathbf{z} \in \mathcal{A}^n$. If dec is a decoding map for $\mathcal{C}$, then the map $\mathbf{x} \mapsto \mathbf{z} \star \mathrm{dec}(\mathbf{x} \star \mathbf{z}^{-1})$ is a decoding map for $\mathbf{z} \star \mathcal{C}$.*

Proof. Let $\mathbf{x} \in \mathcal{A}^n$. Since $\mathbf{z} \star \mathrm{dec}_\mathcal{C}(\mathbf{x} \star \mathbf{z}^{-1}) \in \mathbf{z} \star \mathcal{C}$ the map is well defined. Let $\mathbf{y} \in \mathbf{z} \star \mathcal{C}$, $\mathbf{y} = \mathbf{z} \star \mathbf{c}$ for some $\mathbf{c} \in \mathcal{C}$. If $d(\mathbf{x}, \mathbf{y}) < d(\mathbf{x}, \mathbf{z} \star \mathrm{dec}(\mathbf{x} \star \mathbf{z}^{-1}))$ then, by Lemma 2.1, we deduce that $d(\mathbf{x} \star \mathbf{z}^{-1}, \mathbf{c}) < d(\mathbf{x} \star \mathbf{z}^{-1}, \mathrm{dec}(\mathbf{x} \star \mathbf{z}^{-1}))$, which contradicts the fact that dec is a decoding map for $\mathcal{C}$. $\square$

In order to obtain a stegoscheme we need something more. Let us remember that the family of codes must be indexed by $\mathcal{A}^k$. To get it consider a surjective (and hence bijective) map $r : \mathcal{R} \to \mathcal{A}^k$ and, after identifying $\mathcal{R}$ with $\mathcal{A}^n/\mathcal{C}$, extend it to a map rec $: \mathcal{A}^n \to \mathcal{A}^k$, rec $= r \circ \pi$ where $\pi : \mathcal{A}^n \to \mathcal{A}^n/\mathcal{C}$ is the canonical projection. In the language of the previous Paragraph, for every $\mathbf{m} \in \mathcal{A}^k$ we have $\mathcal{C}_\mathbf{m} = r^{-1}(\mathbf{m}) \star \mathcal{C}$. Thus, by applying the results stated there, we obtain an $[n, k]$ stegoscheme $\mathcal{S}$ whose recovering and embedding maps are rec and $\mathrm{emb}(\mathbf{c}, \mathbf{m}) = \mathrm{dec}_\mathbf{m}(\mathbf{c}) = \mathbf{z} \star \mathrm{dec}(\mathbf{c} \star \mathbf{z}^{-1})$ respectively, where $\mathbf{z} = r^{-1}(\mathbf{m})$ and dec is a decoding map for $\mathcal{C}$.

Example 2.8. Let $\mathcal{A} = \mathbb{F}_q$ a finite field. A linear code $\mathcal{C}$ in $\mathbb{F}^n$ is a group code as it is a subgroup of the additive group $(\mathbb{F}_q^n, +)$. We shall study stegoschemes from linear codes in deeper detail in Paragraphs 2.4.3 and 2.4.4

The parameters of the stegoscheme $\mathcal{S}$ can be deduced from the properties of the code $\mathcal{C}$. Let us introduce a new definition. For a set $\mathcal{Y} \subseteq \mathcal{A}^n$, the *average radius* of $\mathcal{Y}$ is the average distance from a vector of $\mathcal{A}^n$ to $\mathcal{Y}$, namely

$$\tilde{\rho}(\mathcal{Y}) = \frac{1}{q^n} \sum_{\mathbf{x} \in \mathcal{A}^n} d(\mathbf{x}, \mathcal{Y})$$

Clearly $\tilde{\rho} \leq \rho$, where ρ is the covering radius of $\mathcal{Y}$.

Lemma 2.2. *If $\mathcal{C}$ is a group code, then all cosets $\mathbf{z} \star \mathcal{C}$ have the same covering radius and the same average radius.*

Proof. According to the invariance properties stated in Lemma 2.1, we have $d(\mathbf{x}, \mathcal{C}) = d(\mathbf{z} \star \mathbf{x}, \mathbf{z} \star \mathcal{C})$. □

Proposition 2.10. *Let $\mathcal{C} \in \mathcal{A}^n$ be a group code with q^{n-k} elements and let $\mathcal{S}$ be the stegoscheme obtained from $\mathcal{C}$. Then*

(1) the relative payload of $\mathcal{S}$ is $\alpha = k/n$;
(2) the embedding radius of $\mathcal{S}$ is the covering radius of $\mathcal{C}$;
(3) the average number of embedding changes of $\mathcal{S}$ is the average radius of $\mathcal{C}$, $R_a = \tilde{\rho}$.

Proof. (1) is already known. To see (2) and (3) note that by definition of embedding and decoding maps we have

$$d(\mathbf{c}, \text{emb}(\mathbf{c}, \mathbf{m})) = d(\mathbf{c} \star \mathbf{z}^{-1}, \text{dec}(\mathbf{c} \star \mathbf{z}^{-1})) = d(\mathbf{c} \star \mathbf{z}^{-1}, \mathcal{C})$$

for $\mathbf{c} \in \mathcal{A}^n, \mathbf{m} \in \mathcal{A}^k$ and $\text{rec}(\mathbf{z}) = \mathbf{m}$. Lemma 2.2 implies the results. □

Remark 2.1. There are translation invariant codes which are not group codes, see Ref. 33. The extension of our results to these codes and the study of their steganographic applications, are left to the reader as a research problem.

2.4.3. Linear stegoschemes over rings $\mathbb{Z}_q$

If $\#\mathcal{A} = q$ and no other restriction is made on $\mathcal{A}$, we can identify $\mathcal{A} = \mathbb{Z}_q$, which is an abelian group for the sum. Then the theory developed in the previous Paragraph can be applied. Let rec : $\mathbb{Z}_q^n \to \mathbb{Z}_q^k$ be a surjective morphism of $\mathbb{Z}_q$-modules. Since $\mathbb{Z}_q^n$ and $\mathbb{Z}_q^k$ are free modules of finite rank, then rec is given by a matrix H, that is $\text{rec}(\mathbf{x}) = H\mathbf{x}^t$ (see Ref. 29, Chapter II, Section B). The kernel $\mathcal{C}$ of rec is a submodule of $\mathbb{Z}_q^n$ and $\mathbb{Z}_q^n/\mathcal{C} \cong \mathbb{Z}_q^k$. Thus $\#\mathcal{C} = q^{n-k}$. $\mathcal{C}$ is said to be a *linear* code over $\mathbb{Z}_q$ and the stegoscheme derived from rec (or equivalently from $\mathcal{C}$) is said to be a *linear* stegoscheme.

The theory of linear codes over rings $\mathbb{Z}_q$ is quite similar to the theory of linear codes over fields, although the former are much more difficult to handle. The interested reader is addressed to Refs. 7, 8 and 31.

As in the case of fields, cosets in $\mathbb{Z}_q^n/\mathcal{C}$ can be managed in terms of the map rec: two vectors $\mathbf{x}, \mathbf{y} \in \mathbb{Z}_q^n$ belong to the same coset if and only if $\text{rec}(\mathbf{x}) = \text{rec}(\mathbf{y})$. The *leader* of a coset $\mathbf{x} + \mathcal{C}$ is a vector $\mathbf{l} \in \mathbf{x} + \mathcal{C}$ whose weight is minimal among the weights of all elements in $\mathbf{x} + \mathcal{C}$ (so some cosets might have more than one leader). Let $\mathcal{R} = \{\mathbf{l}_1, \ldots, \mathbf{l}_{q^k}\}$ be a set of representatives of all cosets in $\mathbb{Z}_q^n/\mathcal{C}$ such that every $\mathbf{l}_i$ is a leader of its corresponding coset. We have a bijection between $\mathcal{R}$ and $\mathbb{Z}_q^n/\mathcal{C}$, hence rec induces a bijective map $\mathcal{R} \to \mathbb{Z}_q^k$. Denote by cl the inverse of this map. Thus for any vector $\mathbf{x} \in \mathbb{Z}_q^n$, $\text{cl}(\text{rec}(\mathbf{x}))$ is a leader of $\mathbf{x} + \mathcal{C}$. Knowing $\mathcal{R}$ and the map cl, leads to a decoding procedure for $\mathcal{C}$, called *coset leader algorithm*.

Proposition 2.11. *The map* $\text{dec} : \mathbb{Z}_q^n \to \mathcal{C}$ *given by* $\text{dec}(\mathbf{x}) = \mathbf{x} - \text{cl}(\text{rec}(\mathbf{x}))$ *is a decoding map for* $\mathcal{C}$.

Proof. Since rec is a homomorphism, we have $\text{rec}(\mathbf{x} - \text{cl}(\text{rec}(\mathbf{x}))) = \text{rec}(\mathbf{x}) - \text{rec}(\text{cl}(\text{rec}(\mathbf{x}))) = \mathbf{0}$, hence $\text{dec}(\mathbf{x}) \in \mathcal{C}$. If there exists $\mathbf{c} \in \mathcal{C}$ such that $d(\mathbf{x}, \mathbf{c}) < d(\mathbf{x}, \text{dec}(\mathbf{x})) = d(\mathbf{x}, \mathbf{x} - \text{cl}(\text{rec}(\mathbf{x})))$, then $\text{wt}(\mathbf{x} - \mathbf{c}) < \text{wt}(\text{cl}(\text{rec}(\mathbf{x})))$. Since both $\mathbf{x} - \mathbf{c}$ and $\text{cl}(\text{rec}(\mathbf{x}))$ belong to the coset $\mathbf{x} + \mathcal{C}$, this is impossible. □

The computation of the map cl is usually made by using a table containing all elements of $\mathbb{Z}_q^k$ and the corresponding leaders. Such a table is called a *syndrome-leader* table. Propositions 2.9 and 2.11 provide a way to obtain an $[n, k]$ stegoscheme $\mathcal{S}$ from $\mathcal{C}$. $\mathcal{C}$ is the *main code* associated to $\mathcal{S}$.

Proposition 2.12. *Let* rec *be the map defined above and let* $\text{emb} : \mathbb{Z}_q^n \times \mathbb{Z}_q^k \to \mathbb{Z}_q^n$ *given by* $\text{emb}(\mathbf{c}, \mathbf{m}) = \mathbf{c} - \text{cl}(\text{rec}(\mathbf{c}) - \mathbf{m})$. *Then the pair* (emb, rec) *is an* $[n, k]$ *stegoscheme over* $\mathbb{Z}_q$.

Proof. The only thing to do is to deduce the expression of emb. From the discussion after Proposition 2.9, we have $\text{emb}(\mathbf{c}, \mathbf{m}) = \mathbf{z} + \text{dec}(\mathbf{c} - \mathbf{z})$, with $\mathbf{z} = \text{cl}(\mathbf{m})$. Then

$$\begin{aligned}\text{emb}(\mathbf{c}, \mathbf{m}) &= \text{cl}(\mathbf{m}) + \text{dec}(\mathbf{c} - \text{cl}(\mathbf{m})) \\ &= \mathbf{c} - \text{cl}(\text{rec}(\mathbf{c} - \text{cl}(\mathbf{m}))) \\ &= \mathbf{c} - \text{cl}(\text{rec}(\mathbf{c}) - \text{rec}(\text{cl}(\mathbf{m}))) \\ &= \mathbf{c} - \text{cl}(\text{rec}(\mathbf{c}) - \mathbf{m}). \end{aligned}$$
□

Proposition 2.13. *Let $\mathcal{S} = (\text{emb}, \text{rec})$ be a $\mathbb{Z}_q$-linear stegoscheme of type $[n, k]$ and let $\mathcal{C}$ be main code associated to $\mathcal{S}$. Then*

(1) The embedding radius of $\mathcal{S}$ equals the covering radius of $\mathcal{C}$.
(2) The average number of embedding changes of $\mathcal{S}$ is

$$R_a = \frac{1}{q^k} \sum_{\mathbf{l} \in \mathcal{R}} \text{wt}(\mathbf{l}).$$

Proof. (1) is a particular case of Proposition 2.10, item (2). Let us prove (2). According to Proposition 2.12, the number of changes made by the embedding process $\text{emb}(\mathbf{c}, \mathbf{m})$ is $\text{wt}(\text{cl}(\text{rec}(\mathbf{c}) - \mathbf{m})) = \text{wt}(\mathbf{l})$, when $\mathbf{l}$ is the leader of the corresponding coset. Since all cosets in $\mathbb{Z}_q^n/\mathcal{C}$ have the same cardinality and we are assuming all possible messages $\mathbf{m}$ and covers $\mathbf{c}$ equally probable, then all leaders are also equally probable and the statement (1) holds. □

The sequence $(\text{wt}(\mathbf{l}) : \mathbf{l} \in \mathcal{R})$ is called the *coset leader weight distribution* of $\mathcal{C}$. The computation of this sequence is a classical problem in Coding Theory.

Research Problem 2.9. (Research Problem 1.1 of Ref. 28). Find an effective method for computing the coset leader weight distribution of a linear code $\mathcal{C}$.

The reader should have noted that the theory of linear codes over rings $\mathbb{Z}_q$ is very similar to the theory of linear codes over finite fields $\mathbb{F}_q$. This is because most of the properties of linear codes defined over a field $\mathbb{F}_q$ depend on the additive structure of $\mathbb{F}_q$.

2.4.4. *Linear stegoschemes over fields*

If $q = p^s$, a prime power, then it is well known that we have an isomorphism $\mathbb{F}_q \cong \mathbb{Z}_p^s$ of additive groups, and then the theory we have developed can also be applied. Thus let us assume now that the alphabet $\mathcal{A}$ is a finite field with q elements, $\mathcal{A} = \mathbb{F}_q$. Let rec : $\mathbb{F}_q^n \to \mathbb{F}_q^k$ be a surjective linear map. There exists a $k \times n$ matrix H of full rank k such that $r(\mathbf{x}) = H\mathbf{x}^t$. Then H can be seen as a parity check matrix of the linear $[n, n-k]$ code $\mathcal{C} = \ker(\text{rec})$. The stegoscheme arising from $\mathcal{C}$ is of type $[n, k]$, it has rec as a recovering map and its embedding map is the one given by Proposition 2.12, namely

$$\text{emb}(\mathbf{c}, \mathbf{m}) = \mathbf{c} - \text{cl}(H\mathbf{c}^t - \mathbf{m}).$$

Exercise 2.10. Study the stegoscheme $\mathcal{S}_s$ obtained from the Hamming code $\mathcal{H}_2(s)$, see Paragraph 2.3.3. Compare with Example 2.7. Show that the parameters of $\mathcal{S}_s$ are as follows: $n_s = 2^s - 1; k_s = s; \rho_s = 1; R_s = (2^s - 1)/2^s; \alpha_s = s/(2^s - 1); c_s = 1/2^s$ and $e_s = s2^s/(2^s - 1)$. Study the performance of these parameters when s tends to infinity.

Research Problem 2.11. Study the average radius $\tilde{\rho}$ of a linear code. Find bounds on $\tilde{\rho}$.

Research Problem 2.12. Find codes providing good stegoschemes.

2.5. Bounds

The next step in our study is to determine bounds on the parameters of a stegoscheme. These bounds will allow us to characterize good schemes.

2.5.1. *The domain of stegoschemes*

We are interested in questions like: what is the maximum amount of information we can embed in the cover for a given distortion? Or what is the maximum possible embedding efficiency for a given payload? Next we shall give several answers to these questions. We again consider stegoschemes over an alphabet $\mathcal{A}$ with q symbols.

A stegoscheme $\mathcal{S} = (\text{emb}, \text{rec})$ of type $[n, k, R_a]$ over $\mathcal{A}$ can be represented by a point $p_{\mathcal{S}} = (R_a/n, k/n)$ in the square $[01] \times [0, 1]$. The set $\mathcal{D}_\mathcal{A} = \{p_{\mathcal{S}} : \mathcal{S} \text{ is a stegoscheme over } \mathcal{A}\}$, is the *domain of stegoschemes*. Similarly to what we do in Coding Theory, we can consider the number

$$\alpha_\mathcal{A}(n, R) = \max\{k/n : \text{ there exists an } [n, k, \leq R] \text{ stegoscheme over } \mathcal{A}\}$$

where n is a positive integer and $R \leq n$. A stegoscheme $\mathcal{S}$ of type $[n, k, R_a]$ such that $k/n = \alpha_\mathcal{A}(n, R_a)$ is called *optimal*.

Example 2.9. The embedding map of naive LSB stegoschemes is $\mathrm{emb}(\mathbf{c}, \mathbf{m}) = \mathbf{m}$. Then they are of type $[n, n]$ and the average number of embedding changes is $R_a = n(q-1)/q$. Thus its relative payload is $\alpha = 1$ and they are optimal.

Let $\theta = (q-1)/q$. According to Example 2.9, it suffices to consider stegoschemes having change rate $c \leq \theta$ because with them we can achieve the maximum possible payload, $\alpha = 1$. Thus for $0 \leq c \leq \theta$ we define

$$\alpha_\mathcal{A}(c) = \limsup_{n \to \infty} \alpha_\mathcal{A}(n, cn).$$

The functions $\alpha_\mathcal{A}(n, R)$ and $\alpha_\mathcal{A}(c)$ are not known. Clearly $\alpha_\mathcal{A}(c)$ is not decreasing, $\alpha_\mathcal{A}(0) = 0$ and $\alpha_\mathcal{A}(\theta) = 1$. In the next Paragraph we shall study an upper bound for $\alpha_\mathcal{A}(c)$.

Research Problem 2.13. Study $\alpha_\mathcal{A}(n, R)$ and $\alpha_\mathcal{A}(c)$. Is $\alpha_\mathcal{A}(c)$ a strictly increasing function?

2.5.2. Balls and entropy

Let $t \in \mathbb{R}$. Remember that the cardinality of a ball $B(\mathbf{c}, t) = \{\mathbf{x} \in \mathcal{A}^n : d(\mathbf{c}, \mathbf{x}) \leq t\}$ is denoted $V_q(n, t)$

$$V_q(n, t) = \#B(\mathbf{c}, t) = \sum_{i=0}^{\lfloor t \rfloor} \binom{n}{i} (q-1)^i.$$

For $t \leq n\theta$, the number $V_q(n, t)$ can be estimated in terms of the entropy function. Let us remember that the binary entropy $H(x)$ is defined in the interval $[0, 1]$ by $H(0) = H(1) = 0$ and

$$H(x) = -x \log_2(x) - (1-x) \log_2(1-x)$$

for $x \in (0, 1)$. This function is ubiquitous in Information Theory and it has a lot of applications. A complete treatment can be found in Ref. 23. In general, for arbitrary $q \geq 2$ we define the q-ary entropy in $[0, 1)$ by $H_q(0) = 0$ and

$$H_q(x) = x \log_q(q-1) - x \log_q(x) - (1-x) \log_q(1-x)$$

for $x \in (0, 1)$, see Ref. 41. See also Figure 2.1 for a representation of H_2.

Exercise 2.14. Prove that $H_q(x)$ increases from 0 to 1 as x runs from 0 to θ.

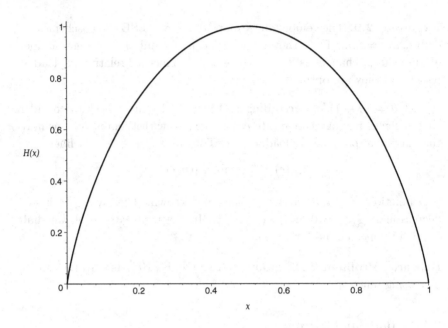

Fig. 2.1. Binary entropy

Proposition 2.14. *Let $q \geq 2, n$, be two integers and $\delta \in (0, \theta]$. Then*

$$\log_q(V_q(n, \delta n)) \leq n H_q(\delta).$$

Proof. By definition of entropy we have

$$q^{-nH_q(\delta)} = \left(\frac{\delta}{q-1}\right)^{\delta n} (1-\delta)^{(1-\delta)n}.$$

Consider now the following sequence of relations

$$1 = (\delta + (1-\delta))^n$$

$$\geq \sum_{i=0}^{\lfloor \delta n \rfloor} \binom{n}{i} \delta^i (1-\delta)^{n-i}$$

$$= \sum_{i=0}^{\lfloor \delta n \rfloor} \binom{n}{i} (q-1)^i \left(\frac{\delta}{q-1}\right)^i (1-\delta)^{n-i}$$

$$= \sum_{i=0}^{\lfloor \delta n \rfloor} \binom{n}{i} (q-1)^i (1-\delta)^n \left(\frac{\delta}{(q-1)(1-\delta)}\right)^i$$

$$\geq \sum_{i=0}^{\lfloor \delta n \rfloor} \binom{n}{i} (q-1)^i (1-\delta)^n \left(\frac{\delta}{(q-1)(1-\delta)}\right)^{\delta n}$$

$$= \sum_{i=0}^{\lfloor \delta n \rfloor} \binom{n}{i} (q-1)^i \left(\frac{\delta}{q-1}\right)^{\delta n} (1-\delta)^{(1-\delta)n}$$

$$= q^{-nH_q(\delta)} \sum_{i=0}^{\lfloor \delta n \rfloor} \binom{n}{i} (q-1)^i$$

where the last inequality comes from the fact that $\delta \leq \theta$ and hence $\delta \leq (q-1)(1-\delta)$. Then

$$1 \geq q^{-nH_q(\delta)} V_q(n, \delta n)$$

which proves the result. □

Exercise 2.15. By using a similar argument as in Proposition 2.14, prove that for $\delta \in (0, \theta]$ and n large enough we have $\log_q(V_q(n, \delta n)) \geq nH_q(\delta) - o(n)$. As a consequence,

$$\lim_{n \to \infty} n^{-1} \log_q(V_q(n, \delta n)) = H_q(\delta).$$

This fact shows that the estimate of $V_q(n, \delta n)$ given by the entropy is quite good for large n. See Ref. 41.

2.5.3. A Hamming-like bound

Proposition 2.15. *Let $\mathcal{S} = (\text{emb}, \text{rec})$ be a steganographic scheme of type $[n, k]$ and embedding radius ρ over $\mathcal{A}$. Then $q^k \leq V_q(n, \rho)$.*

Proof. Let $\mathbf{c} \in \mathcal{A}^n$. By definition of covering radius, for every $\mathbf{m} \in \mathcal{A}^k$ we have $\mathrm{emb}(\mathbf{c}, \mathbf{m}) \in B(\mathbf{c}, \rho)$. For fixed $\mathbf{c}$ the map $\mathrm{emb}(\mathbf{c}, -)$ is injective hence we have $q^k = \#\mathcal{A}^k \leq V_q(n, \rho)$. □

This bound is analogous to the Hamming bound in Coding Theory (Proposition 2.4). Thus, we call it the *Hamming bound for steganographic schemes*. In the same way, schemes reaching equality are called *perfect* (other authors call them *maximum length embeddable* schemes, see Ref. 49).

Proposition 2.16. *Let $\mathcal{S} = (\mathrm{emb}, \mathrm{rec})$ be a stegoscheme arising from a group code $\mathcal{C}$. Then $\mathcal{S}$ is perfect if and only if so is $\mathcal{C}$.*

Proof. Since the embedding radius of $\mathcal{S}$ equals the covering radius of $\mathcal{C}$, the Hamming bound for Error-Correcting Codes (Proposition 2.4) and the steganographic Hamming bound (Proposition 2.15) coincide. □

By taking logarithms in Proposition 2.15 we can deduce a bound on the relative payload.

Corollary 2.1. *Let $\mathcal{S} = (\mathrm{emb}, \mathrm{rec})$ be a q-ary steganographic scheme of type $[n, k]$ and embedding radius ρ. Then its relative payload verifies*

$$\alpha \leq \frac{\log_q(V_q(n, \rho))}{n}.$$

Corollary 2.2. *Let $\mathcal{S} = (\mathrm{emb}, \mathrm{rec})$ be a q-ary steganographic scheme of type $[n, k]$ and embedding radius $\rho \leq \theta n$. Then*

(1) its relative payload verifies $\alpha \leq H_q(\rho/n)$.
(2) its lower embedding efficiency verifies $\underline{e} \leq \alpha/H_q^{-1}(\alpha)$.

Proof. By applying the bound of Proposition 2.14 to the result stated in Corollary 2.1 we obtain the first statement. For the second one, let H_q^{-1} be the inverse function of H_q restricted to the interval $[0, \theta]$. Since H_q is increasing then so is H_q^{-1}, hence $H_q^{-1}(\alpha) \leq \rho/n$ and thus $\alpha/H_q^{-1}(\alpha) \geq \alpha n/\rho = \underline{e}$. □

Research Problem 2.16. Prove, or disprove, the following statement: if $\#\mathcal{A} = q$ and $R/n \leq \theta$ then $\alpha_\mathcal{A}(n, R) \leq H_q(R/n)$.

2.5.4. *Asymptotic bounds*

The same idea used in Proposition 2.15 can be applied to obtain a bound based on the average number of embedding changes R_a (instead of the embedding radius ρ). For a set $B = \{\mathbf{x}_1, \ldots, \mathbf{x}_m\} \subseteq \mathcal{A}^n$ and a vector $\mathbf{c} \in \mathcal{A}^n$, let $\tilde{d}(\mathbf{c}, B)$ be the average distance from $\mathbf{c}$ to the elements of B,

$$\tilde{d}(\mathbf{c}, B) = \frac{1}{m} \sum_{i=1}^{m} d(\mathbf{c}, \mathbf{x}_i).$$

Furthermore, for a nonnegative $t \leq n$, let $b_q(n, t)$ be the number

$$b_q(n, t) = \frac{1}{V_q(n, t)} \sum_{i=1}^{\lfloor t \rfloor} i \binom{n}{i} (q-1)^i = n(q-1) \frac{V_q(n-1, t-1)}{V_q(n, t)}$$

where the right-hand equality is obtained taking into account that

$$i \binom{n}{i} = n \binom{n-1}{i-1}.$$

Lemma 2.3. *Let $B = \{\mathbf{x}_1, \ldots, \mathbf{x}_m\} \subseteq \mathcal{A}^n$ be a set of m distinct elements, $\mathbf{c} \in \mathcal{A}^n$ a vector and $t \leq n$ a nonnegative real number.*

(1) If $B = B(\mathbf{c}, t)$ then $\tilde{d}(\mathbf{c}, B) = b_q(n, t)$.
(2) If there exists τ such that $B(\mathbf{c}, \tau - 1) \subseteq B \subset B(\mathbf{c}, \tau)$ then $b_q(n, \tau - 1) \leq \tilde{d}(\mathbf{c}, B) < b_q(n, \tau)$. In particular the sequence $b_q(n, 0), \ldots, b_q(n, n)$, is strictly increasing.
(3) If $\tilde{d}(\mathbf{c}, B) < b_q(n, t)$ then $\#B < V_q(n, t)$.

Proof. (1) and (2) are clear. (3) Let $B_0 = B$ and for $i = 1, \ldots, m$, iteratively construct the sets $B_i = (B_{i-1} \setminus \{\mathbf{x}_i\}) \cup \{\mathbf{x}'_i\}$, where $\mathbf{x}'_i$ is a vector in $(\mathcal{A}^n \setminus B_i) \cup \{\mathbf{x}_i\}$ whose distance to $\mathbf{c}$ is minimal among all elements of $(\mathcal{A}^n \setminus B_i) \cup \{\mathbf{x}_i\}$. Then $\tilde{d}(\mathbf{c}, B_i) \leq \tilde{d}(\mathbf{c}, B_{i-1})$ for $1 \leq i \leq m$. Consequently $\tilde{d}(\mathbf{c}, B_m) \leq \tilde{d}(\mathbf{c}, B)$. Furthermore, by construction, there exists an integer τ such that $B(\mathbf{c}, \tau - 1) \subseteq B_m \subset B(\mathbf{c}, \tau)$. Hence $b_q(n, \tau - 1) \leq \tilde{d}(\mathbf{c}, B^*) < b_q(n, t)$. Since the sequence $b_q(n, 0), \ldots, b_q(n, n)$, is strictly increasing, we deduce $\tau \leq t$ and then $\#B = \#B' < V_q(n, \tau) \leq V_q(n, t)$. □

Proposition 2.17. *Let $\mathcal{S} = (\text{emb}, \text{rec})$ be a steganographic scheme of type $[n, k]$ and average number of embedding changes R_a, defined over an alphabet $\mathcal{A}$ with q symbols. Let $t \leq n$ be a nonnegative real number. If $R_a < b_q(n, t)$ then $q^k < V_q(n, t)$.*

Proof. For $\mathbf{c} \in \mathcal{A}^n$ let $B_\mathbf{c} = \{\text{emb}(\mathbf{c}, \mathbf{m}) : \mathbf{m} \in \mathcal{A}^k\}$. Then $\#B_\mathbf{c} = q^k$ and R_a is the average of the numbers $\tilde{d}(\mathbf{c}, B_\mathbf{c})$, $\mathbf{c} \in \mathcal{A}^n$. Thus there exists $\mathbf{c}^* \in \mathcal{A}^n$ such that $\tilde{d}(\mathbf{c}^*, B_{\mathbf{c}^*}) \leq R_a$. If $R_a < b_q(n, t)$ then $\tilde{d}(\mathbf{c}^*, B_\mathbf{c}^*) < b_q(n, t)$ and the previous Lemma implies that $q^k = \#B_{\mathbf{c}^*} < V_q(n, t)$. □

Proposition 2.17 allows us to state an asymptotic version of Corollary 2.2.

Lemma 2.4. *Given $\delta \in [0, \theta)$, we have*
$$\lim_{n \to \infty} \frac{b_q(n, \delta n)}{n} = \delta.$$

Proof. If $\delta = 0$ then $b_q(n, 0) = 0$ and the result is clear. If $\delta > 0$ then
$$\lim_{n \to \infty} \frac{b_q(n, \delta n)}{n} = \lim_{n \to \infty} \frac{\binom{n-1}{\lfloor \delta n \rfloor - 1}}{\binom{n}{\lfloor \delta n \rfloor}} = \lim_{n \to \infty} \frac{\lfloor \delta n \rfloor}{n} = \delta.$$
□

Proposition 2.18. *Let $N \subseteq \mathbb{N}$ be an infinite set. For $n \in N$ let S_n be a q-ary $[n, k_n, R_n]$ stegoscheme with $R_n/n \leq \theta$. Let $\alpha = \limsup_{n \to \infty} k_n/n$ and $c = \limsup_{n \to \infty} R_n/n$. Then $\alpha \leq H_q(c)$.*

Proof. First note that $c \leq \theta$. If $c = \theta$ then $H_q(c) = 1$ and the result is clear, see Exercise 2.10. Assume $c < \theta$. Let $\varepsilon \in \mathbb{R}$ be such that $c < c + \varepsilon < \theta$. According to Lemma 2.4 it holds that
$$\lim_{n \to \infty} \frac{b_q(n, (c+\varepsilon)n)}{n} = c + \varepsilon$$
hence there exists $n_0 \in N$ such that for all $n \in N$, $n \geq n_0$, we have $b_q(n, (c+\varepsilon)n) > (c + (\varepsilon/2))n > R_n$. Now Proposition 2.17 implies
$$q^{k_n} < V_q(n, (c+\varepsilon)n)$$
hence
$$k_n < \log_q(V_q(n, (c+\varepsilon)n)) \leq n H_q(c+\varepsilon)$$
or equivalently $k_n/n < H_q(c+\varepsilon)$. Thus $\alpha \leq H_q(c)$. □

From this result we obtain a bound on the function $\alpha_{\mathcal{A}}(c)$.

Corollary 2.3. *If the alphabet $\mathcal{A}$ has q symbols, then for $0 \leq c \leq \theta$ we have $\alpha_{\mathcal{A}}(c) \leq H_q(c)$.* □

Example 2.10. Let us consider the family of stegoschemes S_s obtained from the Hamming codes $\mathcal{H}_2(s)$. Their parameters have been obtained in Exercise 2.9. A simple computation shows that
$$\lim_{s \to \infty} \frac{H_2(c_s)}{\alpha_s} = 1.$$
This fact strongly suggests that for c small we have $\alpha_2(c) \sim H_2(c)$.

We can also derive bounds on the embedding efficiency as a function of the change rate and relative payload respectively.

Corollary 2.4. *Let $N \subseteq \mathbb{N}$ be an infinite set. For $n \in N$ let $\mathcal{S}_n$ be a q-ary $[n, k_n, R_n]$ stegoscheme with $R_n/n \leq \theta$. Let e_n be the embedding efficiency of $\mathcal{S}_n$, $\alpha = \limsup_{n \to \infty} k_n/n$, $c = \lim_{n \to \infty} R_n/n$ and $e = \limsup_{n \to \infty} e_n$. Then $e = \alpha/c$ and*

(1) $e \leq H_q(c)/c$;
(2) $e \leq \alpha/H_q^{-1}(\alpha)$.

Proof. Firstly we note that

$$e = \limsup_{n \to \infty}(e_n) = \limsup_{n \to \infty}(k_n/R_n) = \limsup_{n \to \infty}(\frac{k_n/n}{R_n/n}) = \alpha/c.$$

(1) From Corollary 2.18 we have $\alpha \leq H_q(c)$ hence $e = \alpha/c \leq H_q(c)/c$. (2) Let H_q^{-1} be the inverse function of H restricted to the interval $[0, \theta]$. Since H is increasing then so is H_q^{-1}, hence $H_q^{-1}(\alpha) \leq c$ and thus

$$\frac{\alpha}{H_q^{-1}(\alpha)} \geq \frac{\alpha}{c} = e.$$

□

Graphical representations of the bounds given by this Corollary are shown in Figures 2.2 and 2.3.

As noted before, in the above bounds we have restricted to stegoschemes with $c \leq \theta$. Observe that $c = \theta$ is precisely the change rate of a q-ary stegoscheme $\mathcal{S}$ for which $\text{emb}(\mathbf{c}, \mathbf{m})$ is chosen at random. Also we have the following result.

Exercise 2.17. Let $\mathcal{S}$ be a binary $[n, k]$ stegoscheme with average number of embedding changes $R_a > n/2$. Prove that there exists a stegoscheme $\mathcal{S}'$ with the same parameters n, k, and average number of embedding changes $R'_a = n - R_a$. Deduce that the bound stated in Proposition 2.18 is also true for $c > 1/2$.

It is natural to ask whether the bounds we have developed can be achieved. It is also interesting to find stegoschemes (or codes providing them) whose parameters achieve or approach these bounds. Unfortunately not much can be said in general about these problems. For instance, even if we restrict to stegoschemes arising from group codes, our main asymptotic results, Proposition 2.18 and Corollaries 2.3 and 2.4, give bounds which translated to the language of codes, lead to (or come from) a bound on the

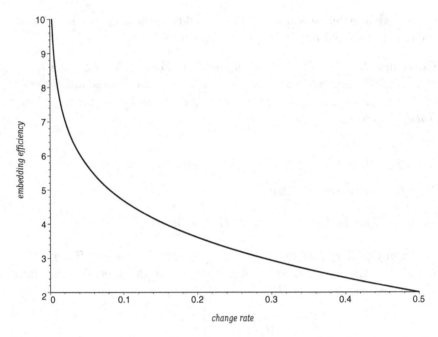

Fig. 2.2. A bound on the embedding efficiency as a function of the change rate (in the binary case).

average radius of a code. But this concept has never been studied so far in Coding Theory. We shall return to this problem in Paragraph 2.5.6.

2.5.5. *Perfect stegoschemes*

If we restrict to stegoschemes over fields arising from group codes, the bound of Corollary 2.1 is attained by perfect stegoschemes, which correspond to perfect codes (not necessarily linear). Recall that there are three main types of nontrivial perfect codes (see Ref. 28, Chapter 6, Section 10):

- the (linear) single-error-correcting Hamming codes over $\mathbb{F}_q$, with parameters $[(q^s - 1)/(q - 1), n - s, 3]$, $s = 2, 3, \ldots$;
- the (linear) binary $[23, 12, 7]$ Golay code; and
- the (linear) ternary $[11, 6, 5]$ Golay code

(definition and properties of Golay codes can be found in Ref. 28 or Ref. 41 for example). The parameters of Golay codes determine them uniquely up

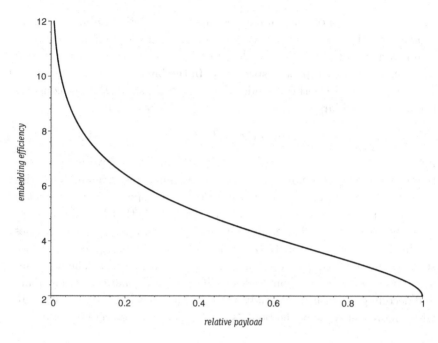

Fig. 2.3. A bound on the embedding efficiency as a function of the relative payload (in the binary case).

to equivalence. On the contrary many nonequivalent single-error-correcting perfect codes are known. However we have the following result (Ref. 28, Theorem 33), which determines the parameters of perfect stegoschemes arising from group codes over fields.

Theorem 2.1 (Tietäväinen and van Lint). *A nontrivial perfect code over a field $\mathbb{F}_q$ must have the same parameters n, M (number of codewords) and d as one of the Hamming or Golay codes.* □

We can ask for a similar result for general stegoschemes over arbitrary alphabets $\mathcal{A}$.

Research Problem 2.19. Determine all possible parameters for a perfect stegoscheme over an arbitrary alphabet $\mathcal{A}$.

With regard to the bounds given in Corollary 2.2, the situation is completely different. Since they are related to the radius, its study leads to the classic problem of determining the minimum covering radius among all (not necessarily linear) (n, M) codes over $\mathbb{F}_q$, and the minimum covering radius among all linear $[n, k]$ codes over $\mathbb{F}_q$. These minima are denoted by $\rho_q(n, M)$ and by $\rho_q[n, k]$ respectively. In the binary case, an asymptotic version of the ball-covering bound, $MV_2(n, \rho) \geq 2^n$, shows that for constant t, $0 < t < 1$, and large n we have (see Ref. 5, Theorem 15)

$$\rho_2[n, nt] \sim nH_2^{-1}(1 - t).$$

Furthermore, for n large enough most $[n, nt]$ codes have covering radius of that order of magnitude, $\rho \sim nH_2^{-1}(1-t)$. A similar result holds for nonlinear codes. Translating this equivalence to the corresponding stegoschemes, we obtain just the statement (2) in Corollary 2.2. So that bound is asymptotically attained by almost all sequences of binary $[n, k_n]$ stegoschemes coming from group codes with $\alpha_n \to \alpha$ fixed. This fact suggests to consider stegosystems derived from random linear codes with large n. The problematic part when using random codes is the computational cost of encoding and decoding algorithms. A study of this subject has been done in Ref. 20, where moreover some of the bounds seen in this Section were introduced.

2.5.6. *Another new problem for coding theory*

We have seen how the study of the embedding efficiency of a steganographic scheme $\mathcal{S}$ arising from a group code $\mathcal{C}$ leads to the computation of the average radius of $\mathcal{C}$. If $\mathcal{C}$ is an (n, M) code over $\mathcal{A}$,

$$\tilde{\rho}(\mathcal{C}) = \frac{1}{q^n} \sum_{\mathbf{x} \in \mathcal{A}^n} d(\mathbf{x}, \mathcal{C}).$$

Besides, if $\mathcal{C}$ is a $[n, k]$ linear code over $\mathbb{F}_q$, then

$$\tilde{\rho}(\mathcal{C}) = \frac{1}{q^{n-k}} \sum_{\mathbf{l} \in \mathcal{R}} \operatorname{wt}(\mathbf{l})$$

where $\mathcal{R}$ is a complete set of coset leaders for $\mathcal{C}$ (see Exercise 2.4 and Research Problem 2.9). The tools we developed in Paragraphs 2.5.3 and 2.5.4 to give a bound on the embedding efficiency may be applied now to study the average radius.

Proposition 2.19. *Let $\mathcal{C}$ be a (not necessarily linear) code of length n, error correcting capacity $t = (d - 1)/2$ and covering radius ρ. Then $b_q(n, t) \leq \tilde{\rho} < \rho$. If $\mathcal{C}$ is perfect then $\tilde{\rho} = b_q(n, t)$.*

Proof. Let $\mathbf{x} \in \mathcal{A}^n$. Since the balls of radius t centered at the codewords of $\mathcal{C}$ are disjoint, $\mathbf{x}$ belongs to at most one of them, $B(\mathbf{c}, t)$. In this case $d(\mathbf{x}, \mathcal{C}) = d(\mathbf{x}, \mathbf{c})$ and the average value of $d(\mathbf{x}, \mathcal{C})$ is $b_q(n, t)$. If none of these balls contains $\mathbf{x}$ then $d(\mathbf{x}, \mathcal{C}) > b_q(n, t)$. This proves the left-hand inequality. That $\tilde{\rho} < \rho$ follows from the definition of covering radius. If $\mathcal{C}$ is perfect then $t = \rho$ and no vector lies outside the balls $B(\mathbf{c}, t)$, $\mathbf{c} \in \mathcal{C}$. □

We are naturally led to the following definition: for given n, M and k let

$$\tilde{\rho}_\mathcal{A}(n, M) = \min\{\tilde{\rho}(\mathcal{C}) : \mathcal{C} \text{ is a } (n, M) \text{ code over } \mathcal{A}\}$$
$$\tilde{\rho}_q[n, k] = \min\{\tilde{\rho}(\mathcal{C}) : \mathcal{C} \text{ is a linear } [n, k] \text{ code over } \mathbb{F}_q\}.$$

Research Problem 2.20. Find the value of $\tilde{\rho}_q[n, k]$. Find $[n, k]$ linear codes having average radius $\tilde{\rho}_q[n, k]$.

The same questions can be proposed for nonlinear codes, but note that the obtained results have steganographic significance only for group codes.

Not much is known about $\tilde{\rho}_q[n, k]$. In Ref. 24, $\tilde{\rho}_q[n, 1]$ and $\tilde{\rho}_q[n, 2]$ are computed. Furthermore we have the following.

Proposition 2.20. *Let q be a primer power. If there exists an $[n, k]$ perfect code $\mathcal{C}$ over $\mathbb{F}_q$, then $\tilde{\rho}_q[n, k] = \tilde{\rho}(\mathcal{C})$.*

Proof. If $\mathcal{C}$ is a perfect code of covering radius ρ then the coset leaders are precisely the vectors with weight at most ρ. □

Thus

- $\tilde{\rho}_q[(q^s-1)/(q-1), (q^s-1)/(q-1)-s-1] = (q^s-1)/q^s$ for all $s = 2, 3, \ldots$;
- $\tilde{\rho}_2[23, 12] = b_2(23, 3) \approx 2.852$;
- $\tilde{\rho}_3[11, 6] = b_3(11, 2) \approx 1.901$.

2.6. Nonshared selection rules

Hitherto we have assumed that sender and receiver share a particular selection rule that allow them to unambiguously determine the placement of symbols that can be altered in the cover object. However in some situations sender and receiver cannot agree on such a particular selection rule to be used in an information transmission. In these cases the only option to obtain a cover sequence $\mathbf{c}$ from the cover object is to use generic rules such as "use the t least significant bits of all pixels". These rules cannot include criteria to select the placement of embedding changes. As a result not all symbols in the cover sequence **c** can be changed. We speak about *nonshared selection rules*.

To give a concrete example of this situation, imagine that using an image as cover object, we select the pixels that can be altered based on some visual and statistical criteria, comparing each pixel with its neighborhood. But any change alters the image, so it is possible that, despite using the same criteria, the receiver does not identify the same pixels as carriers of the secret message.

Then the difference between the methods used with shared and nonshared selection rules is significant. In both cases, the sender and the receiver are able to agree on a cover sequence $\mathbf{c} \in \mathcal{A}^n$. But while in the shared case all coordinates of **c** can be altered to store the secret message, when using nonshared rules only some of them (called *dry* coordinates) may be altered, and the others (*wet* coordinates) cannot be. Furthermore, the receiver is unable to distinguish dry and wet coordinates.

Stegosystems used for nonshared selection rules are commonly known as *wet paper codes*. This name comes from the article [14] where such codes were studied. To explain this denomination suppose the cover image has been exposed to rain, and thus the sender can only modify the dry pixels. During transmission the cover image dries out, so that the receiver cannot determine the pixels used by the sender. Wet paper codes are closely related to Error-Correcting Codes used for the erasure channel: dry symbols may be seen as erasures and wet symbols as symbols transmitted correctly. Decoding of these codes is based on solving a system of linear equations. Thus in this Section we shall restrict to the case in which $\mathcal{A}$ is a finite field $\mathbb{F}_q$.

2.6.1. *Wet paper codes*

We want to embed a message $\mathbf{m} = (m_1, \ldots, m_k) \in \mathbb{F}_q^k$ into a cover vector $\mathbf{c} = (c_1, \ldots, c_n) \in \mathbb{F}_q^n$. However, not all of coordinates of **c** can be used for this purpose: there is a set $\mathcal{D} \subseteq \{1, \ldots, n\}$ of u *dry* coordinates ($u \geq k$) that may be modified by the sender, while the other $n-u$ coordinates are *wet* and cannot be altered during the embedding process. Let $\mathcal{W} = \{1, \ldots, n\} \setminus \mathcal{D}$. The sets $\mathcal{D}$ and $\mathcal{W}$ are known to the sender but not to the receiver.

We will try to exploit the same idea we used in the case of shared selection rules. Remember that this idea is based on designing first the recovering map. If we refer to stegoschemes arising from linear codes, the recovering is obtained as a syndrome. Thus we set $\mathrm{emb}(\mathbf{c}, \mathbf{m}) = \mathbf{x}$ with

$$H\mathbf{x}^t = \mathbf{m},$$
$$x_i = c_i \text{ if } i \in \mathcal{W}$$

where H is a $k \times n$ matrix of full rank k constructed following some method previously agreed by the sender and the receiver. The sender solves the system and sends $\mathbf{x}$ to the receiver. The receiver has to determine first the matrix H. To that end s/he knows the number of columns n but not the number of rows k. This information must be sent at the beginning of the message $\mathbf{m}$. Then H is constructed row by row. Once the first rows are determined, k is also known to the receiver. Then s/he constructs H and recovers $\mathbf{m}$.

The requirement that only dry coordinates can be altered is written as $\mathbf{x} = \mathbf{c} + \mathbf{u}$, with $\mathrm{supp}(\mathbf{u}) \subseteq \mathcal{D}$. Then $H\mathbf{u}^t = \mathbf{m} - H\mathbf{c}^t$ and, as in the case of shared selection rules, the process of embedding is equivalent to finding a vector $\mathbf{u}$ with known syndrome, $\mathbf{m} - H\mathbf{c}^t$. The difference is that here we do not seek for a vector of minimum weight, but for a vector whose support is contained in $\mathcal{D}$. In short, the embedding process is equivalent to solving a system of linear equations. Since $\mathbf{u}$ has $n - u$ zero coordinates, we can simplify this system, by deleting these coordinates of $\mathbf{u}$ and the corresponding $n - u$ columns in H. Thus we arrive at a system

$$D\mathbf{u}^t = \mathbf{m} - H\mathbf{c}^t$$

where D is the matrix obtained from H after deleting columns with indices in $\mathcal{W}$ and we keep the notation $\mathbf{u}$ for the vector obtained after deleting zeros (that is, we consider $\mathbf{u} \in \mathbb{F}_q^u$). At this moment, the embedding problem is similar to the already studied of shared selection rules except for two important differences: firstly, the receiver does not know D (which is not relevant, because the secret message $\mathbf{m}$ may be obtained from H); and secondly, the sender cannot choose the matrix D, which depends on the set $\mathcal{D}$, that in turn depends on the cover object (through the selection rule used).

2.6.2. Solvability and the weight hierarchy of codes

We have transformed the embedding problem into another of solving a system $D\mathbf{u}^t = \mathbf{m} - H\mathbf{c}^t$ of k linear equations with u unknowns, $u \geq k$. Let us study the solvability of the linear equation $D\mathbf{u}^t = \mathbf{z}$. By Linear Algebra, this system has a solution for general $\mathbf{z}$ if and only if the matrix D has full rank k. Then the minimum number of dry coordinates necessary to embed k message symbols, by using the matrix H, is

$$\text{dry}(H) = \min \ \{ \ u : \text{every submatrix of } H \text{ formed} \\ \text{by } u \text{ columns has full rank } k\}.$$

Consequently the matrix H must have full rank. Assuming this condition let us see the relation of our problem with the theory of codes. Given a code $\mathcal{C}$, *shortening* $\mathcal{C}$ is the process of keeping only those codewords that have a given symbol (e.g. a 0 if the alphabet $\mathcal{A}$ is a numerical set) in a given position i, and then deleting this position. If $\mathcal{C}$ is a linear $[n,k]$ code and the given symbol is 0 then the shortened code is linear of type $[n-1, k']$, with $k-1 \leq k' \leq k$.

Now let $\mathcal{C}$ be the $[n, n-k]$ linear code whose parity check matrix is H. Let H_i be the matrix obtained by deleting the column i of H, and $\mathcal{C}_i$ the code with parity check matrix H_i. Then $\mathcal{C}_i$ is obtained from $\mathcal{C}$ by shortening at the position i. Thus H_i has full rank if and only if $\dim(\mathcal{C}_i) < \dim(\mathcal{C})$. This shows that $\text{dry}(H)$ depends on the code $\mathcal{C}$. We can write $\text{dry}(\mathcal{C}) = \text{dry}(H)$ for any parity check matrix of $\mathcal{C}$.

Proposition 2.21. *Keeping the above notations, the matrix H_i has full rank if and only if $i \in \text{supp}(\mathcal{C})$.*

Proof. H_i has full rank if and only if $\dim(\mathcal{C}_i) < \dim(\mathcal{C})$, that is if and only if $\mathcal{C} \cap \{x_i = 0\} \neq \mathcal{C}$ which is equivalent to $i \in \text{supp}(\mathcal{C})$. □

Thus, if $\mathcal{C}$ is a degenerate code then $\text{dry}(\mathcal{C}) = n$.

Proposition 2.22. *Let $\mathcal{C}$ be an $[n, n-k]$ linear code with parity check matrix H. Then $\text{dry}(\mathcal{C}) \leq n - d(\mathcal{C}^\perp) + 1$, where $\mathcal{C}^\perp$ is the dual of $\mathcal{C}$.*

Proof. If $\mathcal{C}$ is degenerate then $d(\mathcal{C}^\perp) \leq 1$ and $\text{dry}(\mathcal{C}) = n$. Assume that $\mathcal{C}$ has full support. If $\mathcal{C}$ has MDS rank r (see Paragraph 2.3.4), then for all $s \geq r$ we have $d_s(\mathcal{C}) = k + s$ hence for every subspace $L \subseteq \mathcal{C}$ of dimension s we have $\#\text{supp}(L) \geq k+s$. Thus we can shorten $\mathcal{C}$ at least $n-k-r+1$ times consecutively, obtaining in all cases codes of full support. Consequently the matrix obtained from H by deleting $n-k-r+1$ columns has maximum rank, according to Proposition 2.21. Therefore $\text{dry}(\mathcal{C}) \leq n - (n-k-r+1) = k + r - 1$. In Proposition 2.6 we showed that $r = n - k - d(\mathcal{C}^\perp) + 2$ hence $\text{dry}(\mathcal{C}) \leq n - d(\mathcal{C}^\perp) + 1$. □

Corollary 2.5. *Keeping the above notations, using a parity check matrix of an $[n, n-k]$ code $\mathcal{C}$, at most $n - d(\mathcal{C}^\perp) + 1$ dry symbols are needed to embed k information symbols.* □

Example 2.11. By using a parity check matrix of the binary Hamming code $\mathcal{H}_2(s)$ we can embed s bits of information into a cover vector of length $n = 2^s - 1$ with $2^{s-1} \approx n/2$ dry positions. To see the equality here, note that when deleting the last 2^{s-1} columns of the parity check matrix H given in Paragraph 2.3.3, we obtain a matrix whose last row is 0.

2.6.3. The rank of random matrices

Looking in detail at the above Example we note that, despite requiring 2^{s-1} dry positions to ensure that all submatrices of H (the parity check matrix of $\mathcal{H}_2(s)$) have full rank, almost all $s \times s$ submatrices of H have full rank. Thus, *on average* s dry symbols are enough to transmit s information symbols.

The rank properties of random matrices over finite fields have been widely investigated as a problem in Graph Theory and Coding Theory, related to codes for the erasure channel. In this Paragraph we shall describe –without proof– some of the main results in this line of research, in connection with the problem of wet paper codes. The interested reader is addressed to Refs. 6, 25 and 38.

Let M be a $k \times k$ matrix over the finite field with q elements $\mathbb{F}_q$. If the first j columns of M are linearly independent, then they span a vector space of size q^j. The probability that the next column avoids this space is $(q^k - q^j)/q^k$, hence

$$\text{prob}(M \text{ is nonsingular}) = \prod_{j=1}^{k} \left(1 - \frac{1}{q^j}\right).$$

Let k, m be nonnegative integers and $M_{k,k+m}$ be a random $k \times (k+m)$ matrix over $\mathbb{F}_q$ with $m \geq 0$. The following result can be proved by a recursive argument (see Refs. 6 and 38).

Theorem 2.2. *Let $M_{k,k+m}$ be a matrix where all elements of $\mathbb{F}_q$ are equally likely. Then*

$$\lim_{k \to \infty} \text{prob}\left(\text{rank}(M_{k,k+m}) = k\right) = \prod_{j=m+1}^{\infty} \left(1 - \frac{1}{q^j}\right).$$

It can be shown that this formula is very accurate even for k small. For $m = 0, 1, \ldots$, let

$$Q_m = \prod_{j=m+1}^{\infty} \left(1 - \frac{1}{q^j}\right).$$

From these expressions we deduce the probability that exactly m extra columns beyond k are needed to obtain a $k \times (k+m)$ random matrix of full rank over $\mathbb{F}_q$: $Q_m - Q_{m-1}$. Note that $Q_{m-1} = ((q^m - 1)/q^m)Q_m$ hence

$$Q_m - Q_{m-1} = \frac{Q_m}{q^m}.$$

Then the average number of extra columns needed to have full rank is

$$\tilde{m}(q) = \sum_{m=1}^{\infty} m(Q_m - Q_{m-1}) = \sum_{m=1}^{\infty} \frac{m}{q^m} Q_m.$$

To see the convergence of this series, let us remember the following result from elementary Calculus.

Exercise 2.21. Let $t \in \mathbb{R}$ with $|t| < 1$. Prove that

$$\sum_{m=1}^{\infty} mt^m = \frac{t}{(1-t)^2}.$$

(Hint: Starting from the equality $1 + t + t^2 + \cdots = 1/(1-t)$, differentiate both sides with respect to t and then multiply by t.)

Then the series giving $\tilde{m}(q)$ is convergent as it is upper-bounded by a convergent arithmetic-geometric series. According to Exercise 2.21,

$$\tilde{m}(q) = \sum_{m=1}^{\infty} \frac{m}{q^m} Q_m < \sum_{m=1}^{\infty} \frac{m}{q^m} = \frac{q}{(q-1)^2}.$$

It follows that the average number of extra columns needed to have full rank depends only on q. Furthermore this number is really small: for $q \geq 3$ we have $\tilde{m}(q) < 1$ and $\tilde{m}(q)$ decreases when q increases. In the binary case, a direct computation shows that a square random matrix is nonsingular with probability

$$Q_0 = \prod_{j=1}^{\infty} \left(1 - \frac{1}{2^j}\right) \approx 0.2887$$

as k tends to infinity. Besides $\tilde{m}(2) \approx 1.6067$.

Let us return to our original problem of computing the number of extra dry symbols needed on average to transmit k information symbols by using wet paper codes (the *overhead* of the system). According to the previous results this number is constant for constant q. Furthermore it is very small. In practice we can consider an overhead in the range of 5 to 10 in the binary case. Thus the number of dry symbols needed *on average* to transmit k information symbols by using wet paper codes is $u \sim k$.

2.7. The ZZW embedding construction

Wet paper codes can be used together with stegoschemes for shared selection rules in order to improve the performance of these schemes. This idea has been exploited by W. Zhang, X. Zhang and S. Wang, Ref. 50. Let us see the simplest case of this construction.

2.7.1. *Description of the method*

Starting from an $[n, k]$ binary stegoscheme $\mathcal{S}_0$ with average number of embedding changes R_a, we can construct a sequence of stegoschemes $(\mathcal{S}_\ell)_{\ell=1,2,\ldots}$, where $\mathcal{S}_\ell$ is binary of type $[n2^\ell, k + \ell R_a]$ and average number of embedding changes R_a. We shall describe the embedding process of $\mathcal{S}_\ell$. The recovering is straightforward.

Write the cover sequence as a matrix

$$C = \begin{pmatrix} c_{1,1} & \cdots & c_{1,n} \\ \vdots & & \vdots \\ c_{2^\ell,1} & \cdots & c_{2^\ell,n} \end{pmatrix}.$$

Let $\mathbf{c}_1^*, \ldots, \mathbf{c}_n^* \in \mathbb{F}_2^{2^\ell}$ be the columns of C and $\mathbf{c}_1, \ldots, \mathbf{c}_n \in \mathbb{F}_2^{2^\ell-1}$ be the vectors obtained from these columns by deleting the last coordinate. Let $\mathbf{v} \in \mathbb{F}_2^n$ be the vector obtained by adding all rows of C, that is $v_i = c_{1,i} + \cdots + c_{2^\ell,i}$. First we embed k bits of message into $\mathbf{v}$ by using the stegoscheme $\mathcal{S}_0$. On average this process requires R_a changes, say at coordinates $j_1, \ldots, j_r$ (so $r \approx R_a$). To change a coordinate v_j one bit must be changed in the column $\mathbf{c}_j$. The position of this change in $\mathbf{c}_j$ is arbitrary and its choice allows us to embed ℓ more information bits, to complete the $k + \ell r \approx k + \ell R_a$ bits of embedding capacity of $\mathcal{S}_\ell$. Furthermore the number of embedding changes in C is exactly $r \approx R_a$.

Let us describe how to embed these additional ℓr bits. The receiver does not know the positions $j_1, \ldots, j_r$ where the changes must be made.

Otherwise a Hamming $[2^\ell - 1, \ell]$ stegoscheme is enough to embed these ℓr bits. Just write them as r vectors $\mathbf{m}_1, \ldots, \mathbf{m}_r \in \mathbb{F}_2^\ell$. Let H be the $\ell \times (2^\ell - 1)$ parity check matrix of the Hamming code $\mathcal{H}_2(\ell)$. Since the covering radius of Hamming codes is 1, for $i = 1, \ldots, r$, there exists $\mathbf{x}_i \in \mathbb{F}_2^{2^\ell - 1}$ such that $H\mathbf{x}_i^t = \mathbf{m}_i$ and $d(\mathbf{x}_i, \mathbf{c}_{j_i}) \leq 1$. Then substitute $\mathbf{c}_{j_i}$ by $\mathbf{x}_i$ in C. Finally, in case $\mathbf{x}_i = \mathbf{c}_{j_i}$, to change v_{j_i} we just flip the bit c_{2^ℓ, j_i}.

As we noted below, the problem of this approach is that the receiver does not know the positions $j_1, \ldots, j_r$. To solve this problem we shall use wet paper codes. Compute the syndromes $\mathbf{s}_i = H\mathbf{c}_i^t \in \mathbb{F}_2^\ell$, $i = 1, \ldots, n$, and then concatenate all these syndromes to one vector $\mathbf{s} \in \mathbb{F}_2^{n\ell}$. We shall use $\mathbf{s}$ as a cover sequence for wet paper codes. Label the bits corresponding to syndromes $j_1, \ldots, j_r$ as dry, and the remainder bits as wet. As explained before, if the wet paper code requires that the dry syndrome $\mathbf{s}_i$ has to be changed, we change one position in the corresponding vector $\mathbf{c}_i$. If a dry syndrome remains unchanged, we change the last bit of that column. As seen in the previous Section, ℓr dry bits allow to communicate about $\ell r \approx \ell R_a$ information bits on average.

2.7.2. *Asymptotic behavior*

Let us study the asymptotic behavior of the ZZW construction. For $\ell = 0, 1, \ldots$, let α_ℓ, c_ℓ and e_ℓ be the relative payload, change rate and embedding efficiency respectively of the stegoscheme $\mathcal{S}_\ell$. Note that $e_\ell = e_0 + \ell$ whereas $\alpha_\ell \to 0$ and $c_\ell \to 0$ as $\ell \to \infty$. A simple computation shows that

$$\lim_{\ell \to \infty} \frac{H_2(c_\ell)}{\alpha_\ell} = \frac{1}{\ln(2)}.$$

This result seems to be worst than the corresponding for stegoschemes coming from Hamming codes, see Example 2.10. On the other hand we can also study the asymptotic embedding efficiency in terms of the change rate. Let us remember that, according to Corollary 2.4, we have $e_\ell \leq H_2(c_\ell)/c_\ell$ for ℓ large enough.

Proposition 2.23. *With the above notations we have*

$$\lim_{\ell \to \infty} \left(\frac{H_2(c_\ell)}{c_\ell} - e_\ell \right) = \frac{1}{\ln(2)} - \log_2(c_0) - e_0.$$

Proof. By using the continuous variable $x \in \mathbb{R}$ instead of the discrete variable ℓ, the limit takes the form

$$\lim_{x \to \infty} \left(\frac{n 2^x H_2 \left(\frac{R_a}{n 2^x} \right)}{R_a} - e_0 - x \right).$$

A straightforward computation shows that the value of this limit is

$$\frac{-\ln(R_a/n) + 1}{\ln(2)} - e_0 = \log_2 \left(\frac{n}{R_a} \right) + \frac{1}{\ln(2)} - e_0$$

which implies the result. □

2.8. Bibliographical notes and further reading

Steganography is today an active growing field to the point that, as noted in the Introduction, almost 90 % of publications have been written in the last five years. In order to guide the reader, along the text we have included numerous bibliographic citations in which s/he can extend the information provided herein. Below we provide some other indications for further reading.

In this Chapter we have focused our attention on the relationship between Steganography and Error-Correcting Codes and therefore on steganographic schemes. Other relevant parts of Steganography have been succinctly cited or even ignored at all. For a more general treatment of Steganography (including Watermarking) the reader is addressed to the books [2, 11]. Another interesting reading is the Wayner's book, Ref. 44. The reader interested on mimic functions can also see Ref. 43.

The stegosystem F5 was introduced by D. Westfeld, Ref. 45. An exhaustive steganalysis of this method can be found in Ref. 12.

Group codes and codes over $\mathbb{Z}_q$ have been studied by several authors, see Refs. 7, 8, 31 and the references therein. See also Ref. 33 for translation-invariant codes which are not commutative groups. Stegoschemes arising from some particular Error-Correcting Codes have been studied in Refs. 26, 45 (Hamming and Golay codes), Refs. 9, 30, 36 (BCH and Reed-Solomon codes), Ref. 34 (product perfect codes), Ref. 35 ($\mathbb{Z}_2\mathbb{Z}_4$-linear codes), etc.

More information about wet paper codes can be found in Refs. 13–17, 53. Codes for the erasure channel are based on sparse matrices, introduced by Gallager, see Ref. 21. These include LT codes, Ref. 27, raptor codes, Ref. 37, and window codes, Ref. 38.

The ZZW construction was introduced in Ref. 50 in its original form and later generalized in Ref. 52. The embedding efficiency of this construction was studied in Ref. 10.

Acknowledgments

The author would like to thank Dr. Stanislav Bulygin and Prof. David Joyner, who carefully read the Chapter and whose suggestions have helped to improve the final presentation.

References

[1] A. Ashikhmin and A. Barg, Minimal vectors in linear codes, *IEEE Transactions on Information Theory* **44**, 2010–2017 (1998).
[2] I. Cox, M. Miller, J. Bloom, J. Fridrich and T. Kalker, *Digital watermarking and Steganography* (Elsevier-Morgan Kaufmann, Burlington, 2008).
[3] E. R. Berlekamp, R. J. McEliece, and H. C. A. Van Tilborg, On the inherent intractability of certain coding problems, *IEEE Transactions on Information Theory* **24**, 384–386 (1978).
[4] J. Bruck and M. Naor, The hardness of decoding linear codes with preprocessing, *IEEE Transactions on Information Theory* **36**, 381–385 (1990).
[5] G. Cohen, M. Karpovsky, H.F. Mattson and J. Schatz, Covering Radius.- Survey and Recent Results, *IEEE Transactions on Information Theory* **31**, 328-343 (1985).
[6] C. Cooper, On the rank of random matrices, *Random Structures and Algorithms* **16**, 209–232 (2000).
[7] S.T. Dougherty, T.A. Gulliver and Y. H. Park, Optimal linear codes over $\mathbb{Z}_m$, *J. Korean Mathematical Society* **44**, 1139–1162 (2007).
[8] S.T. Dougherty and K. Shiromoto, MDR codes over $\mathbb{Z}_k$, *IEEE Transactions on Information Theory* **46**, 265–269 (2000).
[9] C. Fontaine and F. Galand, How Reed-Solomon Codes Can Improve Steganographic Schemes, *EURASIP Journal on Information Security* **2009**. doi:10.1155/2009/274845.
[10] J. Fridrich, Asymptotic behavior of the ZZW embedding construction, *IEEE Transactions on Information Forensics and Security* **4**, 151–154 (2009).
[11] J. Fridrich, *Steganography in Digital Media: Principles, Algorithms, and Applications* (Cambridge University Press, Cambridge, 2010).
[12] J. Fridrich, M. Goljan and D. Hogea, Steganalysis of JPEG images: Breaking the F5 algorithm. In *Proceedings of 5th Information Hiding Workshop*, Noordwijkerhout, The Netherlands, 310–323, 2002.
[13] J. Fridrich, M. Goljan and D. Soukal, Perturbed quantization Steganography using wet paper codes. In *Proceedings of the 2004 Workshop on Multimedia and Security*, J. Dittman and J. Fridrich, eds. ACM Press, 2004.

[14] J. Fridrich, M. Goljan, P. Lisonek and D. Soukal, Writing on wet paper, *IEEE Transactions on Signal Processing* **53**, 3923–3935 (2005).
[15] J. Fridrich, M. Goljan and D. Soukal, Efficient wet paper codes. In *Proceedings of Information Hiding. IHW 2005*, Springer Verlag, LNCS, 2005.
[16] J. Fridrich, M. Goljan and D. Soukal, Wet paper codes with improved embedding efficiency, *IEEE Transactions on Information Forensics and Security* **1**, 102–110 (2006).
[17] J. Fridrich, M. Goljan and D. Soukal, Steganography via codes for memory with defective cells. In *Proceedings of the Forty-Third Annual Allerton Conference On Communication, Control, and Computing* 1521–1538, 2005
[18] J. Fridrich and P. Lisoněk, Grid colorings in Steganography, *IEEE Transactions on Information Theory* **53**, 1547–1549 (2007).
[19] J. Fridrich, P. Lisonek, and D. Soukal, On steganographic embedding efficiency. In *Proceedings of 8th Int. Workshop Information Hiding*, J. L. Camenisch, C. S. Collberg, N. F. Johnson, and P. Sallee, Eds., 282–296, New York 2006.
[20] J. Friddrich and D. Soukal, Matrix embedding for large payloads, *IEEE Transactions on Information Security and Forensics* **1**, 390–394 (2006).
[21] R.G. Gallager, *Low density parity check codes* (MIT Press, Cambridge, 1963).
[22] M. R. Garey and David S. Johnson, *Computers and Intractability: A Guide to the Theory of NP-Completeness* (W. H. Freeman and Co., New York, 1979).
[23] M. Gray, *Entropy and Information Theory* (Springer, New York, 1990).
[24] M. Khatirinejad and P. Lisoněk, Linear codes for high payload Steganography *Discrete Applied Mathematics* **157**, 971–981 (2009).
[25] V.F. Kolchin, *Random Graphs* (Cambridge University Press, Cambridge, 1999).
[26] C.-Q. Liu, X.-J. Ping, T. Zhang, L. Zhou and Y. Wang, A Research on Steganography Method Based on Error-Correcting Codes, In *International Conference on Intelligent Information Hiding and Multimedia Signal Processing*, 377–380, 2006.
[27] M. Luby, LT codes, The 43rd Annual IEEE Symposium on Foundations of Computer Science, 2002.
[28] F.J. MacWilliams and N. Sloane, *The Theory of Error-Correcting Codes* (North-Holland, Amsterdam, 1977).
[29] B. R. McDonald, *Linear Algebra over Commutative Rings* (Marcel Dekker, New York, 1984).
[30] C. Munuera, Steganography and Error-Correcting Codes, *Signal Processing* **87**, 1528–1533 (2007).
[31] Y. H. Park, Modular independence and generator matrices for codes over $\mathbb{Z}_m$, *Designs, codes and Cryptography* **50**, 147–162 (2009).
[32] W. B. Pennebaker and J. L. Mitchell, *JPEG Still Image Data Compression Standard* (Springer, New York, 1993).
[33] J. Rifà and J. Pujol, Translation-invariant properlinear codes, *IEEE Transactions on Information Theory* **43**, 590–598 (1997).

[34] H. Rifà and J. Rifà, Product perfect codes and Steganography, *Digital Signal Processing* **19**, 764–769 (2009).
[35] H. Rifà, J. Rifà and L. Ronquillo, Perfect $\mathbb{Z}_2\mathbb{Z}_4$-linear codes in Steganography, arXiv:1002.0026v1 (2010).
[36] D. Schönfeld and A. Winkler, Embedding with syndrome coding based on BCH codes. In *Proceedings ACM 8th Workshop on Multimedia Security*, 214–23 (2006).
[37] A. Shokrollahi, Raptor codes, *IEEE Transactions on Information Theory* **52**, 2552–2567 (2006).
[38] C. Studholme and I.F. Blake, Random matrices and codes for the erasure channel, *Algoritmica* **56**, 605–620 (2010).
[39] J. Trithemius, *Steganographia: Ars per occultam scripturam animi sui voluntatem absentibus aperiendi certu*. (*Circa* 1500. First printed edition in Frankfurt, 1606).
[40] Y.C. Tseng and H.K. Pan, Data hiding in 2-color images, *IEEE Transactions on Computers* **51**, 873–890 (2002).
[41] J.H. van Lint, *Introduction to Coding Theory* (Springer-Verlag, GTM-86, New York, 1982).
[42] C. Wang, N. Wu and C. Tsai, High payload image steganographic method based on human visual perception, preprint (2007).
[43] P. Wayner, Mimic functions, *Cryptologia* **16**, 193–214 (1992).
[44] P. Wayner, *Disappearing Cryptography* (Morgan Kaufmann Publishers, San Francisco, 2002).
[45] D. Westfeld, F5. High capacity despite better steganalysis. In *Lecture Notes in Computer Science*, vol. 2137, 289–302, Springer, New York (2001).
[46] F.M.J. Willems and M. van Dijk, Capacity and codes for embedding information in grayscale signals, *IEEE Transactions on Information Theory* **51**, 1209–1214 (2005).
[47] V.K. Wei, Generalized Hamming weights for linear codes, *IEEE Transactions on Information Theory* **37**, 1412–1418 (1991).
[48] N-I. Wu1 and M-S. Hwang, Data Hiding: Current Status and Key Issues, *International Journal of Network Security* **4**, 19 (2007).
[49] W. Zhang and S. Li, Steganographic Codes. A new problem of coding theory, arXiv:cs/0505072 (2005).
[50] W. Zhang, X. Zhang and S. Wang, Maximizing steganographic embedding efficiency by combining Hamming codes and wet paper codes. In *10th International Workshop on Information Hiding*, K. Solani, K. Sullivan and U. Madhow, Editors. LNCS, 60–71, Springer, New York (2008).
[51] W. Zhang, X. Zhang and S. Wang, Near-Optimal Codes for Information Embedding in Gray-Scale Signals, *IEEE Transactions on Information Theory* **56**, 1262–1269 (2010).
[52] W. Zhang and X. Wang, Generalization of the ZZW embedding construction for Steganography, *IEEE Transactions on Information Forensics and Security* **4**, 564–569 (2009).
[53] W. Zhang and X. Zhu Improving the embedding efficiency of wet paper codes by paper folding, *IEEE Signal Processing Letters* **16**, 794–797 (2009).

Chapter 3

An Introduction to LDPC Codes

I. Márquez-Corbella[*] and E. Martínez-Moro[†]

*Institute of Mathematics IMUVa, University of Valladolid
Valladolid, Castilla, Spain
[*]iremarquez@agt.uva.es, [†]edgar@maf.uva.es*

The focus of this chapter is on low-density parity-check (LDPC) codes and iterative decoding methods. Low density parity check codes were originally introduced by Gallager in his doctoral thesis in 1963 [19]. Gallager discussed several iterative decoding algorithms for these codes that in general have very good error-correcting capabilities. Later, Tanner [32] in 1981 gave a graphical interpretation to these codes as a bipartite graph whose incidence matrix is the parity check matrix of the LDPC code, and showed how to view the iterative decoding algorithms as message-passing algorithms on these graphs.

Contents

3.1 Introduction 129
3.2 Representation for LDPC codes 130
    3.2.1 Tanner graph 130
3.3 Communication channels 134
3.4 Decoding algorithms 136
    3.4.1 Maximum-Likelihood Decoding 136
    3.4.2 Iterative decoding 137
    3.4.3 Linear Programming Decoding 145
3.5 Connections between LP and ML decoding 148
3.6 Pseudocodewords 155
3.7 Connections between LP and iterative decoding 158
References 162

3.1. Introduction

After the discovery of turbo codes in 1993 by Berrou et al. [6], LDPC codes were rediscovered by Mackay and Neal [27] in 1995. Both classes have excellent performances in terms of error correction close to the Shannon limit.

It is well known, that any linear code $\mathcal{C} \subseteq \mathbb{F}_q^n$ can be defined as the nullspace of a parity-check matrix $H_\mathcal{C}$ for $\mathcal{C}$, that is, $\mathcal{C} = \{\mathbf{c} \in \mathbb{F}_q^n \mid H_\mathcal{C}\, \mathbf{c}^t = 0\}$. A (d_r, d_c)-regular LDPC code, as originally defined by Gallager, is a binary linear code determined by the condition that every codeword bit participates in exactly d_c parity-check equations and that every such check equation involves exactly d_r codewords bits, where d_c and d_r are parameters that can be chosen freely.

Definition 3.1. Low-density parity-check (LDPC) codes are a class of linear block codes with a sparse parity-check matrix. That is, their parity-check matrix contains few non-zero entries in comparison to the amount of 0's.

In other words, an $m \times n$ matrix will be called *c-sparse* if as $m, n \to \infty$, the number of non-zero entries in this matrix is always less than $c \times \max(m, n)$. We can define two numbers describing these matrices: $d_r << n$ for the number of non-zero entries in each row and $d_c << m$ for the columns. An LDPC matrix is called (j, k)-regular if it contains j non-zero entries in every column (i.e. $d_r = j$) and k non-zero entries in every row (i.e. $d_c = k$).

3.2. Representation for LDPC codes

To represent LDPC codes we have basically two different possibilities: we can use its parity check matrix which are low-density or we can use the Tanner graph associated to it.

3.2.1. *Tanner graph*

Let V be a finite set, and denote by $E(V) = \{\{u, v\} \mid u, v \in V\}$ the subsets of two elements of V.

Definition 3.2. A graph G is a pair (V_G, E_G), where V_G is a nonempty set of elements called vertices and $E_G \subseteq E(V)$ is a (possibly empty) set of elements called edges, such that each edge $e \in E_G$ is assigned an unordered pair of vertices $\{u, v\}$ called endpoints of e. The graph G is finite if V_G is a finite set.

A *directed graph* or *digraph* is a graph where the edges have direction, in this case $\{u, v\} \neq \{v, u\}$. An edge of a graph which joins a vertex to itself is a *graph loop*. *Multiple edges* are two or more edges that are incident to the same two vertices. A *simple graph* is an unweighted, undirected graph containing no graph loops or multiple edges.

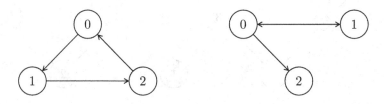

Fig. 3.1. An example of a directed graph and a graph with a multiple edge.

Definition 3.3. Let $G = (V_G, E_G)$ be a simple graph. For $v \in V_G$, the neighborhood of v is the set $N(v)$ of vertices $u \in V_G$ such that $\{u, v\} \in E_G$. Elements of $N(v)$ are called neighbors of v, and the degree of v is the number of neighbors it has. We say that G is d-regular if every vertex in G has degree d.

A *path* in $G = (V_G, E_G)$ is a finite sequence of vertices $v_0, \ldots v_k \in V_G$ such that $\{v_{i-1}, v_i\} \in E_G$ for $1 \leq i \leq k$. A *cycle* is a path such that $v_0 = v_k$.

A graph $G = (V_G, E_G)$ is said to be connected if for any two vertices $u, v \in V_G$ there is a path from u to v in G.

Definition 3.4. We say that $G = (V_G, E_G)$ is a tree if G is connected and has no cycles. A forest is a disjoint union of trees.

Definition 3.5. We say that $G = (V_G, E_G)$ is a bipartite graph if there is a partition of $V_G = U \cup W$ into two nonempty disjoint sets such that no two vertices within the same set are adjacent.

In other words, we say that $G = (V_G, E_G)$ is a bipartite graph and we denoted by $G = (U, W; E_G)$ if $V_G = U \cup W$ with $U \cap V = \emptyset$ and any $e \in E_G$ is of the form $e = \{u, v\}$ with $u \in U$ and $v \in V$.

In 1981, Tanner introduced a graphical representation for LDPC codes which not only provide a complete representation of the code but also allows efficient iterative message passing algorithms. The Tanner graph associated to an LDPC code is simply a bipartite graph whose incidence matrix is the parity check matrix of the LDPC code.

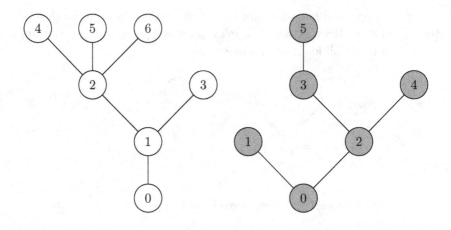

Fig. 3.2. An example of a forest with two trees.

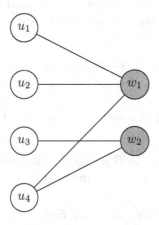

Fig. 3.3. An example of a bipartite graph.

Definition 3.6. A Tanner graph is a finite bipartite graph $T = (V, C; E_T)$. We call V the set of variable nodes of T and C the set of check nodes of T.

Given a binary linear code $\mathcal{C} \subseteq \mathbb{F}_2^n$, there is a one-to-one correspondence between Tanner graphs for $\mathcal{C}$ and parity check matrices for $\mathcal{C}$. Indeed if $H = (h_{ji})$ is an $m \times n$ binary matrix, then we can associate a Tanner graph $T = (V, C; E_T)$ to the matrix H by setting the variable nodes to the columns of H (i.e. $V = \{v_1, \ldots, v_n\}$) and the check nodes to the rows of H (i.e.

$C = \{c_1, \ldots, c_m\}$), and defining any edge of T as $e = \{v_i, c_j\}$ with $v_i \in V$, $c_j \in C$ if and only if $h_{ji} = 1$. Conversely, if $T = (V = \{v_1, \ldots, v_n\}, C = \{c_1, \ldots, c_m\}; E_T)$ is a Tanner graph, then we associate a binary $m \times n$ matrix $H = (h_{ji})$ to T, where $h_{ji} = 1$ if and only if $(v_i, c_j) \in E_T$.

For LDPC codes, the sparsity of H means that the corresponding graph is sparse in the number of edges. Therefore using a Tanner graph the LDPC code associated can be viewed as the set of all binary sequences assigned to the variable nodes such that all of the constraints imposed by the parity-check nodes are satisfied. That is, the codewords are those vectors $(c_1, \ldots, c_n)$ such that for all check nodes the sum of the neighboring positions among the variable nodes is zero.

Example 3.1. Let $H_\mathcal{C}$ be a (2,4)-regular sparse matrix, i.e. $H_\mathcal{C}$ contains 2 non-zero entries in every column and 4 non-zero entries in every row.

$$H_\mathcal{C} = \begin{pmatrix} 0 & 1 & 1 & 1 & 0 & 1 & 0 & 0 & 0 & 0 \\ 1 & 0 & 0 & 0 & 0 & 0 & 1 & 0 & 1 & 1 \\ 1 & 1 & 1 & 0 & 0 & 0 & 1 & 0 & 0 & 0 \\ 0 & 0 & 0 & 0 & 1 & 1 & 0 & 1 & 1 & 0 \\ 0 & 0 & 0 & 1 & 1 & 0 & 0 & 1 & 0 & 1 \end{pmatrix} \in \mathbb{F}_2^{5 \times 10}$$

We can associate to this matrix a binary linear code of parameters $[10, 6, 2]$ with generator matrix $G_\mathcal{C}$.

$$G_\mathcal{C} = \begin{pmatrix} 1 & 0 & 0 & 0 & 0 & 0 & 1 & 0 & 0 & 0 \\ 0 & 1 & 0 & 0 & 0 & 0 & 1 & 1 & 0 & 1 & 0 \\ 0 & 0 & 1 & 0 & 0 & 1 & 1 & 0 & 1 & 0 \\ 0 & 0 & 0 & 1 & 0 & 1 & 0 & 0 & 1 & 1 \\ 0 & 0 & 0 & 0 & 1 & 0 & 0 & 0 & 1 & 1 \\ 0 & 0 & 0 & 0 & 0 & 0 & 0 & 1 & 1 & 1 \end{pmatrix} \in \mathbb{F}_2^{6 \times 10}$$

To describe these elements in Magma we proceed as follows:

```
> A:=SparseMatrix(GF(2), 5, 10, [
    4, 2,1, 3,1, 4,1,  6,1,
    4, 1,1, 7,1, 9,1, 10,1,
    4, 1,1, 2,1, 3,1,  7,1,
    4, 5,1, 6,1, 8,1,  9,1,
    4, 4,1, 5,1, 8,1, 10,1 ]);
> Matrix(A);
[0 1 1 1 0 1 0 0 0 0]
[1 0 0 0 0 0 1 0 1 1]
```

```
[1 1 1 0 0 0 1 0 0 0]
[0 0 0 0 1 1 0 1 1 0]
[0 0 0 1 1 0 0 1 0 1]

> C := LDPCCode(A);
> C;
[10, 6, 2] Linear Code over GF(2)
Generator matrix:
[1 0 0 0 0 0 1 0 0 0]
[0 1 0 0 0 1 1 0 1 0]
[0 0 1 0 0 1 1 0 1 0]
[0 0 0 1 0 1 0 0 1 1]
[0 0 0 0 1 0 0 0 1 1]
[0 0 0 0 0 0 0 1 1 1]
```

Figure 3.4 shows the Tanner graph of the parity check matrix H_C where white nodes are *variable nodes* and gray nodes are *check nodes*.

3.3. Communication channels

In information theory a channel refers to a theoretical *channel model* with certain error characteristics. In this section, we will describe three basic communication channel models. In all three cases the input alphabet is binary, and the elements of the input alphabet are called bits.

- The *Binary Erasure Channel* (BEC) of communication was introduced by Elias [13] in 1995, but it was regarded as a rather theoretical channel model until the expansion of Internet. In this model, a transmitter sends a bit (zero or one) and the receiver either receives the bit or it receives a message that the bit was erased. Thus the output alphabet consists of 0, 1 and an additional element denoted by e and called *erasure*. Each bit is either transmitted correctly (with probability $1 - p$), or it is erased (with probability p). The main advantage of this channel is that the receiver can identify the location of all errors since any position that has a 1 or a 0 in the received word is correct.
- The *Binary Symmetric Channel* (BSC) of communication is defined by a binary input, a binary output and a probability of error p. That is, the transmission is not perfect and occasionally the receiver receives a bit which can be flipped (changed from 1 to 0, or vice

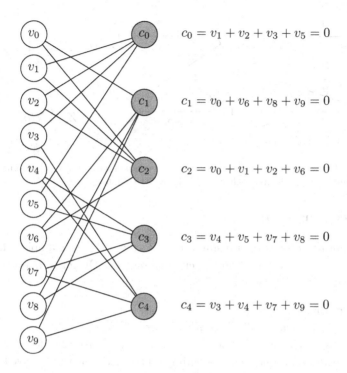

$c_0 = v_1 + v_2 + v_3 + v_5 = 0$

$c_1 = v_0 + v_6 + v_8 + v_9 = 0$

$c_2 = v_0 + v_1 + v_2 + v_6 = 0$

$c_3 = v_4 + v_5 + v_7 + v_8 = 0$

$c_4 = v_3 + v_4 + v_7 + v_9 = 0$

Fig. 3.4. Tanner graph of the parity check matrix H_C.

versa) with probability p. In this case, the location of errors is unknown by the receiver.
- The *Additive White Gaussian Noise* (AWGN) channel which adds white noise with a constant spectral density and a Gaussian distribution of amplitude to a binary-input information. Therefore, let X_i the binary-inputs and Z_i the noise, where Z_i is independent and identically-distributed and drawn from a zero-mean normal distribution with variance σ^2, then the AWGN channel is represented by a series of continuous outputs $Y_i = X_i + Z_i$. Wideband Gaussian noise comes from many natural sources, such as the thermal vibration of atoms in conductors, shot noise, black body radiation from the earth and other warm objects, and from celestial sources such as the sun.

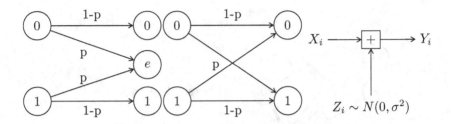

Fig. 3.5. Three examples of channels: The BEC channel with eraure probability p, the BSC channel with error probability p and the AWGN channel.

3.4. Decoding algorithms

The aim of *decoding* is to transmit a codeword $\mathbf{c} \in \mathcal{C}$ across a noisy channel and then estimate $\bar{\mathbf{c}}$ based on the channel output. During this process we can obtain three results:

(1) $\bar{\mathbf{c}} = \mathbf{c}$, this is a *decoding success*.
(2) $\bar{\mathbf{c}} = \mathbf{c}' \neq \mathbf{c}$, this is called *decoding error*.
(3) $\bar{\mathbf{c}} = \mathbf{r} \notin \mathcal{C}$, this case is known as a *decoding failure*, the decoder has a *noncodeword output* and, depending on the decoder, $\mathbf{r}$ may have binary, rational or real entries.

3.4.1. *Maximum-Likelihood Decoding*

The *Maximum-Likelihood Decoding* (MLD) is one of the most fundamental and significant topics in coding theory. In MLD, the decoders minimize the message error probability when each codeword is transmitted with equal probability, that is given a received message $\mathbf{y} \in \mathbb{F}_q^n$ the decoder tries to estimate the most likely transmitted message by finding the codeword $\mathbf{c} \in \mathcal{C}$ that maximizes the probability of receiving the sequence $\mathbf{y}$ from the channel, given that $\mathbf{c}$ was transmitted. Thus, the ML decoder estimates the codeword as:

$$\mathbf{c}_{ML} = \max_{\mathbf{c} \in \mathcal{C}} \mathrm{P}\left(\mathbf{y} \text{ received} \mid \mathbf{c} \text{ sent}\right).$$

Where the quantity $\mathrm{P}\left(\mathbf{y} \text{ received} \mid \mathbf{c} \text{ sent}\right)$ represents the probability of receiving the sequence $\mathbf{y} \in \mathbb{F}_q^n$ from the channel given that the codeword $\mathbf{c}$ was transmitted. This quantity depends on the type of errors the channel introduces in the transmitted message, the channel model.

In the case that all codewords are equally likely to be sent, then by Bayes Theorem we have:

$$\mathrm{P}\left(\mathbf{y} \text{ received} \mid \mathbf{c} \text{ sent}\right) = \frac{\mathrm{P}\left(\mathbf{y} \text{ received, } \mathbf{c} \text{ sent}\right)}{\mathrm{P}(\mathbf{c} \text{ sent})}$$

$$= \mathrm{P}\left(\mathbf{c} \text{ sent} \mid \mathbf{y} \text{ received}\right) \cdot \frac{\mathrm{P}(\mathbf{y} \text{ received})}{\mathrm{P}(\mathbf{c} \text{ sent})}$$

Since P(**y** received) is fixed if **y** is fixed and P(**c** sent) is constant by hypothesis, then P (**y** received | **c** sent) is maximized when P (**c** sent | **y** received) is maximized.

Note as well that if the probability of error on a discrete memoryless channel p is strictly less than one half, then the *minimum distance decoding* is equivalent to the *maximum likelihood decoding*. In the problem of minimum distance decoding, also known as *nearest neighbor decoding*, the decoders try to calculate the codeword that differs in less position with the received vector, i.e. the codeword having minimum hamming distance with the received vector. Hence if we denote by $d = d_H(\mathbf{c}, \mathbf{y}) = \sharp\{i \mid 1 \leq i \leq n, c_i \neq y_i\}$ to the Hamming distance between the received vector $\mathbf{y} \in \mathbb{F}_q^n$ and the sent codeword $\mathbf{c} \in \mathcal{C}$, then we have:

$$\mathrm{P}\left(\mathbf{y} \text{ received} \mid \mathbf{c} \text{ sent}\right) = (1-p)^{n-d} \cdot p^d = (1-p)^n \cdot \left(\frac{p}{1-p}\right)^d.$$

Since $p < \frac{1}{2}$, maximizing the above formula would be equivalent to minimizing the parameter d.

Although the ML decoding problem has a simple description its complexity increases exponentially with the code length for decoding general linear codes, since there is an exponentially large number of codewords in terms of the code length that are compared with the received vector. The computational complexity of ML decoding of general linear codes has been extensively studied. In [4], [5] it is shown that the problem is NP-hard even if the code is known in advance and we allow data preprocessing [8].

3.4.2. *Iterative decoding*

One of the features that makes LDPC codes attractive is the existence of various iterative message-passing algorithms which achieve excellent decoding performance. Iterative decoding algorithms are based on correspondences between graph theory and probability theory [35, 38] and have applications outside of coding theory in various applied statistical and computation fields. These applications include Bayesian networks [29], computer vision [20], statistical physics [12], statistical signal, image processing, artificial intelligence, etc.

The most common ones are: *sum-product* algorithm (also known as belief propagation) and *min-sum* algorithm which is widely studied due to its simplicity but it does not perform as well as the previous one (we refer the reader to [17], [26], [36]). These two algorithms are graph-based message passing algorithms which either converge to a solution that may be or not the ML solution (they do not guarantee the ML certificate property) or they do not converge at all.

At each round of a message-passing algorithm, messages are exchanged along the edges of the code's constraint graph (i.e. messages are passed from variable nodes to check nodes and vice versa). The message sent must not take into account the message sent in the previous round.

Example 3.2. Let us consider a binary LDPC code $\mathcal{C}$ whose transmission takes place over a BSC channel and let $H_\mathcal{C}$ be a parity check matrix of $\mathcal{C}$.

$$H_\mathcal{C} = \begin{pmatrix} 0 & 1 & 0 & 1 & 1 & 0 & 0 & 1 \\ 1 & 1 & 1 & 0 & 0 & 1 & 0 & 0 \\ 0 & 0 & 1 & 0 & 0 & 1 & 1 & 1 \\ 1 & 0 & 0 & 1 & 1 & 0 & 1 & 0 \end{pmatrix}.$$

Let $T = (V, C; E_T)$ be the Tanner graph corresponding to the code we have defined where $V = \{v_0, \ldots, v_7\}$ represents the variable nodes and $C = \{c_0, \ldots, c_3\}$ represent the check nodes. Figure 3.6 shows such Tanner graph.

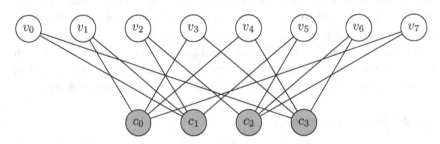

Fig. 3.6. Tanner graph of the parity check matrix $H_\mathcal{C}$.

Let $\mathbf{c} = (10010101) \in \mathcal{C}$ be the transmitted codeword and $\mathbf{y} = (11010101) \in \mathbb{F}_2^8$ the received vector. That is, $\mathbf{e} = (01000000)$ is the transmission error caused by the channel noise. We will explain with this simple example an iterative decoding algorithm which works with hard decision.

(1) In the first step all variable nodes v_i send a message to their check nodes c_j containing the bit they think is the correct one. The only information that the variable nodes have at this step is the received vector, thus the sent message is exactly the vector **y**. That means that if $(v_i, c_j) \in E_T$ then v_i sent the bit y_i to c_j.

c_0	received:	$v_1 \to 1$	$v_3 \to 1$	$v_4 \to 0$	$v_7 \to 1$
c_1	received:	$v_0 \to 1$	$v_1 \to 1$	$v_2 \to 0$	$v_5 \to 1$
c_2	received:	$v_2 \to 0$	$v_5 \to 1$	$v_6 \to 0$	$v_7 \to 1$
c_3	received:	$v_0 \to 1$	$v_3 \to 1$	$v_4 \to 0$	$v_6 \to 0$

(2) In the second step all check nodes, following the information received and the equations of the parity check matrix, calculated a response to every connected variable node with the bit they think is correct. That is, if V_c are the set of variable nodes incident to the check node c then, for each $v \in V_c$, c looks at the message received from the variable nodes $V_c \setminus \{v\}$ and calculates the bit that v should have in order to fulfill the parity check equation.

c_0	sent:	$0 \to v_1$	$0 \to v_3$	$1 \to v_4$	$0 \to v_7$
c_1	sent:	$0 \to v_0$	$0 \to v_1$	$1 \to v_2$	$0 \to v_5$
c_2	sent:	$0 \to v_2$	$1 \to v_5$	$0 \to v_6$	$1 \to v_7$
c_3	sent:	$1 \to v_0$	$1 \to v_3$	$0 \to v_4$	$0 \to v_6$

(3) **Can we estimate the transmitted codeword?**
The variable nodes use the information received from the check nodes to determine whether the original bit is correct. A simple way to do this is using majority vote between the original bit and the received messages.
- If the chosen vector is equal to the vector sent in the second step then the algorithm ends and we can deduce an estimate of the transmitted codeword.
- Otherwise, variable nodes send another message with the chosen vector to the check nodes and we return to step 2.

Table 3.3 shows step 3 of the described decoding algorithm applied to our example.

variable node	y_i received	received messages	decision
v_0	1	$c_1 \to 0$ $c_3 \to 1$	1
v_1	1	$c_0 \to 0$ $c_1 \to 0$	0
v_2	0	$c_1 \to 1$ $c_2 \to 0$	0
v_3	1	$c_0 \to 0$ $c_3 \to 1$	1
v_4	0	$c_0 \to 1$ $c_3 \to 0$	0
v_5	1	$c_1 \to 0$ $c_2 \to 1$	1
v_6	0	$c_2 \to 0$ $c_3 \to 0$	0
v_7	1	$c_0 \to 1$ $c_2 \to 1$	1

Note that the original received vector $\mathbf{y} = (11010101)$ does not coincide with the chosen vector $\hat{\mathbf{y}} = (10010101)$, hence we have to repeat step 2 using now the chosen vector $\hat{\mathbf{y}}$.

c_0	received:	$v_1 \to 0$	$v_3 \to 1$	$v_4 \to 0$	$v_7 \to 1$
	sent:	$0 \to v_1$	$1 \to v_3$	$0 \to v_4$	$1 \to v_7$
c_1	received:	$v_0 \to 1$	$v_1 \to 0$	$v_2 \to 0$	$v_5 \to 1$
	sent:	$1 \to v_0$	$0 \to v_1$	$0 \to v_2$	$1 \to v_5$
c_2	received:	$v_2 \to 0$	$v_5 \to 1$	$v_6 \to 0$	$v_7 \to 1$
	sent:	$0 \to v_2$	$1 \to v_5$	$0 \to v_6$	$1 \to v_7$
c_3	received:	$v_0 \to 1$	$v_3 \to 1$	$v_4 \to 0$	$v_6 \to 0$
	sent:	$1 \to v_0$	$1 \to v_3$	$0 \to v_4$	$0 \to v_6$

variable node	$\hat{y}_i$ received	received messages	decision
v_0	1	$c_1 \to 1$ $c_3 \to 1$	1
v_1	0	$c_0 \to 0$ $c_1 \to 0$	0
v_2	0	$c_1 \to 0$ $c_2 \to 0$	0
v_3	1	$c_0 \to 1$ $c_3 \to 1$	1
v_4	0	$c_0 \to 0$ $c_3 \to 0$	0
v_5	1	$c_1 \to 1$ $c_2 \to 1$	1
v_6	0	$c_2 \to 0$ $c_3 \to 0$	0
v_7	1	$c_0 \to 1$ $c_2 \to 1$	1

In this case the vector $\hat{\mathbf{y}}$ and the chosen vector are equal so we have corrected the transmission error. The estimation given by this algorithm of the sent codeword is (10010101).

For binary LDPC codes, the variable nodes assume the values one or zero. Hence, one important subclass of message passing algorithms is the *belief propagation algorithm* which is present in Gallager's work [19] and it is also used in Artificial Intelligence (see [27]). Let us call $\mathbf{y} \in \mathbb{F}_2^n$ the received vector and $\mathbf{c} \in \mathcal{C}$ the transmitted codeword. In this subclass a message is represented:

- Either as a probability vector p_0 or p_1 where:

$$\begin{cases} p_0 = \mathrm{P}(x_i = 0|\ y_i) \text{ is the probability that the variable node } x_i \\ \qquad \text{assumes the value 0.} \\ p_1 = \mathrm{P}(x_i = 1|\ y_i) \text{ is the probability that the variable node } x_i \\ \qquad \text{assumes the value 1.} \end{cases}$$

- Or as *log-likelihood ratios* $\ln \frac{\mathrm{P}(x_i=0|\ y_i)}{\mathrm{P}(x_i=1|y_i)} = \ln \frac{p_0}{p_1}$ denoted by $\ln \mathrm{L}(x_i|\ y_i)$.

Assuming variable independency we can easily derive formulas for these probabilities:

(1) Let x be an equiprobable random variable then by Bayes' theorem we have

$$\ln \mathrm{L}(x|\ y) = \ln \mathrm{L}(y|\ x).$$

(2) If $y_1, \ldots, y_k$ are independent random variables we have :

$$\ln \mathrm{L}(x|\ y_1, \ldots, y_k) = \sum_{i=1}^{k} \ln \mathrm{L}(x|\ y_i).$$

(3) If $x_1, \ldots, x_k$ are binary random variables and $y_1, \ldots, y_k$ are random variables:

$$\ln \mathrm{L}(x_1 \oplus \ldots \oplus x_k|\ y_1, \ldots, y_k) = \ln \frac{1 + \prod_{i=1}^{l} \tan(\frac{l_i}{2})}{1 - \prod_{i=1}^{l} \tan(\frac{l_i}{2})},$$

where $l_i = \ln \mathrm{L}(x_i|y_i)$ and $\oplus$ denotes addition over $\mathbb{F}_2$.

Before detailing the characteristics of min-sum and sum-product algorithms, we will fix some notation.

- Each iteration consists of two parts, first variable nodes send messages to its neighboring check nodes and then check nodes send messages to the corresponding variable nodes.
- Let $m_{vc}^{(l)}$ be the message passed from a variable node v to a check node c at the l-th round of the algorithm. This message is the result of applying a function whose input parameters are the messages received at v from neighboring check nodes other than c.
- Similarly, let $m_{cv}^{(l)}$ be the message passed from a check node c to a variable node v at the l-th round of the algorithm. This message is the result of a function whose input parameters are the messages received at c from neighboring variable nodes other than v.
- Let C_v denote the set of check nodes incident to a variable node v and V_c denote the set of variable nodes incident to check node c. Therefore $C_w \setminus \{u\}$ or $V_w \setminus \{u\}$ denote the set of neighbors of w excluding node u.

One important aspect of belief propagation is running time. Since these algorithms traverse edges in a sparse graph (i.e. the number of edges is small) the time needed to implement these algorithms is not high. Moreover the number of operations performed is linear in the number of variables nodes.

Another important property is that these algorithms are channel independent, whereas the messages passed during the algorithm are completely dependent on the channel.

3.4.2.1. Min-Sum Decoding

The *Min-Sum* (MS) algorithm (See [37] [16], [36]) for decoding LDPC codes is a parallel iterative soft decoding algorithm. At each iteration, first messages are sent from the variable nodes to the check nodes, then from check nodes back to the variable nodes. Later a probability for each bit is computed and finally a hard decoding decision is made for each bit based on the previous probability.

At iteration l, the updated equations for the messages can be described as:

- At a variable node:

$$m_{vc}^{(l)} = \begin{cases} \ln L(v^{(0)}) & \text{if } l = 0 \\ \ln L(v^{(0)}) + \sum_{c' \in C_v \setminus \{c\}} m_{c'v}^{(l-1)} & \text{if } l \geq 1 \end{cases}$$

where $\ln L(v^{(0)})$ represents the log-likelihood ratio for the input data bit v and $m_{vc}^{(l)}$ denotes the message sent from variable node v to check node c at iteration l. If we assume that v corresponds to the i-th bit of the input data $\mathbf{x}$ then m_{vc}^l is the log-likelihood ratio that the ith bit of the input data has the value 0 versus 1, given the information obtained via the check nodes other than check node c.

- At a check node:

$$m_{cv}^{(l)} = \prod_{v' \in V_c \setminus \{v\}} \text{sgn}(m_{v'c}^{(l)}) \cdot \min_{v' \in V_c \setminus \{v\}} |m_{v'c}^{(l)}|,$$

where $m_{cv}^{(l)}$ denotes the message sent from check node c to variable node v and $\text{sgn}(v)$ denotes the sign of v. If we assume that v corresponds to the i-th bit of the input data then $m_{cv}^{(l)}$ is the log-likelihood ratio that the check node c is satisfied when the input data bit i is fixed to value 0 versus value 1, and the other bits are independent with log-likelihood ratios $m_{v'c}^{(l)}$ for each $v' \in V_c \setminus \{v\}$.

The pseudo-code for the min-sum algorithm is given in Algorithm 1.

3.4.2.2. Sum-Product Decoding

The sum-product algorithm, also known as the belief propagation algorithm, is the most powerful iterative soft decoding algorithm for LDPC codes. This algorithm is known to perform very well in practice [10]. However it presents two problems: it does not always converge and it does not have the ML certificate property (i.e. even if the algorithm outputs a codeword it is not guaranteed to be an ML codeword).

This algorithm uses the local probabilities of each bit v_i which are based only on the received symbol y_i and the channel model used. For example, in the BSC, we have:

$$P(y_i|\ v_i = 0) = \begin{cases} p & \text{if } y_i = 1 \\ 1 - p & \text{if } y_i = 0 \end{cases}$$

The messages are pairs (m^0, m^1) that represent an estimate of the distribution on the marginal probabilities $P(y_i|v)$ over the settings of the received word $\mathbf{y}$. At iteration l, the updated equations for the messages can be described as:

- At a variable node the message $m_{vc}^{(l)} = (m_{vc}^0, m_{vc}^1)^{(l)}$ is performed by multiplying all its incoming messages (besides the one from c) together with the local probability. Thus the update is as follows:

$$(m_{vc}^0)^{(l)} = P(y_i| \ v = 0) \prod_{c' \in C_v \setminus \{c\}} (m_{c'v}^0)^{(l-1)}$$

$$(m_{vc}^1)^{(l)} = P(y_i| \ v = 1) \prod_{c' \in C_v \setminus \{c\}} (m_{c'v}^1)^{(l-1)}$$

- At a check node the message $m_{cv}^{(l)} = (m_{cv}^0, m_{cv}^1)^{(l)}$ is performed by adding all the probabilities of its local configuration for each setting of the bit v based on the incoming messages other than v. Thus the update is as follows:

$$(m_{cv}^0)^{(l)} = \sum_{\{S \in E_c : v \notin S\}} \prod_{v' \in S} (m_{v'c}^1)^{(l)} \prod_{v' \notin S \cup \{v\}} (m_{v'c}^0)^{(l)}$$

$$(m_{cv}^1)^{(l)} = \sum_{\{S \in E_c : v \in S\}} \prod_{v' \in S \setminus \{v\}} (m_{v'c}^1)^{(l)} \prod_{v' \notin S} (m_{v'c}^0)^{(l)}$$

Where S represents all possible configurations of the variable nodes. That is, S is the subset of the set of variable nodes $N(c)$ that are neighbors of a particular check node $c \in C$ that contain an even number of variable nodes, i.e.

$$S \subseteq E_c = \{S \subseteq N(c) \mid |S| \text{ even}\}.$$

Each subset defines a local codeword set, where v is set to 0 if $v \notin S$ and v is set to 1 otherwise.

Finally, at each bit a hard decision is made by multiplying all the incoming messages to a variable node along with the local probability. Thus we compute:

$$\alpha_i^0 = P(y_i|v = 0) \prod_{c \in C_v} (m_{vc}^0)^{(l)}$$

$$\alpha_i^1 = P(y_i|v = 1) \prod_{c \in C_v} (m_{vc}^1)^{(l)}$$

Then, if $\alpha_i^0 > \alpha_i^1$ we set $y_i = 0$, otherwise we set $y_i = 1$.

The pseudo-code for the sum-product algorithm is given in Algorithm 2.

3.4.3. Linear Programming Decoding

An *integer linear programming problem* is a technique which tries to find a non-negative integral vector that minimizes (or maximizes) a linear objective function subject to linear constraint equations with integral coefficients on the vector.

Definition 3.7. Given an integral matrix $A \in \mathbb{Z}^{m \times n}$, which is known as the *matrix of coefficients*, and the vectors $\mathbf{b} \in \mathbb{Z}^m$ and $\mathbf{w} \in \mathbb{R}^n$, the vector $\mathbf{w}$ is called the *cost vector*. The *integer linear programming problem*, denoted by $\text{IP}_{A,\mathbf{w}}(\mathbf{b})$ consists of three parts: a *linear function* to be minimized (or maximized), a *problem constraint* and *non-negative variables*. Therefore if we express the problem in matrix form it becomes:

$$\text{IP}_{A,\mathbf{w}}(\mathbf{b}) = \begin{cases} \text{minimize } \mathbf{w} \cdot \mathbf{u} \\ \text{subject to } \begin{cases} A\mathbf{u}^t = \mathbf{b} \\ \mathbf{u} \in \mathbb{Z}^n_{\geq 0} \end{cases} \end{cases}$$

A solution $\mathbf{u} \in \mathbb{Z}^n_{\geq 0}$ which satisfies $A\mathbf{u}^t = \mathbf{b}$ is called *optimal* if $\mathbf{u}$ minimizes the inner product $\mathbf{w} \cdot \mathbf{u}$.

Although an integer program differs from a linear program only in the requirement that solutions are integral instead of real, the general integer program is NP-complete while linear programs can be solved in polynomial time. The first general algorithm to solve an integer program was *Gomory's cutting plane method*. Then further methods around *branching and bounding* integer linear programs were designed. In [9], Conti and Traverso introduced a Gröbner basis based algorithm to solve this type of problem. For other algorithms and further reading on both linear and integer programming see [31].

We define the following *characteristic crossing* functions:

$$\blacktriangledown : \mathbb{Z}^s \to \mathbb{Z}^s_q \quad \text{and} \quad \blacktriangle : \mathbb{Z}^s_q \to \mathbb{Z}^s$$

Where s is determined by context and the spaces may also be matrix spaces. The map $\blacktriangledown$ is reduction modulo q, but the map $\blacktriangle$ replaces the class of $0, 1, \ldots, q-1$ by the same symbols regarded as integers, both maps act coordinate-wise. These maps will be used with matrices and vectors, themselves regarded as maps, acting on the right.

Definition 3.8. Similar to the previous definition, for an integer $q \geq 2$ and considering the matrix $A \in \mathbb{Z}^{m \times n}_q$ and the vectors $\mathbf{b} \in \mathbb{Z}^n_q$, $\mathbf{w} \in \mathbb{R}^n$, we define a *modular integer program*, denoted by $\text{IP}_{A,\mathbf{w},q}(\mathbf{b})$ as the problem of finding a vector $\mathbf{w} \in \mathbb{Z}_q^n$ that minimizes the inner product $\mathbf{w} \cdot \blacktriangle \mathbf{u}$ subject to $A\mathbf{u}^t \equiv \mathbf{b} \mod q$. If we express the problem in matrix form it becomes:

$$\mathrm{IP}_{A,\mathbf{w},q}(\mathbf{b}) = \begin{cases} \text{minimize } \mathbf{w} \cdot \blacktriangle \mathbf{u} \\ \text{subject to } \begin{cases} A\mathbf{u}^t \equiv \mathbf{b} \mod q \\ \mathbf{u} \in \mathbb{Z}_q^n \end{cases} \end{cases}$$

Note that the constraint conditions are modular ones but the weight minimizing condition is over the reals.

The ML decoding problem can be regarded as an integer program as follows. Let $q = 2$ and $\mathcal{C}$ be a binary linear block code of length n with parity check matrix $H_\mathcal{C}$. Then solving the modular program $\mathrm{IP}_{H_\mathcal{C},1,2}(\mathbf{b})$, where $\mathbf{1} = (1, 1, \ldots, 1) \in \mathbb{R}^n$, is equivalent to complete decoding $\mathbf{b}$.

The *support* of a vector $\mathbf{y} \in \mathbb{F}_2^n$ is the set of its non-zero positions, i.e.

$$\mathrm{supp}(\mathbf{c}) = \{i \mid c_i \neq 0\}.$$

Definition 3.9. A codeword $\mathbf{m}$ in the code $\mathcal{C}$ is said to be *minimal* if there is no other codeword $\mathbf{c} \in \mathcal{C}$ such that $\mathrm{supp}(\mathbf{c}) \subseteq \mathrm{supp}(\mathbf{m})$.

We will denote by $\mathcal{M}_\mathcal{C}$ the set of all the minimal codewords of $\mathcal{C}$.

Knowing the minimal codewords of the code $\mathcal{C}$ is sufficient in order to assess its ML decoding performance since the sets of minimal support codewords in linear codes had been considered to be related to *gradient-like decoding* algorithms [4, 21]. It is quite difficult to describe the set of minimal codewords for an arbitrary linear code even in the binary case since it is related with ML decoding problem. The problem of determining the set of minimal codewords has been solved for q-ary Hamming codes and for the second order binary Reed-Muller codes, and there have been attempts to characterize for other classes of codes like BCH codes and the rth-order binary Reed-Muller code, see [7] and the references therein. In [28] we describe the set of codewords of minimal support of codes defined on $\mathbb{Z}_q^n$ using the Graver basis associated to a modular integer programming.

It is important to recall that the modular integer programming approach does not allow the performance of complete decoding in a straightforward way for $q > 2$, but the description of the Graver basis of the modular problem provides a description of the minimal support codewords set of linear codes over $\mathbb{Z}_q^n$.

In recent years LP decoding for LDPC codes has received increased attention because of its practical performance, close to that of iterative decoding algorithms. The method introduced by Feldman et al. in [18], that we will detail below, show how to analyze LP decoding using properties of Tanner graphs.

Let's call $\mathbf{y} \in \mathbb{F}_2^n$ the received vector and $\mathbf{c} \in \mathcal{C}$ the transmitted codeword. Thus we can define the *log-likelihood ratios* to be

$$\lambda_i = \frac{P(Y_i = y_i | c_i = 0)}{P(Y_i = y_i | c_i = 1)}, \ i = 1, \ldots, n.$$

Thereby ML decoding estimates the nearest codeword to the received vector $\mathbf{y} \in \mathbb{F}_2^n$ as:

$$\mathbf{c}_{ML} = \min_{\mathbf{c} \in \mathcal{C}} \sum_{i=1}^{n} c_i \lambda_i.$$

For a given code $\mathcal{C}$ we define the codeword polytope to be the convex hull of all possible codewords:

$$\text{poly}(\mathcal{C}) = \{\sum_{\mathbf{c} \in \mathcal{C}} \beta_\mathbf{c} \mathbf{c} \mid \beta_\mathbf{c} \geq 0 \text{ and } \sum_{\mathbf{c} \in \mathcal{C}} \beta_\mathbf{c} = 1\}.$$

Then we can define ML decoding as the problem of

$$\begin{cases} \text{minimize } \sum \lambda_i c_i \\ \text{subject to } \mathbf{c} \in \text{poly}(\mathcal{C}). \end{cases}$$

This formulation is a linear program, since it involve minimizing a linear cost function over a polytope. Consequently, the optimum will always be attained at a vertex of poly($\mathcal{C}$) and these vertices are in one-to-one correspondence with codewords.

Let's remember that if the probability of error on a discrete memoryless channel p is strictly less than one half, then the *maximum likelihood decoding* for a code $\mathcal{C} \subseteq \mathbb{F}_2^n$ is equivalent to the *minimum distance decoding* where, given a vector $\mathbf{y} \in \mathbb{F}_2^n$, the goal is to find a codeword $\mathbf{x} \in \mathcal{C}$ so as to minimize the Hamming distance $d_H(\mathbf{x}, \mathbf{y}) = ||\mathbf{x} - \mathbf{y}||_1$. So with appropriate assumptions the problem can be reformulated as follows:

$$\mathbf{c}_{ML} = \min_{\mathbf{c} \in \text{poly}(\mathcal{C})} ||\mathbf{c} - \mathbf{y}||_1.$$

However, for most of the codes, the complexity of poly($\mathcal{C}$) grows exponentially in the block length and therefore finding the minimum of the above expressions using linear programming over long codes is impractical. To improve efficiency Feldman et al. in [14, 18] replaced the minimization over poly($\mathcal{C}$) by a minimization over some easily describable polytope $\mathcal{P}$ called *fundamental polytope* which is a relaxation of poly($\mathcal{C}$).

3.5. Connections between LP and ML decoding

Let $\mathcal{I} = \{1, \ldots, n\}$ and $\mathcal{J} = \{1, \ldots, m\}$ be a set of indices. We consider a binary linear code $\mathcal{C}$ represented by a Tanner graph $T = (V, C; E_T)$ where $V = \{v_i \mid i \in \mathcal{I}\}$ and $C = \{c_j \mid j \in \mathcal{J}\}$ defines the variable nodes and the check nodes respectively of T. Following the notation established in the previous sections we will denote by $N(c_j)$ the neighborhood of a check node $c_j \in C$ which is the set of variable nodes $v_i \in V$ that are incident to c_j in T. We will use the element $\mathbf{f} = (f_1, \ldots, f_n)$ to denote a set of bits where the value $f_i \in \{0, 1\}$ represent the assignment given to each variable node $v_i \in V$.

Since each check node represent a parity check equation then for each $c_j \in C$ we can define a local code $\mathcal{C}_j$ as:

$$\mathcal{C}_j = \{\mathbf{f} = (f_1, \ldots, f_n) \in \mathbb{F}_2^n \mid \sum_{v_i \in N(c_j)} f_i \equiv 0 \mod 2\}.$$

Furthermore, for each check node $c_j \in C$ if we consider the subsets $S \subseteq N(c_j)$ that contain an even number of variable nodes, i.e. S is in the set

$$E_j = \{S \subseteq N(c_j) \mid |S| \text{ even}\},$$

and we made a local code, $\mathcal{C}(j, S)$, match each of those subsets whose codewords are defined by setting $f_i = 1$ for each $v_i \in S$, $f_i = 0$ for each $v_i \in N(c_j) \setminus S$ and setting all other f_i arbitrarily.

Similarly in terms of polytopes we can define for each check node c_j and for each subset $S \in E_j$ a local codeword polytope, $\text{poly}(\mathcal{C}(j, S))$ as the convex hull of all possible codewords of the local code $\mathcal{C}(j, S)$.

Since we can define the original code $\mathcal{C}$ as the intersection of all the local codes, it seems natural to consider the relaxation polytope as the intersection of all the local codewords polytopes.

Feldman et al. in [18] introduce an auxiliary LP variable $w_{j,S}$ which can be seen as a way to verify that a codeword verifies check c_j using the configuration S, i.e. such codeword belongs to $\text{poly}(\mathcal{C}(j, S))$. Thus the variables $\{w_{j,S}\}$ must satisfy the following constraints:

$$0 \leq w_{j,S} \leq 1 \; \forall S \in E_j \tag{3.1}$$

$$\sum_{S \in E_j} w_{j,S} = 1 \tag{3.2}$$

Furthermore the indicator f_i at each variable node $v_i \in N(c_j)$ must belong to the local codeword polytope associate to the check node c_j, which is the intersection of all poly($\mathcal{C}(j,S)$) with $S \in E_j$. That means that the following equation must hold:

$$f_i = \sum_{S \in E_j : v_i \in S} w_{j,S} \ \forall v_i \in N(c_j) \tag{3.3}$$

With this assumptions, Feldman et al. defined the *fundamental polytope*, $\mathcal{P}$, as the set of points $(\mathbf{f}, \mathbf{w})$ such that Eqs. (3.1), (3.2) and (3.3) hold for all $j \in \mathcal{J}$ and they define a *Linear Code Linear Program* (LCLP) problem to be:

$$\begin{cases} \text{minimize} \sum_{i=1}^n \lambda_i f_i \\ \text{subject to } (\mathbf{f}, \mathbf{w}) \in \mathcal{P}. \end{cases}$$

We define the polytope $\overline{\mathcal{P}} = \{\mathbf{f} \mid \exists \mathbf{w} \text{ s.t. } (\mathbf{f}, \mathbf{w}) \in \mathcal{P}\}$ as the projection of $\mathcal{P}$ onto the $\{f_i\}$ variables. Since LCLP only involves the $\{f_i\}$ variables, optimizing over $\mathcal{P}$ or $\overline{\mathcal{P}}$ will produce the same result. Hence from now on we will refer to $\overline{\mathcal{P}}$ as the fundamental polytope.

If we enforce $f_i \in \{0,1\}$ for all $i \in \mathcal{I}$. Then Jeroslow in [22] proves that the polytope $\overline{\mathcal{P}}$ is exactly the set of points that satisfy:

$$\sum_{v_i \in S} f_i + \sum_{v_i \in (N(c_j) \setminus S)} (1 - f_i) \leq |N(c_j)| - 1 \tag{3.4}$$

for all check nodes $c_j \in C$ and $S \subseteq N(c_j)$ where $|S|$ is odd.

Furthermore we can rewrite Eq. (3.4) as follows:

$$\sum_{v_i \in N(c_j) \setminus S} f_i + \sum_{v_i \in S} (1 - f_i) \geq 1. \tag{3.5}$$

From the previous lines we can obtain the following definition:

Definition 3.10. Let $H \in \mathbb{F}_2^{m \times n}$ be the sparse matrix associated to a LDPC code $\mathcal{C}$ and let us denote by h_i the i-th row of H for $i \in \{1, \ldots, m\}$. Then the fundamental polytope is defined as:

$$\mathcal{P} = \bigcup_{i=1}^m \text{poly}(\mathcal{C}_i) \text{ where } \mathcal{C}_i = \{\mathbf{x} \in \mathbb{F}_2^n \mid h_i \mathbf{x}^t \equiv 0 \mod 2\}.$$

Let $T = (V, C; E_T)$ be the Tanner graph which represents $\mathcal{C}$ where $V = \{v_1, \ldots, v_n\}$ and $C = \{c_1, \ldots, c_m\}$ define the variable nodes and the check nodes respectively for T. Then the above definition can also be formulated as:

$$\mathcal{P} = \bigcup_{c_i \in C} \text{conv}(\mathcal{C}_i) \text{ where } \mathcal{C}_i = \{\mathbf{x} \in \mathbb{F}_2^n \mid \sum_{v_j \in N(c_i)} x_j \equiv 0 \mod 2\}.$$

Recall that $N(c_j)$ denotes the neighborhood of the check node c_j in T.

Note that the fundamental polytope depends on the parity check matrix that describes the code $\mathcal{C}$, so different parity check matrices for the same code might lead to different fundamental polytopes.

With the following proposition Feldman et al. in [18] prove that there is a one-to-one correspondence between codewords and integral solutions to LCLP.

Proposition 3.1. *Let $\mathcal{C}$ be a binary linear code and $\mathcal{P}$ the fundamental polytope associated to $\mathcal{C}$ then $\mathbf{f} = (f_1, \ldots, f_n)$ is a codeword of $\mathcal{C}$ if and only if $\mathbf{f}$ is an integral point in $\mathcal{P}$.*

Proof. Let $T = (V, C; E_T)$ be the Tanner graph which represents the code $\mathcal{C}$ where $V = \{v_i \mid i \in \mathcal{I}\}$ and $C = \{c_j \mid j \in \mathcal{J}\}$ define the variable nodes and the check nodes respectively for T.

First we will show that every codeword $\mathbf{c} \in \mathcal{C}$ is an integral point in the fundamental polytope $\mathcal{P}$. Let us suppose that there exists one codeword $\mathbf{f} \in \mathcal{C}$ which is not a point in $\mathcal{P}$.

Since $\mathbf{f}$ is a codeword, then it belongs to all local codes defined from the check nodes $c_j \in C$. That is, for all $j \in \mathcal{J}$ there exists a subset $S \subseteq E_j$ (i.e. $S \subseteq N(c_j)$ with $|S|$ odd) such that $f_i = 1$ for all $v_i \in S$ and $f_i = 0$ for all $v_i \in N(c_j) \setminus S$.

Now we define $S' = S \setminus \{v\}$ for some $v \in S$ and let us denote by f_v the value assigned in $\mathbf{f}$ to the variable node v. Then:

$$\sum_{v_i \in S'} f_i + \sum_{v_i \in (N(c_j) \setminus S')} (1 - f_i) = \sum_{v_i \in S \setminus \{v\}} f_i + \sum_{v_i \in N(c_j) \setminus S} (1 - f_i) + (1 - f_v)$$
$$= |S| - 1 + |N(c_j)| - |S|$$
$$= |N(c_j)| - 1.$$

Therefore by Eq. (3.4) we have that $\mathbf{f} = (f_1, \ldots, f_n)$ is a point in $\mathcal{P}$. In addition, since $\mathbf{f} \in \mathbb{F}_2^n$ then $\mathbf{f}$ is an integral point in $\mathcal{P}$.

Now we will show that every integral point in $\mathcal{P}$ is a codeword of $\mathcal{C}$. Let $\mathbf{f}$ be an arbitrary integral point of $\mathcal{P}$ such that $\mathbf{f}$ is not a codeword. Then we know that there is some check equation that $\mathbf{f}$ does not verify, i.e. for some check node $c_j \in C$ we have that

$$\sum_{v_i \in N(c_j)} f_i \not\equiv 0 \mod 2.$$

Therefore there exists $S \subseteq N(c_j)$, $|S|$ odd such that $f_i = 1$ for all $v_i \in S$

and $f_i = 0$ for all $v_i \in N(c_j) \setminus S$. Thus we have that:

$$\sum_{v_i \in N(c_j)} f_i = \sum_{v_i \in S} f_i + \sum_{v_i \in N(c_j) \setminus S} f_i = |S| \not\equiv 0 \mod 2.$$

This implies that:

$$\sum_{v_i \in S} f_i + \sum_{v_i \in N(c_j) \setminus S} (1 - f_i) = |N(c_j)| > |N(c_j)| - 1$$

contradicting Eq. (3.4), i.e. contradicting the fact that $\mathbf{f} \in \mathcal{P}$. □

The decoding algorithm based on the LCLP problem consists on applying linear programming LP on this problem and then either outputs the integral solution (if the solution obtained verifies $\mathbf{f} \in \{0, 1\}^n$) or outputs a *decoding failure*.

Proposition 3.2. *The LP decoding has the ML certificate property. That is if the algorithm outputs a codeword, it is guaranteed to be the ML codeword.*

Proof. If the LP algorithm outputs a codeword $\mathbf{f} \in \mathcal{C}$ then, by definition, the cost of $\mathbf{f} \in \mathcal{C}$ is less than or equal to the cost of all points of the fundamental polytope $\mathcal{P}$. From Proposition 3.1 we know that all codewords of $\mathcal{C}$ are points of $\mathcal{P}$. Then the cost of $\mathbf{f} \in \mathcal{C}$ is less than or equal to the cost of the other codewords of the code. □

Theorem 3.1. *Let us suppose that the codeword $\mathbf{c} \in \mathcal{C} \subseteq \mathbb{F}_2^n$ is transmitted. The LCLP decoder will succeed if and only if there is no other feasible solution to LCLP with cost less than the cost of $\mathbf{c} \in \mathcal{C}$.*

Proof. By Proposition 3.1 $\mathbf{c} \in \mathcal{C}$ is a feasible solution to LCLP. Furthermore if $\mathbf{c}$ minimizes the cost function then it is the unique optimal solution to LCLP. Thus the decoder will output $\mathbf{c}$ which is the transmitted codeword and we will obtain a decoding success.

On the contrary, if there is a feasible solution $\mathbf{y} \in \mathbb{F}_2^n$ to LCLP with lower cost than the cost of the transmitted codeword $\mathbf{c} \in \mathcal{C}$, i.e. $\mathbf{y}, \mathbf{c} \in \mathcal{P}$ and $\sum \lambda_i y_i < \sum \lambda_i c_i$. Therefore, the decoder either outputs $\mathbf{c}' \in \mathcal{C} \setminus \{\mathbf{c}\}$ which will be a *decoding error* or it outputs $\mathbf{r} \notin \mathcal{C}$ which will be a *decoding failure*. □

Corollary 3.1. *The LP decoder will fail if and only if there is a non-zero point in $\mathcal{P}$ with cost less than or equal to zero.*

Wainwright et al. in [34] show that if the Tanner graph associated to a linear binary code $\mathcal{C}$ has no cycles then all solutions of the LCLP problem are integral. However if the Tanner graph has cycles then the optimal solution to LCLP may not be integral.

The following example has been extracted from the article [18].

Example 3.3. Let us consider $\mathcal{C}$ the Hamming code of parameters $[7,4,3]$ and let $H_\mathcal{C} \in \mathbb{F}_2^{3\times 7}$ be a parity check matrix of $\mathcal{C}$.

$$H_\mathcal{C} = \begin{pmatrix} 1 & 0 & 1 & 0 & 1 & 0 & 1 \\ 0 & 1 & 1 & 0 & 0 & 1 & 1 \\ 0 & 0 & 0 & 1 & 1 & 1 & 1 \end{pmatrix} \in \mathbb{F}_2^{3\times 7}.$$

Let $T = (V, C; E_T)$ be the Tanner graph corresponding to $H_\mathcal{C}$ where $V = \{v_i \mid i \in \mathcal{I} = \{1, \ldots 6\}\}$ represent the variable nodes and $C = \{c_j \mid j \in \mathcal{J} = \{0, 1, 2\}\}$ represent the check nodes. Figure 3.7 shows two representations of this Tanner graph.

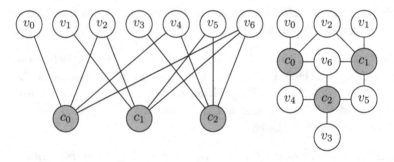

Fig. 3.7. Two representations of the Tanner graph of $H_\mathcal{C}$.

We define the cost vector λ as the vector $\left(-\frac{7}{2}, 1, \ldots, 1\right) \in \mathbb{F}_2^7$. We observe that all codewords have nonnegative cost since any codeword $\mathbf{f} = (f_0, \ldots, f_6) \in \mathcal{C}$ with negative cost would have $f_0 = 1$ and therefore, from the parity check matrix, we would have that at least two other different elements of $\mathbf{f}$ would have value 1 having a total cost greater than $\frac{1}{4}$.

If we set $\mathbf{f} = (1, 0, \frac{1}{2}, \frac{1}{2}, 0, 0, \frac{1}{2})$ and we define the following auxiliary variables of the form $w_{j,S}$:

$$\begin{cases} w_{0,\{0,2\}} = w_{0,\{0,6\}} = \frac{1}{2} \\ w_{1,\{2,6\}} = w_{1,\emptyset} = \frac{1}{2} \\ w_{2,\{3,6\}} = w_{2,\emptyset} = \frac{1}{2} \end{cases}$$

Then it is easy to verify that $(\mathbf{f}, \mathbf{w})$ satisfied all of the LCLP constraint and the cost of $(\mathbf{f}, \mathbf{w})$ is $-\frac{1}{4}$ which is strictly less than the cost of any codeword.

As we have already seen, in general LP decoding is not equivalent to ML decoding. Note that ML decoding only considers codewords as outputs whereas LP decoding considers any vertex of the fundamental polytope which include both codewords and non-trivial LP pseudocodewords. However in the special case of AWGN channel, Axvig et al. in [2] show that this difference is smaller.

Over the AWGN channel, ML decoding is based on squared Euclidean distance between modulated points where the modulation map for any $\mathbf{x} \in \mathbb{R}^n$ is given by $m(\mathbf{x}) = 2\mathbf{x} - \mathbf{1}$ with $\mathbf{1} = (1, \ldots, 1) \in \mathbb{R}^n$.

Let us consider a binary linear code $\mathcal{C} \subseteq \mathbb{F}_2^n$ and a received vector $\mathbf{y} \in \mathbb{R}^n$. Then the output of ML decoding is given by:

$$\hat{\mathbf{c}}_{ML} = m^{-1}\left(\min_{\mathbf{c} \in m(\mathcal{C})} \sum_{i=1}^n (y_i - c_i)^2\right) = m^{-1}\left(\min_{\mathbf{c} \in m(\mathcal{C})} \sum_{i=1}^n (y_i^2 - 2y_i c_i + c_i^2)\right).$$

If we denote by $\mathcal{V}(\mathcal{P})$ the set of vertices of the fundamental polytope (recall that $\mathcal{C} \subseteq \mathcal{V}(\mathcal{P})$), then the ML estimation can be extended to the following rule called *generalized maximum likelihood* (GML) output:

$$\hat{\mathbf{x}}_{GML} = m^{-1}\left(\min_{\mathbf{v} \in m(\mathcal{V}(\mathcal{P}))} \sum_{i=1}^n (y_i^2 - 2y_i c_i + c_i^2)\right).$$

In the following Theorem Axvig et al. in [3] compare ML decoding and GML decoding to LP decoding.

Theorem 3.2. *Let us consider a binary linear code $\mathcal{C} \subseteq \mathbb{F}_2^n$ whose transmission takes place over an AWGN channel and let $H_\mathcal{C}$ be a parity check matrix of $\mathcal{C}$. Then the following statements are equivalent:*

(1) $\mathcal{C}$ has no non-trivial LP pseudocodewords with respect to $H_\mathcal{C}$.

(2) LP decoding for $\mathcal{C}$ with respect to $H_\mathcal{C}$ is equivalent to ML decoding for $\mathcal{C}$.

(3) LP decoding for $\mathcal{C}$ with respect to $H_\mathcal{C}$ is equivalent to GML decoding for $\mathcal{C}$ with respect to $H_\mathcal{C}$.

Proof. Following the notation established above, let $\mathcal{V}(\mathcal{P})$ be the set of vertices of the fundamental polytope and $m(\mathcal{V}(\mathcal{P}))$ be the set of n-dimensional vertices of the modulated fundamental polytope $m(\mathcal{P})$.

Given a received vector $\mathbf{y} \in \mathbb{R}^n$, the log-likelihood ratios for $\mathbf{y}$ on the AWGN channel with noise variance σ^2 are given by:

$$\lambda_i = \ln \frac{\mathrm{P}(y_i|\ x_i = 0)}{\mathrm{P}(y_i|\ x_i = 1)} = \frac{-2y_i}{\sigma^2}.$$

Hence the decision rule for the LP decoding can be formulated as:

$$\hat{\mathbf{x}}_{LP} = \min_{\mathbf{w} \in \mathcal{V}(\mathcal{P})} \sum_{i=1}^{n} \lambda_i w_i = \min_{\mathbf{w} \in \mathcal{V}(\mathcal{P})} \sum_{i=1}^{n} \frac{-2y_i}{\sigma^2} w_i.$$

Since the modulation map is coordinate-wise linear and strictly increasing we have:

$$\hat{\mathbf{x}}_{LP} = m^{-1} \left(\min_{\mathbf{v} \in m(\mathcal{V}(\mathcal{P}))} \sum_{i=1}^{n} \frac{-2y_i}{\sigma^2} v_i \right).$$

As σ is independent of $\mathbf{v} \in m(\mathcal{V}(\mathcal{P}))$ and adding y_i^2 to the ith entry will not change the minimization then:

$$\hat{\mathbf{x}}_{LP} = m^{-1} \left(\min_{\mathbf{v} \in m(\mathcal{V}(\mathcal{P}))} \sum_{i=1}^{n} -2y_i v_i \right) = m^{-1} \left(\min_{\mathbf{v} \in m(\mathcal{V}(\mathcal{P}))} \sum_{i=1}^{n} (y_i^2 - 2y_i v_i) \right)$$

- If we assume that $\mathcal{C}$ has no non-trivial LP pseudocodewords with respect to $H_\mathcal{C}$ then $m(\mathcal{V}(\mathcal{P})) = m(\mathcal{C})$ and $\sum_{i=1}^{n} v_i^2 = n$ since $v_i = \pm 1$ for all $\mathbf{v} \in m(\mathcal{C})$. Thus adding v_i^2 to the ith entry does not change the minimization problem described above. Hence $\hat{\mathbf{x}}_{LP} = \hat{\mathbf{x}}_{ML} = \hat{\mathbf{x}}_{GML}$.
- Conversely, let us suppose that LP decoding for $\mathcal{C}$ with respect to $H_\mathcal{C}$ is equivalent to ML decoding for $\mathcal{C}$ and assume that $\mathcal{C}$ has a non-trivial LP pseudo-codeword $\mathbf{w}$ with respect to $H_\mathcal{C}$. Then, by our assumption, there exists a vector λ of log-likelihood ratios such that

$$\sum_{i=1}^{n} \lambda_i w_i \leq \sum_{i=1}^{n} \lambda_i c_i \text{ for any } \mathbf{c} \in \mathcal{C}.$$

Over the AWGN channel the received vector may be any vector in $\mathbb{R}^n$, so we can construct a received vector $\mathbf{y}$ such that the resulting log-likelihood vector is λ. Thus, if $\mathbf{y}$ is received, LP decoding will return a nontrivial pseudocodeword, whereas ML decoding will always return a codeword. Hence, LP and ML decodings are not equivalent which contradicts the assumption made.
- Furthermore, because $\mathcal{C}$ has a nontrivial LP pseudocodeword with respect to $H_\mathcal{C}$ then $\sum_{i=1}^{n} v_i^2$ is not constant over $\mathbf{v} \in m(\mathcal{V}(\mathcal{P}))$. Hence the LP decision rule differs from the GML decision rule, which contradicts the fact that LP and GML are equivalent.

□

3.6. Pseudocodewords

The set of points of the *fundamental polytope* $\mathcal{P}$ associated with the linear code $\mathcal{C}$ is greater than or equal to the set of points of poly($\mathcal{C}$), so it can happen that $\mathcal{P}$ has vertices that are not codewords of $\mathcal{C}$. This fact leads to the definition of pseudocodeword of a code which is the main point of this section.

Definition 3.11. A linear programming pseudocodeword of a code $\mathcal{C}$ is any vertex $\mathbf{f} \in \mathbb{F}_2^n$ of the fundamental polytope $\mathcal{P}$ associated to one of its parity check matrix $H_\mathcal{C}$. A non-trivial linear programming pseudocodeword is a linear programming pseudocodeword that is not a codeword.

Definition 3.12. Let $\mathcal{I} = \{1, \ldots, n\}$ and $\mathcal{J} = \{1, \ldots, m\}$ be a set of indices and let $T = (V, C; E_T)$ be the Tanner graph representing a binary LDPC code $\mathcal{C}$ where $V = \{v_i \mid i \in \mathcal{I}\}$ and $C = \{c_j \mid j \in \mathcal{J}\}$ denotes the variable nodes and the check nodes respectively of T and $E_T = \{(v, c) \mid v \in V, c \in C\}$. Assume that an iterative message-passing algorithm has been run on T for a total of l iterations, where a single iteration consists of a message passing from the variable nodes to the check nodes and then back to the variable nodes. Then a *computation tree for T of depth l* is formed by enumerating the Tanner graph from an arbitrary node, called the root of the tree, down through the desired number of decoding iteration.

A computation tree for l iterations and having variable node v_i acting as the root node of the tree is denoted by $C_i(T)_l$.

Remark that a *computation tree of depth l*, $C_i(T)_l$, depends on the chosen iterative message-passing algorithm but it will always have $2l + 1$ levels where the 0th level consists of the root variable v_i, each even-numbered level contains only variable nodes and each odd-numbered level contains only check nodes.

A *valid assignment* on a computation tree is an assignment of zeros and ones to the variable nodes such that all check nodes are satisfied. Given a codeword $\mathbf{c} = (c_1, \ldots, c_n) \in \mathcal{C}$ and a computation tree $C_j(T)_l$ for l iterations and having variable node $v_j \in V$ of the Tanner graph $T = (V, C; E_T)$ which represents $\mathcal{C}$, if we assign the value c_i to each copy of the variable node v_i for $i \in \mathcal{I}$ we will obtain a valid assignment on $C_j(T)_l$. However, not all valid assignments of zeros and ones to the variable nodes correspond to a codeword. This motivates the following definitions.

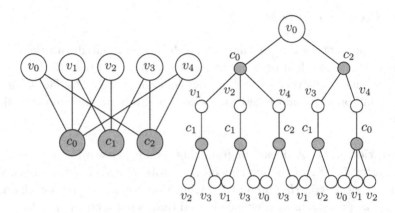

Fig. 3.8. A Tanner graph T and its computation tree for 2 iterations and having variable node v_0, $C_0(T)_2$.

Definition 3.13. Let $C_j(T)_l$ be a computation tree for l iterations and having variable node $v_j \in V$ of the Tanner graph $T = (V, C; E_T)$ which represents the code $\mathcal{C}$. A *codeword* $\mathbf{c} \in \mathcal{C}$ corresponds to a valid assignment on the computation tree such that, for each $i \in \mathcal{I}$, all copies of the variable node $v_i \in V$ in the computation tree are assigned the same value. Otherwise, a valid assignment which does not verify the above property corresponds to a *non-trivial pseudocodeword* of the code.

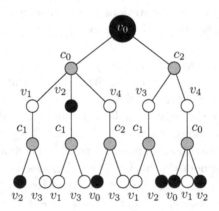

Fig. 3.9. A valid assignment corresponding to a codeword of the computation tree of Figure 3.8, where black nodes represent the assignment of the value 0 and white nodes represent the assignment of the value 1.

An Introduction to LDPC Codes

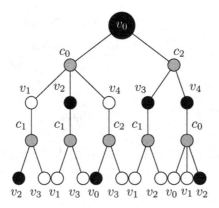

Fig. 3.10. A valid assignment corresponding to a non-trivial pseudocodeword of the computation tree of Figure 3.8, where black nodes represent the assignment of the value 0 and white nodes represent the assignment of the value 1.

The definition of pseudocodewords in terms of computation tree was originally made by Wiber in [36]. Later this work was extended by Frey et al. in [18] and recently, Koetter and Vontobel in [24] introduced another interpretation of pseudocodeword as we will describe below.

In [36] Wiber also gives explicit cost functions for both min-sum and sum-product decoding algorithm such that the i-th entry of the output vector of these algorithms after l iterations coincide with the value assigned to the variable node v_i by the cost function on a computation tree for l iterations and having variable node v_i acting as the root node, i.e. $C_i(T)_l$.

Definition 3.14. A finite degree l cover of a bipartite graph $G = (V, U; E_G)$ is a bipartite graph $\hat{G}$ where for each vertex $x_i \in V \cup U$ we define l vertices $\hat{x}_{i_1}, \ldots \hat{x}_{i_l}$ of $\hat{G}$ with $\deg(\hat{x}_{i_j}) = \deg(x_i)$ for all $1 \leq j \leq l$ and for every $\{x_i, x_j\} \in E_G$ there are l edges connecting each vertex $\hat{x}_{i_r}$ with a vertex $\hat{x}_{j_s}$.

Definition 3.15. A pseudocodeword of G is a vector $\mathbf{p} = (p_1, \ldots, p_n)$ obtained by reducing a codeword

$$\hat{c} = (\hat{c}_{1_1}, \hat{c}_{1_2}, \ldots, \hat{c}_{1_l}, \hat{c}_{2_1}, \ldots, \hat{c}_{2_l}, \ldots, \hat{c}_{n_1}, \ldots, \hat{c}_{n_l})$$

of the Tanner graph $\hat{G}$, where $\hat{G}$ represents a degree l lift of G. The reduction can be performed in two different ways:

(1) In the first case we obtain a pseudocodeword with integer entries and we refer to it as *unscaled pseudocodeword*. In this case the reduction is

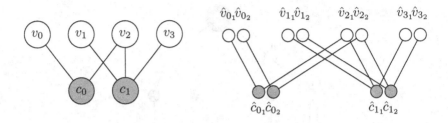

Fig. 3.11. A 2 degree cover of a Tanner graph.

given by:
$$p_i = \hat{c}_{i_1} + \hat{c}_{i_2} + \ldots + \hat{c}_{i_l}, \text{ for all } i = 1, \ldots, n.$$

Where $+$ denotes the real addition.

Note that the element p_i is the number of variables $\hat{c}_{i_j}$ with $1 \leq j \leq l$ that are assigned the value 1 in the degree l lift of G, i.e. for $1 \leq i \leq n$, $p_i = \sharp\{j \mid \hat{c}_{i_j} = 1\}$.

(2) The second case yields to a pseudocodeword with rational entries defined as follows:
$$p_i = \frac{\hat{c}_{i_1} + \hat{c}_{i_2} + \ldots + \hat{c}_{i_l}}{l}$$

Definition 3.16. An unscaled pseudocodeword $\mathbf{p} = (p_1, \ldots, p_n)$ is *irreducible* if it cannot be written as a sum of other unscaled pseudocodewords. Otherwise the pseudocodeword $\mathbf{p}$ is said to be *reducible*.

Definition 3.17. A pseudocodeword $\mathbf{p}$ of a Tanner graph G is said to be *lift-realizable* if $\mathbf{p}$ is obtained by reducing (using Definition 3.15) a valid assignment on some degree lift of G.

There are some relationships between LP decoding and iterative decoding that we will study in more detail in the next section. However, it is noteworthy that the pseudocodewords of the LP decoder are exactly the pseudocodewords arising from a graph cover for the min-sum algorithm.

3.7. Connections between LP and iterative decoding

Comparing the performance of LP decoding and iterative decoding is an open issue. In theory, LP decoding, which runs in polynomial time, is less efficient than iterative decoding, most of which run in linear time for a fixed number of iterations. However in [14] they show that LP decoding is equivalent to iterative decoding in several cases. For example in the binary erasure channel (BEC), they show that the performance of LP decoding is equivalent to belief propagation algorithm.

Definition 3.18. A stopping set S is a subset of the set of variable nodes such that all neighbors of S are connected to S at least twice.

Note, in particular, that the empty set is a stopping set.

Lemma 3.1. *The space of stopping sets is closed under unions.*

Proof. We have to prove that if S_1 and S_2 are both stopping sets then so is $S_1 \cup S_2$. If c is a neighbor of $S_1 \cup S_2$ then it must be a neighbor of at least one $v \in S_1$ or $v \in S_2$. Assume that c is a neighbor of $v \in S_1$, since S_1 is a stopping set, then c has at least two connections to S_1 and therefore at least two connections to $S_1 \cup S_2$. □

We will call *size of a stopping set* S the number of elements of S. If there is no smaller sized nonempty stopping set containing S, then S is said to be *minimal*. The smallest minimal stopping set is called a minimum stopping set which is not necessarily unique. However, each subset of the set of variable nodes contains a unique maximal stopping set (which might be the empty set).

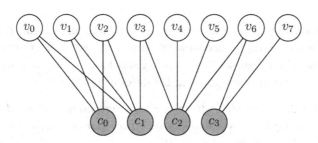

Fig. 3.12. Stopping sets in a Tanner graph. We observe that $\{v_0, v_1, v_2\}$ is a minimum stopping set whereas $\{v_4, v_5, v_6, v_7\}$ is a minimal stopping set.

One useful observation is that the support of a lift-realizable pseudocodeword forms a stopping set in the Tanner graph associated to a code $\mathcal{C}$.

Next Lemma is presented in [11] and shows the importance of stopping sets in the process of iterative decoding of LDPC codes over a binary erasure channel (BEC).

Lemma 3.2. *Belief propagation (sum-product) algorithm over BEC fails if and only if a stopping set exists among the erased bits.*

Proof. Suppose that we use a binary LDPC code $\mathcal{C}$ over the binary erasure channel (BEC) and we want to decode the received word using an iterative decoding algorithm which would be applied until either a codeword is recovered or until the decoder fails. Let $T = (V, C; E_T)$ be the Tanner graph associated to a parity check matrix of $\mathcal{C}$, where $V = \{v_i \mid i \in \mathcal{I}\}$ and $C = \{c_j \mid j \in \mathcal{J}\}$ defines the variable nodes and the check nodes respectively of T. We will denote by $\mathcal{E}$ the set of variable nodes which were erased by the channel.

We have to see that the set of erasures which remains when the decoder stops is equal to the unique maximal stopping set S contained in $\mathcal{E}$. From Definition 3.18 we know that every neighbor of an element of S has at least two connections to the set S. Thus, since $S \subseteq \mathcal{E}$, the iterative decoder cannot determine the variable nodes contained in S, which is the unique maximal stopping set contained in $\mathcal{E}$.

Conversely, suppose that the decoder is unable to determine the value of a set of variable nodes denoted by S. This implies that all neighbor of these variables has at least two connections to the set S, that is, S is a stopping set. Furthermore S must be the maximal stopping set since no erasure contained in a stopping set can be determined by the iterative decoder. □

From Corollary 3.1 we know that the LP decoder will fail if and only if there is a non-zero pseudocodeword with zero cost. Next theorem which was presented in [14] shows that stopping sets for iterative decoding are the special case of pseudocodewords in LP decoder in the BEC.

Theorem 3.3. *In the BEC, there is a non-zero pseudocodeword with zero cost if and only if there is a stopping set. Therefore, the performance of LP and BP decoding are equivalent in BEC.*

Proof. Suppose that we use an LDPC binary code $\mathcal{C}$ over the BEC. Since the BEC is memoryless and symmetric, we can model it with a cost function λ setting $\lambda_i = -1$ if the received bit f_i has the value 1, $\lambda_i = 1$ if the received bit f_i has the value 0 and $\lambda_i = 0$ if the received bit is an erasure. Let $T = (V, C; E_T)$ be the Tanner graph associated to a parity check matrix $H_\mathcal{C}$ of $\mathcal{C}$, where $V = \{v_i \mid i \in \mathcal{I}\}$ and $C = \{c_j \mid j \in \mathcal{J}\}$ defines the variable nodes and the check nodes respectively of T and as before, we will denote by $\mathcal{E}$ to the set of variable nodes which were erased by the channel.

First we will see that if there exists a zero-cost codeword, then there exists a stopping set among the erased bits. Let $\mathbf{f} = (f_1, \ldots, f_n) \in \mathbb{F}_2^n$ be a pseudocodeword with zero cost, i.e.

$$\sum_{i \in \mathcal{I}} \lambda_i f_i = 0. \quad (3.6)$$

By definition of pseudocodeword, for each check node $c_j \in C$ if we consider the subsets $S \subseteq N(c_j)$ that contain an even number of variable nodes, i.e. S is in the set $E_j = \{S \subseteq N(c_j) \mid |S| \text{ even }\}$, then there exist LP variables $\{w_{j,S}\}$ which verify the following constraint:

$$0 \leq w_{j,S} \leq 1 \quad \forall S \in E_j \quad (3.7)$$

$$\sum_{S \in E_j} w_{j,S} = 1 \quad (3.8)$$

$$f_i = \sum_{S \in E_j : v_i \in S} w_{j,S} \; \forall v_i \in N(c_j). \quad (3.9)$$

Let us define the subset $M = \{v_i \in V \mid f_i > 0\}$. Since $\lambda_i \geq 0$ for all $v_i \in V$ we have that $\lambda_i = 0$ for all $v_i \in M$, therefore $M \subseteq \mathcal{E}$. Now suppose that M is not a stopping set, then there exists at least one check node $c_j \in \left(\bigcup_{v_i \in M} N(v_i) \right)$ such that c_j has only one neighbor v_{k_1} in the set M. From Eq. (3.9) we have that:

$$f_i = \sum_{S \in E_j : v_i \in S} w_{j,S} \; \forall v_i \in N(c_j).$$

Since $v_{k_1} \in M$ by definition of M we have that $f_{k_1} > 0$. Thus there must be some $S \in E_j$ such that $v_{k_1} \in S$ and $w_{j,S} > 0$. Since S has even cardinality, there exists at least another variable node $v_{k_2} \in S \subseteq N(c_j)$ which not only is neighbor of the check node c_j but also verifies that $f_{k_2} > 0$. Indeed, from Eq. (3.9) we have that:

$$h_{k_2} = \sum_{S \in E_j : v_i \in S} w_{j,S} \geq w_{j,S} > 0.$$

This implies that $v_{k_1}, v_{k_2} \in N(c_j)$ and $v_{k_1}, v_{k_2} \in M$, which contradicts the fact that c_j has only one neighbor in M.

Conversely, let us see that if there exists a stopping set, then there is a zero-cost pseudocodeword. Let S be a stopping set, then we can build a pseudocodeword $\mathbf{f}$ by setting $f_i = 2$ for all $v_i \in S$ and $f_i = 0$ for all $v_i \notin S$. Since $S \subseteq \mathcal{E}$ we have that $\sum_{i \in \mathcal{I}} \lambda_i f_i = 0$, therefore it suffices to prove that $\mathbf{f}$ is well defined, i.e. we have to see that $\mathbf{f}$ is a vertex of the fundamental polytope $\mathcal{P}$ associated to the parity check matrix $H_\mathcal{C}$.

For all check node, let us define the set $M(c_j) = N(c_j) \cap S$. By definition of stopping set we have that $|M(c_j)| \geq 2$. If $|M(c_j)|$ is even then for all check node $c_j \in \bigcup_{v_i \in S} N(v_i)$ we set $w_{j,M(c_j)} = 2$, whereas if $|M(c_j)|$ is odd, let $I = \{i_1, i_2, i_3\}$ be three arbitrary elements of $M(c_j)$, we set

$$\begin{cases} w_{j,M(c_j)\setminus I} = 2 \\ w_{j,\{i_1,i_2\}} = w_{j,\{i_1,i_3\}} = w_{j,\{i_2,i_3\}} = 1 \end{cases}$$

and the other LP variables we set them as zeros. Thus we have that:

$$f_i = \sum_{M \in E_j : v_i \in M} w_{j,S} = \begin{cases} 2 \ \forall v_i \in S \text{ and } c_j \in N(v_i) \\ 0 \ \forall v_i \notin S \text{ and } c_j \in N(v_i). \end{cases}$$

$\square$

References

[1] S. Arora, C. Daskalakis and D. Steurer. *Message passing algorithms and improved LP decoding*. Proceedings of the 41st annual ACM symposium on Theory of computing, 3-12, New York, 2009.

[2] N. Axvig, E. Price, E. Psota, D. Turk, L. C. Pérez and J. L. Walker. *A universal theory of pseudocodewords*. Proceedings of the 45th Annual Allerton Conference on Communication, Control and Computing, Sept. 2007.

[3] N. Axvig, D. Dreher, K. Morrison, E. Psota, L. C. Perez and J. L. Walker. *Analysis of Connections Between Pseudocodewords*. IEEE Transaction on Information Theory, volume 55, 4099-4107, 2009.

[4] A. Barg. *Complexity issues in coding theory*. Handbook of Coding Theory Elsevier Science, volume 1, 649-754, 1998.

[5] Elwyn R. Berlekamp, Robert J. McEliece and Henk C. A. van Tilborg. *On the inherent intractability of certain coding problems*. Institute of Electrical and Electronics Engineers. Transactions on Information Theory, volume 24, no.3, 384-386, 1978.

[6] C. Berrou, A. Glavieux and P. Thitimajshima. *Near Shannon limit error-correcting coding and decoding: Turbo-codes*. Proceedings of the IEEE International Conference on Communication, 1064-1070, Geneva, Switzerland, 1993.

[7] Y. Borissow and N. Manev. *Minimal codewords in linear codes*. Serdica Math. J., volume 30, 303-324, 2004.

[8] J. Bruck and M. Naor. *The Hardness of Decoding Linear Codes with Preprocessing*. IEEE Trans. Inform. Theory, volume 36, no.2, 1990.

[9] P. Conti and C. Traverso. *Buchberger algorithm and integer programming*. Applied algebra, algebraic algorithms and error-correcting codes, Lecture Notes in Comput. Sci, volume 539, 130-139, 1991.

[10] D. Divsalar, H. Jin and R. McEliece. *Coding theorems for "turbo-like" codes*. Proc. 36th Annual Allerton Conference on Communication, Control and Computing, 201-210, Sept. 1998.

[11] C. Di, D. Proietti, T. Richardson, E. Telatar and R. Urbanke. *Finite Length analysis of low-density parity check nodes.* IEEE Transactions on Information Theory, volume 48(6), 2002.
[12] S. F. Edwards and P. W. Anderson. *J. Phys. F..* 5(965), 1975.
[13] P. Elias. *Coding for two noisy channels.* Proc. 3rd London Symp. Information Theory, 61-76, London U.K., 1963.
[14] J. Feldman. *Decoding Error-Correcting Codes via Linear Programming.* PhD thesis, Massachusetts Institute of Technology. Cambridge, MA, 2000.
[15] J. Feldman, M. Wainwright and D.R. Karger. *Using Linear Programming to Decode Linear Codes.* IEEE Trans. Inform. Theory, volume 51, 954-972, 2005.
[16] M. Fossorier, M. Mihaljevic and H. Imai. *Reduced complexity iterative decoding of low density parity check codes based on belief propagation.* IEEE Transactions on Communications, volume 47, 673-680, 1999.
[17] G. D. Forney, Jr., R. Koetter, F. Kschischang and A. Reznik. *On the effective weights of pseudocodewords for codes defined on graphs with cycles.* The IMA Volumes in Mathematics and its applications, volume 123, chapter 5, 101-112, 2001.
[18] B. Frey, R. Koetter and A. Vardy. *Signal-space characterization of iterative decoding.* IEEE Transactions on Information Theory, IT-47(2), 766-781, 2001.
[19] R. J. Gallager. *Low Density Parity Check Codes.* M.I.T. Press, Cambridge, MA, 1963.
[20] S. Geman and D. Geman. *Stochastic relaxation, gibbs distributions and the bayesian restoration of images.* IEEE Transaction in Pattern Analysis and Machine Intelligence, 6(6): 721-741, 1984.
[21] T. Y. Hwang. *Decoding linear block codes for minimizing word error rate.* IEEE Trans. Inform. Theory, volume 25, 733-737, 1979.
[22] R. G. Jeroslow. *On defining sets of vertices of the hypercube by linear inequalities.* Discr. Math. volume 11, 119-124, 1975.
[23] C. A. Kelley. *Pseudocodewords, expander graphs and the algebraic construction of low-density parity-check codes.* Ph.D dissertation, Notre Dame, Indiana, 2006.
[24] R. Koetter and P. O. Vontobel. *Graph-covers and iterative decoding of finite length codes.* In Proceedings of the IEEE International Symposium on Turbo Codes and Applications, Brest, France, Sept. 2003.
[25] R. Koetter, W. C. W. Li, P. O. Vontobel and J. L. Walker. *Pseudocodewors of cycle codes via zeta functions.* IEEE Inform. Theory Workshop, 7-12, San Antonio, TX, USA, 2004.
[26] F. Kschischang, B. J. Frey and H. A. Loeligar. *Factor graphs and the sum-product algorithm.* IEEE Transaction in Information Theory, IT-47(2), 498-519, 2001.
[27] D. J. C. Mackay and R. M. Neal. *Good codes based on very sparse matrices.* Cryptography and Coding, 5th IMA Conference, 1995.
[28] I. Márquez-Corbella and E. Martínez-Moro. *Combinatorics of minimal*

codewords of some linear codes. Submitted to Advances in Mathematics of Communications, 2010.

[29] J. Pearl. *Probabilistic Reasoning in Intelligent Systems: Networks of Plausible Inference.* Morgan Kaufmann, 1988.

[30] R. Smarandache and P. O. Vontobel. *Pseudo-Codeword Analysis of Tanner Graphs from Projective and Euclidean Planes.* IEEE Transactions on Information Theory, volume 53, 2376-2393, 2007.

[31] A. Schrijver. *Theory of Linear and Integer Programming.* Wiley-Interscience, 1996.

[32] R. M. Tanner. *A recursive approach to low complexity codes.* IEEE Transactions on Information Theory, IT-27(5): 533-547, 1981.

[33] P. O. Vontobel, R. Smarandache, N. Kiyavash, J. Teutsch and D. Vukobratovic. *On the Minimal Pseudo-Codewords of Codes from Finite Geometries.* Proceedings of ISIT 2005 (International Symposium on Information Theory), 980-984, 2005.

[34] M. J. Wainwright, T. S. Jaakkola and A. S. Willsky. *MAP estimation via agreement on (hyper)trees: Message-passing and linear programming approaches*, Proc. 40th Annu. Allerton Conf. Communication, Control and Computing, Monticello, Oct. 2002.

[35] M. J. Wainwright. *Stochastic processes on graphs with cycles: geometric and variational approaches.* PhD thesis, Linkoping University, Sweden, 1996.

[36] N. Wiberg. *Codes and decoding on general graphs.* Ph.D dissertation, Department of Electrical Engineering, Linkoping University, Linkoping, Sweden, 1996.

[37] X. Huang. *Sigle-Scan Min-Sum Algorithms for Fast Decoding of LDPC Codes.* IEEE Information Theory Workshop. ITW'06 Chengdu, 140-143, 2006.

[38] Y. S. Yedidia, W. T. Freeman and Y. Weiss. *Understanding belief propagation and its generalizations.* Technical Report TR2001-22, Mitsubishi Electric Research Labs, January 2002.

Algorithm 1: MS algorithm

Input : A binary LDPC code $\mathcal{C}$ with parity check matrix $H_{\mathcal{C}} \in \mathbb{F}_2^{m \times n}$, a Tanner graph $T = (V, C; E_T)$ associated to $H_{\mathcal{C}}$ where $V = \{v_1, \ldots, v_n\}$ and $C = \{c_1, \ldots, c_m\}$ define the variable nodes and check nodes respectively of T. And a received vector $\mathbf{y} \in \mathbb{F}_2^n$.

Output: The min-sum (MS) decoder estimation $\mathbf{c}_{MS}$ of the original codeword transmitted.

<u>Initialization:</u>
for $v \in V$
  | We define $\ln L(v^{(0)}) = \ln \frac{P(v=0 \mid y_i)}{P(v=1 \mid y_i)}$;
endfor
$l := 0$;
// where l represent the number of iterations performed.
while l do not exceed some cap or $H_{\mathcal{C}} \, \mathbf{c}_{MS}^t = 0$ **do**
  | l:=l+1;
  | **for** $v \in V$
  | | <u>Variable node update rule:</u>
  | | $m_{vc}^{(l)} = \begin{cases} \ln L(v^{(0)}) & \text{if } l = 0 \\ \ln L(v^{(0)}) + \sum_{c' \in C_v \setminus \{c\}} m_{c'v}^{(l-1)} & \text{if } l \geq 1 \end{cases}$;
  | **endfor**
  | **for** $c \in C$
  | | <u>Check node update rule:</u>
  | | $m_{cv}^{(l)} = \prod_{v' \in V_c \setminus \{v\}} \mathrm{sgn}(m_{v'c}^{(l)}) \cdot \min_{v' \in V_c \setminus \{v\}} |m_{v'c}^{(l)}|$;
  | **endfor**
  | <u>Decoding:</u>
  | For each bit compute its log-likelihood ratio:
  | $$\ln L(v^{(l+1)}) = \ln L(v^{(0)}) + \sum_{c \in C_v} m_{cv}^l$$
  | Then estimate the original codeword $\mathbf{c}_{MS} = (\hat{c}_1, \ldots, \hat{c}_n)$ where:
  | $$\hat{c}_i = \begin{cases} 0 \text{ if } \ln L(v^{(l+1)}) > 0 \\ 1 \text{ otherwise} \end{cases} \text{ for } i = 1, \ldots, n.$$
end
return $\mathbf{c}_{MS}$.

Algorithm 2: SP algorithm

Input : A binary LDPC code $\mathcal{C}$ with parity check matrix $H_\mathcal{C} \in \mathbb{F}_2^{m \times n}$, a Tanner graph $T = (V, C; E_T)$ associated to $H_\mathcal{C}$ where $V = \{v_1, \ldots, v_n\}$ and $C = \{c_1, \ldots, c_m\}$ define the variable nodes and check nodes respectively of T. And a received vector $\mathbf{y} \in \mathbb{F}_2^n$.

Output : The sum-product (SP) decoder estimation $\mathbf{c}_{SP}$ of the original codeword transmitted.

$l := 0$;
// where l represents the number of iterations performed.
while l do not exceed some cap or $H_\mathcal{C}\, \mathbf{c}_{SP}^t = 0$ **do**
  l:=l+1;
  for $v \in V$
    Variable node update rule:
    $$(m_{vc}^0)^{(l)} = \mathrm{P}(y_i|\ v = 0) \prod_{c' \in C_v \setminus \{c\}} (m_{c'v}^0)^{(l-1)}$$
    $$(m_{vc}^1)^{(l)} = \mathrm{P}(y_i|\ v = 1) \prod_{c' \in C_v \setminus \{c\}} (m_{c'v}^1)^{(l-1)}$$
  endfor
  for $c \in C$
    Check node update rule:
    $$(m_{cv}^0)^{(l)} = \sum_{\{S \in E_c : v \notin S\}} \prod_{v' \in S} (m_{v'c}^1)^{(l)} \prod_{v' \notin S \cup \{v\}} (m_{v'c}^0)^{(l)}$$
    $$(m_{cv}^1)^{(l)} = \sum_{\{S \in E_c : v \in S\}} \prod_{v' \in S \setminus \{v\}} (m_{v'c}^1)^{(l)} \prod_{v' \notin S} (m_{v'c}^0)^{(l)}$$
    // Where $S \subseteq E_c = \{S \subseteq N(c) \mid |S| \text{ even}\}$.
  endfor
  Decoding:
  For each bit we compute:
  $$\alpha_i^0 = \mathrm{P}(y_i|v=0) \prod_{c \in C_v} (m_{vc}^0)^{(l)}$$
  $$\alpha_i^1 = \mathrm{P}(y_i|v=1) \prod_{c \in C_v} (m_{vc}^1)^{(l)}$$
  Then estimate the original codeword $\mathbf{c}_{PS} = (\hat{c}_1, \ldots, \hat{c}_n)$ where:
  $$\hat{c}_i = \begin{cases} 0 \text{ if } \alpha_i^0 > \alpha_i^1 \\ 1 \text{ otherwise} \end{cases} \text{ for } i = 1, \ldots, n.$$
end
return $\mathbf{c}_{SP}$.

Chapter 4

Numerical Semigroups and Codes

Maria Bras-Amorós [*]

Universitat Rovira i Virgili

maria.bras@urv.cat

A numerical semigroup is a subset of $\mathbb{N}$ containing 0, closed under addition and with finite complement in $\mathbb{N}$. An important example of numerical semigroup is given by the Weierstrass semigroup at one point of a curve. In the theory of algebraic geometry codes, Weierstrass semigroups are crucial for defining bounds on the minimum distance as well as for defining improvements on the dimension of codes. We present these applications and some theoretical problems related to classification, characterization and counting of numerical semigroups.

Contents

4.1 Introduction 167
4.2 Numerical semigroups 168
    4.2.1 Paradigmatic example: Weierstrass semigroups on algebraic curves . . 168
    4.2.2 Basic notions and problems 173
    4.2.3 Classification 175
    4.2.4 Characterization 183
    4.2.5 Counting 192
4.3 Numerical semigroups and codes 195
    4.3.1 One-point codes and their decoding 195
    4.3.2 The ν sequence, classical codes, and Feng-Rao improved codes 198
    4.3.3 Generic errors and the τ sequence 208
References 214

4.1. Introduction

Numerical semigroups are probably one of the most simple mathematical objects. However they are involved in very hard (and some very old)

---

[*]This work was partly supported by the Spanish Ministry of Education through projects TSI2007-65406-C03-01 "E-AEGIS" and CONSOLIDER CSD2007-00004 "ARES", and by the Government of Catalonia under grant 2005 SGR 00446.

problems. They can also be found in several applied fields such as error-correcting codes, cryptography, or combinatorial structures for privacy applications.

In this chapter we present numerical semigroups with some of the related classical problems and we explore their importance in the field of algebraic-geometry codes.

The material is divided into two parts. In the first part we give a brief introduction to Weierstrass semigroups as the paradigmatic example of numerical semigroups, we present some classical problems related to general numerical semigroups, we deal with some problems on classification and characterization of numerical semigroups which have an application to coding theory, and we finally present a conjecture on counting numerical semigroups by their genus.

In the second part we present one-point algebraic-geometry codes and we focus on the applications that numerical semigroups have for defining bounds on the minimum distance as well as for defining improvements on the dimension of these codes. Based on the decoding algorithm for one-point codes one can deduce sufficient conditions for decoding, and from these conditions one can define minimal sets of parity checks (and so codes with improved correction capability) either for correcting any kind of error or at least for guaranteeing the correction of the so-called generic errors. The decoding conditions are related to the associated Weierstrass semigroups and so the improvements can be defined in terms of semigroups.

4.2. Numerical semigroups

4.2.1. *Paradigmatic example: Weierstrass semigroups on algebraic curves*

4.2.1.1. *Algebraic curves*

Consider a field K and a bivariate polynomial $f(x, y) \in K[x, y]$. If $\bar{K}$ is the algebraic closure of K, the (plane) *affine curve* associated to f is the set of points in $\bar{K}^2$ at which f vanishes. Now given a *homogeneous* polynomial $F(X, Y, Z) \in K[X, Y, Z]$ the (plane) *projective curve* associated to F is the set of points in $\mathbb{P}^2(\bar{K})$ at which F vanishes. We use the notation $\mathcal{X}_F$ to denote it.

From the affine curve defined by the polynomial $f(x, y)$ of degree d we can obtain a projective curve defined by the *homogenization* of f, that is, $f^*(X, Y, Z) = Z^d f(\frac{X}{Z}, \frac{Y}{Z})$. Conversely, a projective curve defined by a homogeneous polynomial $F(X, Y, Z)$ defines three affine curves with *dehomogenized* polynomials $F(x, y, 1)$, $F(1, u, v)$, $F(w, 1, z)$. The points $(a, b) \in \bar{K}^2$ of the affine curve defined by $f(x, y)$ correspond to the points $(a : b : 1) \in \mathbb{P}^2(\bar{K})$ of $\mathcal{X}_{f^*}$. Conversely, the points $(X : Y : Z)$ with $Z \neq 0$ (resp. $X \neq 0$, $Y \neq 0$) of a projective curve $\mathcal{X}_F$ correspond to the points of the affine curve defined by $F(x, y, 1)$ (resp. $F(1, u, v)$, $F(w, 1, z)$) and so they are called affine points of $F(x, y, 1)$. The points with $Z = 0$ are said to be *at infinity*.

In the case $K = \mathbb{F}_q$, any point of $\mathcal{X}_F$ is in $\mathbb{P}^2(\mathbb{F}_{q^m})$ for some m. If L/K is a field extension we define the *L-rational points* of $\mathcal{X}$ as the points in the set $\mathcal{X}_F(L) = \mathcal{X}_F \cap L^2$.

We will assume that F is irreducible in any field extension of K (i.e. *absolutely irreducible*). Otherwise the curve is a proper union of two curves.

If two polynomials in $K(X, Y, Z)$ differ by a multiple of F, when evaluating them at a point of $\mathcal{X}_F$ we obtain the same value. Thus it makes sense to consider $K(X, Y, Z)/(F)$. Since F is irreducible, $K(X, Y, Z)/(F)$ is an integral domain and we can construct its field of fractions Q_F. For evaluating one such fraction at a projective point we want the result not to depend on the representative of the projective point. Hence, we require the numerator and the denominator to have one representative each, which is a homogeneous polynomial and both having the same degree. The *function field* of $\mathcal{X}_F$, denoted $K(\mathcal{X}_F)$, is the set of elements of Q_F admitting one such representation. Its elements are the *rational functions* of $\mathcal{X}_F$. We say that a rational function $f \in K(\mathcal{X}_F)$ is *regular in a point* P if there exists a representation of it as a fraction $\frac{G(X,Y,Z)}{H(X,Y,Z)}$ with $H(P) \neq 0$. In this case we define $f(P) = \frac{G(P)}{H(P)}$. The ring of all rational functions regular in P is denoted $\mathcal{O}_P$. Again it is an integral domain and this time its field of fractions is $K(\mathcal{X}_F)$.

Let $P \in \mathcal{X}_F$ be a point. If all the partial derivatives F_X, F_Y, F_Z vanish at P then P is said to be a *singular point*. Otherwise it is said to be a *simple point*. Curves without singular points are called *non-singular, regular* or *smooth* curves.

From now on we will assume that F is absolutely irreducible and that $\mathcal{X}_F$ is smooth.

The *genus* of a smooth plane curve $\mathcal{X}_F$ may be defined as

$$g = \frac{(\deg(F) - 1)(\deg(F) - 2)}{2}.$$

For general curves the genus is defined using differentials on a curve which is out of the purposes of this survey.

4.2.1.2. Weierstrass semigroup

Theorem 4.1. *Consider a point P in the projective curve $\mathcal{X}_F$. There exists $t \in \mathcal{O}_P$ such that for any non-zero $f \in K(\mathcal{X}_F)$ there exists a unique integer $v_P(f)$ with*

$$f = t^{v_P(f)} u$$

for some $u \in \mathcal{O}_P$ with $u(P) \neq 0$. The value $v_P(f)$ depends only on $\mathcal{X}_F$, P.

If $G(X,Y,Z)$ and $H(X,Y,Z)$ are two homogeneous polynomials of degree 1 such that $G(P) = 0$, $H(P) \neq 0$, and G is not a constant multiple of $F_X(P)X + F_Y(P)Y + F_Z(P)Z$, then we can take t to be the class in $\mathcal{O}_P$ of $\frac{G(X,Y,Z)}{H(X,Y,Z)}$.

An element such as t is called a *local parameter*. If there is no confusion we will write $\frac{G(X,Y,Z)}{H(X,Y,Z)}$ for its class in $\mathcal{O}_P$. The value $v_P(f)$ is called the *valuation* of f at P. The point P is said to be a *zero* of multiplicity m if $v_P(f) = m > 0$ and a *pole* of multiplicity $-m$ if $v_P(f) = m < 0$. The valuation satisfies that $v_P(f) \geqslant 0$ if and only if $f \in \mathcal{O}_P$ and that in this case $v_P(f) > 0$ if and only if $f(P) = 0$.

Lemma 4.1.

(1) $v_P(f) = \infty$ if and only if $f = 0$
(2) $v_P(\lambda f) = v_P(f)$ for all non-zero $\lambda \in K$
(3) $v_P(fg) = v_P(f) + v_P(g)$
(4) $v_P(f+g) \geqslant \min\{v_P(f), v_P(g)\}$ and equality holds if $v_P(f) \neq v_P(g)$
(5) If $v_P(f) = v_P(g) \geqslant 0$ then there exists $\lambda \in K$ such that $v_P(f - \lambda g) > v_P(f)$.

Let $L(mP)$ be the set of rational functions having only poles at P and with pole order at most m. It is a K-vector space and so we can define $l(mP) = \dim_K(L(mP))$. One can prove that $l(mP)$ is either $l((m-1)P)$ or $l((m-1)P) + 1$. There exists a rational function $f \in K(\mathcal{X}_F)$ having only one pole at P with $v_P(f) = -m$ if and only if $l(mP) = l((m-1)P) + 1$.

Let $A = \bigcup_{m \geqslant 0} L(mP)$, that is, A is the ring of rational functions having poles only at P. Define $\Lambda = \{-v_P(f) : f \in A \setminus \{0\}\}$. It is obvious that $\Lambda \subseteq \mathbb{N}_0$, where $\mathbb{N}_0$ denotes the set of all non-negative integers.

Lemma 4.2. *The set $\Lambda \subseteq \mathbb{N}_0$ satisfies*

(1) $0 \in \Lambda$
(2) $m + m' \in \Lambda$ whenever $m, m' \in \Lambda$
(3) $\mathbb{N}_0 \setminus \Lambda$ has a finite number of elements

Proof.

(1) Constant functions $f = a$ have no poles and satisfy $v_P(a) = 0$ for all $P \in \mathcal{X}_F$. Hence, $0 \in \Lambda$.
(2) If $m, m' \in \Lambda$ then there exist $f, g \in A$ with $v_P(f) = -m$, $v_P(g) = -m'$. Now, by Lemma 4.1, $v_P(fg) = -(m+m')$ and so $m + m' \in \Lambda$.
(3) The well-known Riemann-Roch theorem implies that $l(mP) = m+1-g$ if $m \geqslant 2g - 1$. On one hand this means that $m \in \Lambda$ for all $m \geqslant 2g$, and on the other hand, this means that $l(mP) = l((m-1)P)$ only for g different values of m. So, the number of elements in $\mathbb{N}_0$ which are not in Λ is equal to the genus. □

The three properties of a subset of $\mathbb{N}_0$ in the previous lemma will constitute the definition of a *numerical semigroup*. The particular numerical semigroup of the lemma is called the *Weierstrass semigroup* at P and the elements in $\mathbb{N}_0 \setminus \Lambda$ are called the *Weierstrass gaps*.

4.2.1.3. *Examples*

Example 4.1 (Hermitian curve). Let q be a prime power. The Hermitian curve $\mathcal{H}_q$ over $\mathbb{F}_{q^2}$ is defined by the affine equation $x^{q+1} = y^q + y$ and homogeneous equation $X^{q+1} - Y^q Z - Y Z^q = 0$. It is easy to see that its partial derivatives are $F_X = X^q$, $F_Y = -Z^q$, $F_Z = -Y^q$ and so there is no projective point at which $\mathcal{H}_q$ is singular. The point $P_\infty = (0:1:0)$ is the unique point of $\mathcal{H}_q$ at infinity.

We have $F_X(P_\infty)X + F_Y(P_\infty)Y + F_Z(P_\infty)Z = -Z$ and so $t = \frac{X}{Y}$ is a local parameter at P_∞. The rational functions $\frac{X}{Z}$ and $\frac{Y}{Z}$ are regular everywhere except at P_∞. So, they belong to $\cup_{m \geqslant 0} L(mP_\infty)$. One can derive from the homogeneous equation of the curve that $t^{q+1} = \left(\frac{Z}{Y}\right)^q + \frac{Z}{Y}$. So, $v_{P_\infty}(\left(\frac{Z}{Y}\right)^q + \frac{Z}{Y}) = q+1$. By Lemma 4.1 one can deduce that $v_{P_\infty}(\frac{Z}{Y}) = q+1$ and so $v_{P_\infty}(\frac{Y}{Z}) = -(q+1)$. On the other hand, since $\left(\frac{X}{Z}\right)^{q+1} = \left(\frac{Y}{Z}\right)^q + \frac{Y}{Z}$, we have $(q+1)v_{P_\infty}(\frac{X}{Z}) = -q(q+1)$. So, $v_{P_\infty}(\frac{X}{Z}) = -q$.

We have seen that $q, q+1 \in \Lambda$. In this case Λ contains what we will call later the semigroup generated by $q, q+1$ whose complement in $\mathbb{N}_0$ has $\frac{q(q-1)}{2} = g$ elements. Since we know that the complement of Λ in $\mathbb{N}_0$ also has g elements, this means that both semigroups are the same.

For further details on the Hermitian curve see [35, 66].

Example 4.2 (Klein quartic). *The* Klein quartic *over $\mathbb{F}_q$ is defined by the affine equation $x^3y+y^3+x=0$. We shall see that if $\gcd(q,7)=1$ then $\mathcal{K}$ is smooth. Its defining homogeneous polynomial is $F = X^3Y + Y^3Z + Z^3X$ and its partial derivatives are $F_X = 3X^2Y + Z^3$, $F_Y = 3Y^2Z + X^3$, $F_Z = 3Z^2X + Y^3$. If the characteristic of $\mathbb{F}_{q^2}$ is 3 then $F_X = F_Y = F_Z = 0$ implies $X^3 = Y^3 = Z^3 = 0$ and so $X = Y = Z = 0$. Hence there is no projective point $P = (X:Y:Z)$ at which $\mathcal{K}$ is singular. Otherwise, if the characteristic of $\mathbb{F}_{q^2}$ is different from 3 then $F_X = F_Y = F_Z = 0$ implies $X^3Y = -3Y^3Z$ and $Z^3X = -3X^3Y = 9Y^3Z$. Now the equation of the curve translates to $-3Y^3Z + Y^3Z + 9Y^3Z = 7Y^3Z = 0$. By hypothesis $\gcd(q,7)=1$ and so either $Y=0$ or $Z=0$. In the first case, $F_Y=0$ implies $X=0$ and $F_X=0$ implies $Z=0$, a contradiction, and in the second case, $F_Y=0$ implies $X=0$ and $F_Z=0$ implies $Y=0$, another contradiction.*

Let $P_0 = (0:0:1)$. One can easily check that $P_0 \in \mathcal{K}$. We have $F_X(P_0)X + F_Y(P_0)Y + F_Z(P_0)Z = X$ and so $t = \frac{Y}{Z}$ is a local parameter at P_0. From the equation of the curve we get $\left(\frac{X}{Y}\right)^3 + \frac{Z}{Y} + \left(\frac{Z}{Y}\right)^3 \frac{X}{Y} = 0$. So, at least one of the next equalities holds

- $3v_{P_0}(\frac{X}{Y}) = v_{P_0}(\frac{Z}{Y})$
- $3v_{P_0}(\frac{X}{Y}) = 3v_{P_0}(\frac{Z}{Y}) + v_{P_0}(\frac{X}{Y})$
- $v_{P_0}(\frac{Z}{Y}) = 3v_{P_0}(\frac{Z}{Y}) + v_{P_0}(\frac{X}{Y})$

Since $t = \frac{Y}{Z}$ is a local parameter at P_0, $v_{P_0}(\frac{Z}{Y}) = -1$. Now, since $v_{P_0}(\frac{X}{Y})$ is an integer, only the third equality is possible, which leads to the conclusion that $v_{P_0}(\frac{X}{Y}) = 2$. Similarly, $\left(\frac{X}{Z}\right)^3 \frac{Y}{Z} + \left(\frac{Y}{Z}\right)^3 + \frac{X}{Z} = 0$ gives that at least one of the next equalities holds

- $3v_{P_0}(\frac{X}{Z}) + v_{P_0}(\frac{Y}{Z}) = 3v_{P_0}(\frac{Y}{Z})$
- $3v_{P_0}(\frac{X}{Z}) + v_{P_0}(\frac{Y}{Z}) = v_{P_0}(\frac{X}{Z})$
- $3v_{P_0}(\frac{Y}{Z}) = v_{P_0}(\frac{X}{Z})$

Again only the third equality is possible and this leads to $v_{P_0}(\frac{X}{Z}) = 3$.

Now we consider the rational functions $f_{ij} = \frac{Y^i Z^j}{X^{i+j}}$. We have already seen that $v_{P_0}(f_{ij}) = -2i - 3j$ and we want to see under which conditions $f_{ij} \in \cup_{m \geqslant 0} L(mP_0)$. This is equivalent to see when it has no poles rather than P_0. The poles of f_{ij} may only be at points with $X = 0$ and so only at P_0 and $P_1 = (0:1:0)$. Using the symmetries of the curve we get $v_{P_1}(\frac{Y}{X}) = -1$, $v_{P_1}(\frac{Z}{X}) = 2$. So, $v_{P_1}(f_{ij}) = -i + 2j$. Then $f_{ij} \in \cup_{m \geqslant 0} L(mP_0)$ if and only if $-i+2j \geqslant 0$. We get that Λ contains $\{2i+3j : i,j \geqslant 0, 2j \geqslant i\} = \{0,3,5,6,7,8,\dots\}$. This has 3 gaps which is exactly the genus of $\mathcal{K}$. So,

$$\Lambda = \{0,3,5,6,7,8,9,10,\dots\}.$$

It is left as an exercise to prove that all this can be generalized to the curve $\mathcal{K}_m$ with defining polynomial $F = X^m Y + Y^m Z + Z^m X$, provided that $\gcd(1, m^2 - m + 1) = 1$. In this case $v_{P_0}(f_{ij}) = -(m-1)i - mj$ and $f_{ij} \in \cup_{m \geqslant 0} L(mP_0)$ if and only if $-i + (m-1)j \geqslant 0$. Since $(m-1)i + mj = (m-1)i' + mj'$ for some $(i', j') \neq (i, j)$ if and only if $i \geqslant m$ or $j \geqslant m-1$ we deduce that

$$\{-v_{P_0}(f_{ij}) : f_{ij} \in \cup_{m \geqslant 0} L(mP_0)\}$$
$$= \{(m-1)i + mj : (i,j) \neq (1,0), (2,0), \dots, (m-1,0)\}.$$

This set has exactly $\frac{m(m-1)}{2}$ gaps which is the genus of $\mathcal{K}_m$. So it is exactly the Weierstrass semigroup at P_0. For further details on the Klein quartic we refer the reader to [35, 56].

4.2.2. Basic notions and problems

A *numerical semigroup* is a subset Λ of $\mathbb{N}_0$ containing 0, closed under summation and with finite complement in $\mathbb{N}_0$. A general reference on numerical semigroups is [59].

4.2.2.1. Genus, conductor, gaps, non-gaps, enumeration

For a numerical semigroup Λ define the *genus* of Λ as the number $g = \#(\mathbb{N}_0 \setminus \Lambda)$ and the *conductor* of Λ as the unique integer $c \in \Lambda$ such that $c - 1 \notin \Lambda$ and $c + \mathbb{N}_0 \subseteq \Lambda$. The elements in Λ are called the *non-gaps* of Λ while the elements in $\mathbb{N}_0 \setminus \Lambda$ are called the *gaps* of Λ. The *enumeration* of Λ is the unique increasing bijective map $\lambda : \mathbb{N}_0 \longrightarrow \Lambda$. We will use λ_i for $\lambda(i)$.

Lemma 4.3. *Let Λ be a numerical semigroup with conductor c, genus g, and enumeration λ. The following are equivalent.*

(i) $\lambda_i \geqslant c$
(ii) $i \geqslant c - g$
(iii) $\lambda_i = g + i$

Proof. First of all notice that if $g(i)$ is the number of gaps smaller than λ_i, then $\lambda_i = g(i) + i$. To see that (i) and (iii) are equivalent notice that $\lambda_i \geqslant c \iff g(i) = g \iff g(i) + i = g + i \iff \lambda_i = g + i$. Now, from this equivalence we deduce that $c = \lambda_{c-g}$. Since λ is increasing we deduce that $\lambda_i \geqslant c = \lambda_{c-g}$ if and only if $i \geqslant c - g$. □

4.2.2.2. Generators, Apéry set

The *generators* of a numerical semigroup are those non-gaps which cannot be obtained as a sum of two smaller non-gaps. If $a_1, \ldots, a_l$ are the generators of a semigroup Λ then $\Lambda = \{n_1 a_1 + \cdots + n_l a_l : n_1, \ldots, n_l \in \mathbb{N}_0\}$ and so $a_1, \ldots, a_l$ are necessarily coprime. If $a_1, \ldots, a_l$ are coprime, we call $\{n_1 a_1 + \cdots + n_l a_l : n_1, \ldots, n_l \in \mathbb{N}_0\}$ the *semigroup generated* by $a_1, \ldots, a_l$ and denote it by $\langle a_1, \ldots, a_l \rangle$.

The non-gap λ_1 is always a generator. If for each integer i from 0 to $\lambda_1 - 1$ we consider w_i to be the smallest non-gap in Λ that is congruent to i modulo λ_1, then each non-gap of Λ can be expressed as $w_i + k\lambda_1$ for some $i \in \{0, \ldots, \lambda_1 - 1\}$ and some $k \in \mathbb{N}_0$. So, the generators different from λ_1 must be in $\{w_1, \ldots, w_{\lambda_1 - 1}\}$ and this implies that there is always a finite number of generators. The set $\{w_0, w_1, \ldots, w_{\lambda_1 - 1}\}$ is called the *Apéry set* of Λ and denoted $Ap(\Lambda)$. It is easy to check that it equals $\{l \in \Lambda : l - \lambda_1 \notin \Lambda\}$. References related to the Apéry set are [1, 25, 43, 60, 62].

4.2.2.3. Frobenius' coin exchange problem

Frobenius suggested the problem to determine the largest monetary amount that cannot be obtained using only coins of specified denominations. A lot of information on the Frobenius' problem can be found in Ramírez Alfonsín's book [57].

If the different denominations are coprime then the set of amounts that can be obtained forms a numerical semigroup and the question is equivalent to determining the largest gap. This is why the largest gap of a numerical semigroup is called the *Frobenius number* of the numerical semigroup.

If the number of denominations is two and the values of the coins are a, b with a, b coprime, then Sylvester's formula [69] gives that the Frobenius number is

$$ab - a - b.$$

However, when the number of denominations is larger, there is no closed polynomial form as can be derived from the next result due to Curtis [18].

Theorem 4.2. *There is no finite set of polynomials* $\{f_1, \ldots, f_n\}$ *such that for each choice of* $a, b, c \in \mathbb{N}$, *there is some i such that the Frobenius number of a, b, c is $f_i(a, b, c)$.*

4.2.2.4. Hurwitz question

It is usually attributed to Hurwitz the problem of determining whether there exist non-Weierstrass numerical semigroups, to which Buchweitz gave a positive answer, and the problem of characterizing Weierstrass semigroups. For these questions we refer the reader to [38, 41, 71] and all the citations therein.

A related problem is bounding the number of rational points of a curve using Weierstrass semigroups. One can find some bounds in [30, 42, 68].

4.2.2.5. Wilf conjecture

The Wilf conjecture ([19, 73]) states that the number e of generators of a numerical semigroup of genus g and conductor c satisfies

$$e \geqslant \frac{c}{c-g}.$$

It is easy to check it when the numerical semigroup is symmetric, that is, when $c = 2g$. In [19] the inequality is proved for many other cases. In [8] it was proved by brute approach that any numerical semigroup of genus at most 50 also satisfies the conjecture.

4.2.3. Classification

4.2.3.1. Symmetric and pseudo-symmetric numerical semigroups

Definition 4.1. A numerical semigroup Λ with genus g and conductor c is said to be *symmetric* if $c = 2g$.

Symmetric numerical semigroups have been widely studied. For instance in [12, 16, 35, 39].

Example 4.3. Semigroups *generated by two integers* are the semigroups of the form

$$\Lambda = \{ma + nb : a, b \in \mathbb{N}_0\}$$

for some integers a and b. For Λ having finite complement in $\mathbb{N}_0$ it is necessary that a and b are coprime integers. Semigroups generated by two coprime integers are symmetric [35, 39].

Geil introduces in [29] the *norm-trace curve* over $\mathbb{F}_{q^r}$ defined by the affine equation

$$x^{(q^r-1)/(q-1)} = y^{q^{r-1}} + y^{q^{r-2}} + \cdots + y$$

where q is a prime power. It has a single rational point at infinity and the Weierstrass semigroup at the rational point at infinity is generated by the two coprime integers $(q^r - 1)/(q - 1)$ and q^{r-1}. So, it is an example of a symmetric numerical semigroup.

Properties on semigroups generated by two coprime integers can be found in [39]. For instance, the semigroup generated by a and b, has conductor equal to $(a-1)(b-1)$, and any element $l \in \Lambda$ can be written uniquely as $l = ma + nb$ with m, n integers such that $0 \leqslant m < b$.

From the results in [35, Section 3.2] one can get, for any numerical semigroup Λ generated by two integers, the equation of a curve having a point whose Weierstrass semigroup is Λ.

Let us state now a lemma related to symmetric numerical semigroups.

Lemma 4.4. *A numerical semigroup Λ with conductor c is symmetric if and only if for any non-negative integer i, if i is a gap, then $c - 1 - i$ is a non-gap.*

The proof can be found in [39, Remark 4.2] and [35, Proposition 5.7]. It follows by counting the number of gaps and non-gaps smaller than the conductor and the fact that if i is a non-gap then $c - 1 - i$ must be a gap because otherwise $c - 1$ would also be a non-gap.

Definition 4.2. A numerical semigroup Λ with genus g and conductor c is said to be *pseudo-symmetric* if $c = 2g - 1$.

Notice that a symmetric numerical semigroup cannot be pseudo-symmetric. Next lemma as well as its proof is analogous to Lemma 4.4.

Lemma 4.5. *A numerical semigroup Λ with odd conductor c is pseudo-symmetric if and only if for any non-negative integer i different from $(c - 1)/2$, if i is a gap, then $c - 1 - i$ is a non-gap.*

Example 4.4. The Weierstrass semigroup at P_0 of the Klein quartic of Example 4.2 is $\Lambda = \{0, 3\} \cup \{i \in \mathbb{N}_0 : i \geqslant 5\}$. In this case $c = 5$ and the only gaps different from $(c - 1)/2$ are $l = 1$ and $l = 4$. In both cases we have $c - 1 - l \in \Lambda$. This proves that Λ is pseudo-symmetric.

In [58] the authors prove that the set of irreducible semigroups, that is, the semigroups that cannot be expressed as a proper intersection of two numerical semigroups, is the union of the set of symmetric semigroups and the set of pseudo-symmetric semigroups.

4.2.3.2. Arf numerical semigroups

Definition 4.3. A numerical semigroup Λ with enumeration λ is called an *Arf numerical semigroup* if $\lambda_i + \lambda_j - \lambda_k \in \Lambda$ for every $i, j, k \in \mathbb{N}_0$ with $i \geqslant j \geqslant k$ [17].

For further work on Arf numerical semigroups and generalizations we refer the reader to [2, 13, 46, 61]. For results on Arf semigroups related to coding theory, see [4, 17].

Example 4.5. It is easy to check that the Weierstrass semigroup in Example 4.2 is Arf.

Let us state now two results on Arf numerical semigroups that will be used later.

Lemma 4.6. *Suppose Λ is Arf. If $i, i + j \in \Lambda$ for some $i, j \in \mathbb{N}_0$, then $i + kj \in \Lambda$ for all $k \in \mathbb{N}_0$. Consequently, if Λ is Arf and $i, i + 1 \in \Lambda$, then $i \geqslant c$.*

Proof. Let us prove this by induction on k. It is obvious for $k = 0$ and $k = 1$. If $k > 0$ and $i, i+j, i+kj \in \Lambda$ then $(i+j)+(i+kj)-i = i+(k+1)j \in \Lambda$. □

Consequently, Arf semigroups are *sparse semigroups* [46], that is, there are no two non-gaps in a row smaller than the conductor.

Let us give the definition of inductive numerical semigroups. They are an example of Arf numerical semigroups.

Definition 4.4. A sequence (H_n) of numerical semigroups is called *inductive* if there exist sequences (a_n) and (b_n) of positive integers such that $H_1 = \mathbb{N}_0$ and for $n > 1$, $H_n = a_n H_{n-1} \cup \{m \in \mathbb{N}_0 : m \geqslant a_n b_{n-1}\}$. A numerical semigroup is called *inductive* if it is a member of an inductive sequence [55, Definition 2.13].

One can see that inductive numerical semigroups are Arf [17].

Example 4.6. Pellikaan, Stichtenoth and Torres proved in [54] that the numerical semigroups for the codes over $\mathbb{F}_{q^2}$ associated to the second tower of Garcia-Stichtenoth attaining the Drinfeld-Vlăduţ bound [27] are given recursively by $\Lambda_1 = \mathbb{N}_0$ and, for $m > 0$,

$$\Lambda_m = q \cdot \Lambda_{m-1} \cup \{i \in \mathbb{N}_0 : i \geqslant q^m - q^{\lfloor (m+1)/2 \rfloor}\}.$$

They are examples of inductive numerical semigroups and hence, examples of Arf numerical semigroups.

Example 4.7. *Hyperelliptic numerical semigroups.* These are the numerical semigroups generated by 2 and an odd integer. They are of the form
$$\Lambda = \{0, 2, 4, \ldots, 2k-2, 2k, 2k+1, 2k+2, 2k+3, \ldots\}$$
for some positive integer k.

The next lemma is proved in [17].

Lemma 4.7. *The only Arf symmetric semigroups are hyperelliptic semigroups.*

In order to show which are the only Arf pseudo-symmetric semigroups we need the Apéry set that was previously defined.

Lemma 4.8. *Let Λ be a pseudo-symmetric numerical semigroup. For any $l \in Ap(\Lambda)$ different from $\lambda_1 + (c-1)/2$, $\lambda_1 + c - 1 - l \in Ap(\Lambda)$.*

Proof. Let us prove first that $\lambda_1 + c - 1 - l \in \Lambda$. Since $l \in Ap(\Lambda)$, $l - \lambda_1 \notin \Lambda$ and it is different from $(c-1)/2$ by hypothesis. Thus $\lambda_1 + c - 1 - l = c - 1 - (l - \lambda_1) \in \Lambda$ because Λ is pseudo-symmetric.

Now, $\lambda_1 + c - 1 - l - \lambda_1 = c - 1 - l \notin \Lambda$ because otherwise $c - 1 \in \Lambda$. So $\lambda_1 + c - 1 - l$ must belong to $Ap(\Lambda)$. □

Lemma 4.9. *The only Arf pseudo-symmetric semigroups are $\{0, 3, 4, 5, 6, \ldots\}$ and $\{0, 3, 5, 6, 7, \ldots\}$ (corresponding to the Klein quartic).*

Proof. Let Λ be an Arf pseudo-symmetric numerical semigroup. Let us show first that $Ap(\Lambda) = \{0, \lambda_1 + (c-1)/2, \lambda_1 + c - 1\}$. The inclusion $\supseteq$ is obvious. In order to prove the opposite inclusion suppose $l \in Ap(\Lambda)$, $l \notin \{0, \lambda_1 + (c-1)/2, \lambda_1 + c - 1\}$. By Lemma 4.8, $\lambda_1 + c - 1 - l \in \Lambda$ and since $l \neq \lambda_1 + c - 1$, $\lambda_1 + c - 1 - l \geqslant \lambda_1$. On the other hand, if $l \neq 0$ then $l \geqslant \lambda_1$. Now, by the Arf condition, $\lambda_1 + c - 1 - l + l - \lambda_1 = c - 1 \in \Lambda$, which is a contradiction.

Now, if $\#Ap(\Lambda) = 1$ then $\lambda_1 = 1$ and $\Lambda = \mathbb{N}_0$. But $\mathbb{N}_0$ is not pseudo-symmetric.

If $\#Ap(\Lambda) = 2$ then $\lambda_1 = 2$. But then Λ must be hyperelliptic and so Λ is not pseudo-symmetric.

So $\#Ap(\Lambda)$ must be 3. This implies that $\lambda_1 = 3$ and that 1 and 2 are gaps. If $1 = (c-1)/2$ then $c = 3$ and this gives $\Lambda = \{0, 3, 4, 5, 6, \ldots\}$. Else if $2 = (c-1)/2$ then $c = 5$ and this gives $\Lambda = \{0, 3, 5, 6, 7, \ldots\}$. Finally, if $1 \neq (c-1)/2$ and $2 \neq (c-1)/2$, since Λ is pseudo-symmetric, $c-2, c-3 \in \Lambda$. But this contradicts Lemma 4.6. □

The next two lemmas are two characterizations of Arf numerical semigroups. The first one is proved in [17, Proposition 1].

Lemma 4.10. *The numerical semigroup Λ with enumeration λ is Arf if and only if for every two positive integers i, j with $i \geqslant j$, $2\lambda_i - \lambda_j \in \Lambda$.*

Lemma 4.11. *The numerical semigroup Λ is Arf if and only if for any $l \in \Lambda$, the set $S(l) = \{l' - l : l' \in \Lambda, l' \geqslant l\}$ is a numerical semigroup.*

Proof. Suppose Λ is Arf. Then $0 \in S(l)$ and if $m_1 = l' - l$, $m_2 = l'' - l$ with $l', l'' \in \Lambda$ and $l' \geqslant l$, $l'' \geqslant l$, then $m_1 + m_2 = l' + l'' - l - l$. Since Λ is Arf, $l' + l'' - l \in \Lambda$ and it is larger than or equal to l. Thus, $m_1 + m_2 \in S(l)$. The finiteness of the complement of $S(l)$ is a consequence of the finiteness of the complement of Λ.

On the other hand, if Λ is such that $S(l)$ is a numerical semigroup for any $l \in \Lambda$ then, if $\lambda_i \geqslant \lambda_j \geqslant \lambda_k$ are in Λ, we will have $\lambda_i - \lambda_k \in S(\lambda_k)$, $\lambda_j - \lambda_k \in S(\lambda_k)$, $\lambda_i + \lambda_j - \lambda_k - \lambda_k \in S(\lambda_k)$ and therefore $\lambda_i + \lambda_j - \lambda_k \in \Lambda$. □

4.2.3.3. *Numerical semigroups generated by an interval*

A numerical semigroup Λ is *generated by an interval* $\{i, i+1, \ldots, j\}$ with $i, j \in \mathbb{N}_0$, $i \leqslant j$ if

$$\Lambda = \{n_i i + n_{i+1}(i+1) + \cdots + n_j j : n_i, n_{i+1}, \ldots, n_j \in \mathbb{N}_0\}.$$

A study of semigroups generated by intervals was carried out by García-Sánchez and Rosales in [28].

Example 4.8. The Weierstrass semigroup at the rational point at infinity of the Hermitian curve (Example 4.1) is generated by q and $q+1$. So, it is an example of numerical semigroup generated by an interval.

Lemma 4.12. *The semigroup $\Lambda_{\{i,\ldots,j\}}$ generated by the interval $\{i, i+1, \ldots, j\}$ satisfies*

$$\Lambda_{\{i,\ldots,j\}} = \bigcup_{k \geqslant 0} \{ki, ki+1, ki+2, \ldots, kj\}.$$

This lemma is a reformulation of [28, Lemma 1]. In the same reference we can find the next result [28, Theorem 6].

Lemma 4.13. $\Lambda_{\{i,\dots,j\}}$ is symmetric if and only if $i \equiv 2 \mod j - i$.

Lemma 4.14. *The only numerical semigroups which are generated by an interval and Arf, are the semigroups which are equal to* $\{0\} \cup \{i \in \mathbb{N}_0 : i \geqslant c\}$ *for some non-negative integer c.*

Proof. It is a consequence of Lemmas 4.6 and 4.12. □

Lemma 4.15. *The unique numerical semigroup which is pseudo-symmetric and generated by an interval is* $\{0, 3, 4, 5, 6, \dots\}$.

Proof. By Lemma 4.12, for the non-trivial semigroup $\Lambda_{\{i,\dots,j\}}$ generated by the interval $\{i, \dots, j\}$, the intervals of gaps between λ_0 and the conductor satisfy that the length of each interval is equal to the length of the previous interval minus $j - i$. On the other hand, the intervals of non-gaps between 1 and $c - 1$ satisfy that the length of each interval is equal to the length of the previous interval plus $j - i$.

Now, by Lemma 4.5, $(c - 1)/2$ must be the first gap or the last gap of an interval of gaps. Suppose that it is the first gap of an interval of n gaps. If it is equal to 1 then $c = 3$ and $\Lambda = \{0, 3, 4, 5, 6, \dots\}$. Otherwise $(c - 1)/2 > \lambda_1$. Then, if Λ is pseudo-symmetric, the previous interval of non-gaps has length $n - 1$. Since Λ is generated by an interval, the first interval of non-gaps after $(c-1)/2$ must have length $n - 1 + j - i$ and since Λ is pseudo-symmetric the interval of gaps before $(c - 1)/2$ must have the same length. But since Λ is generated by an interval, the interval of gaps previous to $(c-1)/2$ must have length $n+j-i$. This is a contradiction. The same argument proves that $(c-1)/2$ cannot be the last gap of an interval of gaps. So, the only possibility for a pseudo-symmetric semigroup generated by an interval is when $(c-1)/2 = 1$, that is, when $\Lambda = \{0, 3, 4, 5, 6, \dots\}$. □

4.2.3.4. *Acute numerical semigroups*

Definition 4.5. We say that a numerical semigroup is *ordinary* if it is equal to

$$\{0\} \cup \{i \in \mathbb{N}_0 : i \geqslant c\}$$

for some non-negative integer c.

Almost all rational points on a curve of genus g over an algebraically closed field have Weierstrass semigroup of the form $\{0\} \cup \{i \in \mathbb{N}_0 : i \geqslant g + 1\}$. Such points are said to be *ordinary*. This is why we call these numerical semigroups ordinary [21, 32, 67]. Caution must be taken when the characteristic of the ground field is $p > 0$, since there exist curves with infinitely many non-ordinary points [68].

Notice that $\mathbb{N}_0$ is an ordinary numerical semigroup. It will be called the *trivial* numerical semigroup.

Definition 4.6. Let Λ be a numerical semigroup different from $\mathbb{N}_0$ with enumeration λ, genus g and conductor c. The element $\lambda_{\lambda^{-1}(c)-1}$ will be called the *dominant* of the semigroup and will be denoted d. For each $i \in \mathbb{N}_0$ let $g(i)$ be the number of gaps which are smaller than λ_i. In particular, $g(\lambda^{-1}(c)) = g$ and $g(\lambda^{-1}(d)) = g' < g$. If i is the smallest integer for which $g(i) = g'$ then λ_i is called the *subconductor* of Λ and denoted c'.

Remark 4.1. Notice that if $c' > 0$, then $c' - 1 \notin \Lambda$. Otherwise we would have $g(\lambda^{-1}(c'-1)) = g(\lambda^{-1}(c'))$ and $c'-1 < c'$. Notice also that all integers between c' and d are in Λ because otherwise $g(\lambda^{-1}(c')) < g'$.

Remark 4.2. For a numerical semigroup Λ different from $\mathbb{N}_0$ the following are equivalent:

(i) Λ is ordinary,
(ii) the dominant of Λ is 0,
(iii) the subconductor of Λ is 0.

Indeed, (i)$\iff$(ii) and (ii)$\implies$(iii) are obvious. Now, suppose (iii) is satisfied. If the dominant is larger than or equal to 1 it means that 1 is in Λ and so $\Lambda = \mathbb{N}_0$ a contradiction.

Definition 4.7. If Λ is a non-ordinary numerical semigroup with enumeration λ and with subconductor λ_i then the element λ_{i-1} will be called the *subdominant* and denoted d'.

It is well defined because of Remark 4.2.

Definition 4.8. A numerical semigroup Λ is said to be *acute* if Λ is ordinary or if Λ is non-ordinary and its conductor c, its subconductor c', its dominant d and its subdominant d' satisfy $c - d \leqslant c' - d'$.

Roughly speaking, a numerical semigroup is acute if the last interval of gaps before the conductor is smaller than the previous interval of gaps.

Example 4.9. For the Hermitian curve over $\mathbb{F}_{16}$ the Weierstrass semigroup at the unique point at infinity is

$$\{0, 4, 5, 8, 9, 10\} \cup \{i \in \mathbb{N}_0 : i \geqslant 12\}.$$

In this case $c = 12$, $d = 10$, $c' = 8$ and $d' = 5$ and it is easy to check that it is an acute numerical semigroup.

Example 4.10. For the Weierstrass semigroup at the rational point P_0 of the Klein quartic in Example 4.2 we have $c = 5$, $d = c' = 3$ and $d' = 0$. So, it is an example of a non-ordinary acute numerical semigroup.

Lemma 4.16. *Let Λ be a numerical semigroup.*

(1) If Λ is symmetric then it is acute.
(2) If Λ is pseudo-symmetric then it is acute.
(3) If Λ is Arf then it is acute.
(4) If Λ is generated by an interval then it is acute.

Proof. If Λ is ordinary then it is obvious. Let us suppose that Λ is a non-ordinary semigroup with genus g, conductor c, subconductor c', dominant d and subdominant d'.

(1) Suppose that Λ is symmetric. We know by Lemma 4.4 that a numerical semigroup Λ is symmetric if and only if for any non-negative integer i, if i is a gap, then $c - 1 - i \in \Lambda$. If moreover it is not ordinary, then 1 is a gap. So, $c - 2 \in \Lambda$ and it is precisely the dominant. Hence, $c - d = 2$. Since $c' - 1$ is a gap, $c' - d' \geq 2 = c - d$ and so Λ is acute.
(2) Suppose that Λ is pseudo-symmetric. If $1 = (c-1)/2$ then $c = 3$ and $\Lambda = \{0, 3, 4, 5, 6, \dots\}$ which is ordinary. Else if $1 \neq (c-1)/2$ then the proof is equivalent to the one for symmetric semigroups.
(3) Suppose Λ is Arf. Since $d \geq c' > d'$, then $d + c' - d'$ is in Λ and it is strictly larger than the dominant d. Hence it is larger than or equal to c. So, $d + c' - d' \geq c$ and Λ is acute.
(4) Suppose that Λ is generated by the interval $\{i, i+1, \dots, j\}$. Then, by Lemma 4.12, there exists k such that $c = ki$, $c' = (k-1)i$, $d = (k-1)j$ and $d' = (k-2)j$. So, $c - d = k(i-j) + j$ while $c' - d' = k(i-j) - i + 2j$. Hence, Λ is acute. □

In Figure 4.1 we summarize all the relations we have proved between acute semigroups, symmetric and pseudo-symmetric semigroups, Arf semigroups and semigroups generated by an interval.

Remark 4.3. There exist numerical semigroups which are not acute. For instance,
$$\Lambda = \{0, 6, 8, 9\} \cup \{i \in \mathbb{N}_0 : i \geq 12\}.$$

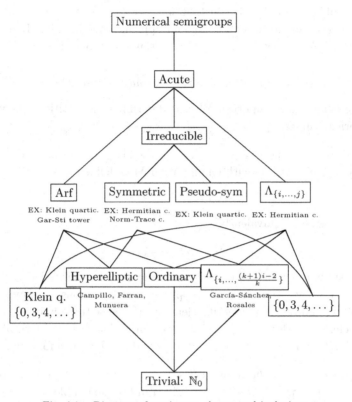

Fig. 4.1. Diagram of semigroup classes and inclusions.

In this case, $c = 12$, $d = 9$, $c' = 8$ and $d' = 6$.

On the other hand there exist numerical semigroups which are acute and which are not symmetric, pseudo-symmetric, Arf or interval-generated. For example,

$$\Lambda = \{0, 10, 11\} \cup \{i \in \mathbb{N}_0 : i \geqslant 15\}.$$

In this case, $c = 15$, $d = 11$, $c' = 10$ and $d' = 0$.

4.2.4. Characterization

4.2.4.1. Homomorphisms of semigroups

Homomorphisms of numerical semigroups, that is, maps f between numerical semigroups such that $f(a + b) = f(a) + f(b)$, are exactly the scale maps $f(a) = ka$ for all a, for some constant $k \geqslant 0$. Indeed, if f is a homomorphism then $\frac{f(a)}{a}$ is constant since $f(ab) = a \cdot f(b) = b \cdot f(a)$. Furthermore, the unique surjective homomorphism is the identity. Indeed, for a semigroup Λ, the set $k\Lambda$ is a numerical semigroup only if $k = 1$.

4.2.4.2. *The $\oplus$ operation, the ν sequence, and the τ sequence*

Next we define three important objects describing the addition behavior of a numerical semigroup.

Definition 4.9. The operation $\oplus_\Lambda : \mathbb{N}_0 \times \mathbb{N}_0 \to \mathbb{N}_0$ associated to the numerical semigroup Λ (with enumeration λ) is defined as

$$i \oplus_\Lambda j = \lambda^{-1}(\lambda_i + \lambda_j)$$

for any $i, j \in \mathbb{N}_0$. Equivalently,

$$\lambda_i + \lambda_j = \lambda_{i \oplus_\Lambda j}.$$

The subindex referring to the semigroup may be omitted if the semigroup is clear by the context. The operation $\oplus$ is obviously commutative, associative, and has 0 as identity element. However there is in general no inverse element. Also, the operation $\oplus$ is compatible with the natural order of $\mathbb{N}_0$. That is, if $a < b$ then $a \oplus c < b \oplus c$ and $c \oplus a < c \oplus b$ for any $c \in \mathbb{N}_0$.

Example 4.11. For the numerical semigroup $\Lambda = \{0, 4, 5, 8, 9, 10, 12, 13, 14, \dots\}$ the first values of $\oplus$ are given in the next table:

$\oplus$	0	1	2	3	4	5	6	7	...
0	0	1	2	3	4	5	6	7	...
1	1	3	4	6	7	8	10	11	...
2	2	4	5	7	8	9	11	12	...
3	3	6	7	10	11	12	14	15	...
4	4	7	8	11	12	13	15	16	...
5	5	8	9	12	13	14	16	17	...
6	6	10	11	14	15	16	18	19	...
7	7	11	12	15	16	17	19	20	...
⋮	⋮	⋮	⋮	⋮	⋮	⋮	⋮	⋮	⋱

Definition 4.10. The partial ordering $\preccurlyeq_\Lambda$ on $\mathbb{N}_0$ associated to the numerical semigroup Λ is defined as follows:

$$i \preccurlyeq_\Lambda j \text{ if and only if } \lambda_j - \lambda_i \in \Lambda.$$

Equivalently

$i \preccurlyeq_\Lambda j$ if and only if there exists k with $i \oplus_\Lambda k = j$.

As before, if Λ is clear by the context then the subindex may be omitted.

Definition 4.11. Given a numerical semigroup Λ, the set N_i is defined by

$$N_i = \{j \in \mathbb{N}_0 : j \preccurlyeq_\Lambda i\} = \{j \in \mathbb{N}_0 : \lambda_i - \lambda_j \in \Lambda\}.$$

The sequence ν_i is defined by $\nu_i = \#N_i$.

Example 4.12. The ν sequence of the trivial semigroup $\Lambda = \mathbb{N}_0$ is $1, 2, 3, 4, \ldots$.

Next lemma states the relationship between $\oplus$ and the ν sequence of a numerical semigroup. Its proof is obvious.

Lemma 4.17. Let Λ be a numerical semigroup and ν the corresponding ν sequence. For all $i \in \mathbb{N}_0$,

$$\nu_i = \#\{(j,k) \in \mathbb{N}_0^2 : j \oplus k = i\}.$$

Example 4.13. If we are given the table in Example 4.11 we can deduce ν_7 by counting the number of occurrences of 7 inside the table. This is exactly 6. Thus, $\nu_7 = 6$. Repeating the same process for all ν_i with $i < 7$ we get $\nu_0 = 1, \nu_1 = 2, \nu_2 = 2, \nu_3 = 3, \nu_4 = 4, \nu_5 = 3, \nu_6 = 4$.

This lemma implies that any finite set in the sequence ν can be determined by a finite set of $\oplus$ values. Indeed, to compute ν_i it is enough to know $\{j \oplus k : 0 \leqslant j, k \leqslant i\}$.

Definition 4.12. Given a numerical semigroup Λ define its τ *sequence* by

$$\tau_i = \max\{j \in \mathbb{N}_0 : \text{ there exists } k \text{ with } j \leqslant k \leqslant i \text{ and } j \oplus_\Lambda k = i\}.$$

Notice that τ_i is the largest element j in N_i with $\lambda_j \leqslant \lambda_i/2$. In particular, if $\lambda_i/2 \in \Lambda$ then $\tau_i = \lambda^{-1}(\lambda_i/2)$. Notice also that τ_i is 0 if and only if λ_i is either 0 or a generator of Λ.

Example 4.14. In Table 4.14 we show the ν sequence and the τ sequence of the numerical semigroups generated by $4, 5$ and generated by $6, 7, 8, 17$.

Table 4.1. ν and τ sequences of the numerical semigroups generated by $4,5$ and by $6,7,8,17$.

(a) $<4,5>$

i	λ_i	ν_i	τ_i
0	0	1	0
1	4	2	0
2	5	2	0
3	8	3	1
4	9	4	1
5	10	3	2
6	12	4	1
7	13	6	2
8	14	6	2
9	15	4	2
10	16	5	3
11	17	8	3
12	18	9	4
13	19	8	4
14	20	9	5
15	21	10	4
16	22	12	5
17	23	12	5
18	24	13	6
19	25	14	6
20	26	15	7
21	27	16	7
22	28	17	8
23	29	18	8
⋮	⋮	⋮	⋮

(b) $<6,7,8,17>$

i	λ_i	ν_i	τ_i
0	0	1	0
1	6	2	0
2	7	2	0
3	8	2	0
4	12	3	1
5	13	4	1
6	14	5	2
7	15	4	2
8	16	3	3
9	17	2	0
10	18	4	1
11	19	6	2
12	20	8	3
13	21	8	3
14	22	8	3
15	23	8	3
16	24	9	4
17	25	10	4
18	26	11	5
19	27	12	5
20	28	13	6
21	29	14	6
22	30	15	7
23	31	16	7
⋮	⋮	⋮	⋮

One difference between the τ sequence and the ν sequence is that, while in the ν sequence not all non-negative integers need to appear, in the τ sequence all of them appear. Notice for instance that 7 does not appear in the ν sequence of the numerical semigroup generated by 4 and 5 nor the numerical semigroup generated by $6, 7, 8, 17$. The reason for which any non-negative integer j appears in the τ sequence is that if $\lambda_i = 2\lambda_j$ then $\tau_i = j$. Furthermore, the smallest i for which $\tau_i = j$ corresponds to $\lambda_i = 2\lambda_j$.

4.2.4.3. Characterization of a numerical semigroup by $\oplus$

The next result was proved in [6, 7].

Lemma 4.18. *The $\oplus$ operation uniquely determines a semigroup.*

Proof. Suppose that two semigroups $\Lambda = \{\lambda_0 < \lambda_1 < \ldots\}$ and $\Lambda' = \{\lambda'_0 < \lambda'_1 < \ldots\}$ have the same associated operation $\oplus$. Define the map $f(\lambda_i) = \lambda'_i$. It is obviously surjective and it is a homomorphism since $f(\lambda_i + \lambda_j) = f(\lambda_{i \oplus j}) = \lambda'_{i \oplus j} = \lambda'_i + \lambda'_j = f(\lambda_i) + f(\lambda_j)$. So, $\Lambda = \Lambda'$. □

Conversely to Lemma 4.18 we next prove that any finite set of $\oplus$ values is shared by an infinite number of semigroups. This was proved in [7].

Lemma 4.19. *Let a, b be positive integers. Let Λ be a numerical semigroup with enumeration λ and let d be an integer with $d \geqslant 2$. Define the numerical semigroup $\Lambda' = d\Lambda \cup \{i \in \mathbb{N} : i \geqslant d\lambda_{a \oplus b}\}$. Then $i \oplus_{\Lambda'} j = i \oplus_\Lambda j$ for all $i \leqslant a$ and all $j \leqslant b$, and $\Lambda' \neq \Lambda$.*

Proof. It is obious that $\Lambda' \neq \Lambda$. Let λ' be the enumeration of Λ'. For all $k \leqslant a \oplus_\Lambda b$, $\lambda'_k = d\lambda_k$. In particular, if $i \leqslant a$ and $j \leqslant b$ then $\lambda'_i = d\lambda_i$ and $\lambda'_j = d\lambda_j$. Hence, $\lambda'_{i \oplus_{\Lambda'} j} = \lambda'_i + \lambda'_j = d\lambda_i + d\lambda_j = d\lambda_{i \oplus_\Lambda j} = \lambda'_{i \oplus_\Lambda j}$. This implies $i \oplus_{\Lambda'} j = i \oplus_\Lambda j$. □

By varying d in Lemma 4.19, we can see that although the values $(i \oplus j)_{0 \leqslant i,j}$ of a numerical semigroup uniquely determine it, any subset $(i \oplus j)_{0 \leqslant i \leqslant a, 0 \leqslant j \leqslant b}$ is exactly the corresponding subset of infinitely many numerical semigroups.

4.2.4.4. Characterization of a numerical semigroup by ν

We will use the following result on the values ν_i. It can be found in [39, Theorem 3.8].

Lemma 4.20. *Let Λ be a numerical semigroup with genus g, conductor c and enumeration λ. Let $g(i)$ be the number of gaps smaller than λ_i and let*

$$D(i) = \{l \in \mathbb{N}_0 \setminus \Lambda : \lambda_i - l \in \mathbb{N}_0 \setminus \Lambda\}.$$

Then, for all $i \in \mathbb{N}_0$,

$$\nu_i = i - g(i) + \#D(i) + 1.$$

In particular, for all $i \geqslant 2c - g - 1$ (or equivalently, for all i such that $\lambda_i \geqslant 2c - 1$), $\nu_i = i - g + 1$.

Proof. The number of gaps smaller than λ_i is $g(i)$ but it is also $\#D(i) + \#\{l \in \mathbb{N}_0 \setminus \Lambda : \lambda_i - l \in \Lambda\}$, so,

$$g(i) = \#D(i) + \#\{l \in \mathbb{N}_0 \setminus \Lambda : \lambda_i - l \in \Lambda\}. \tag{4.1}$$

On the other hand, the number of non-gaps which are at most λ_i is $i+1$ but it is also $\nu_i + \#\{l \in \Lambda : \lambda_i - l \in \mathbb{N}_0 \setminus \Lambda\} = \nu_i + \#\{l \in \mathbb{N}_0 \setminus \Lambda : \lambda_i - l \in \Lambda\}$. So,

$$i + 1 = \nu_i + \#\{l \in \mathbb{N}_0 \setminus \Lambda : \lambda_i - l \in \Lambda\}. \tag{4.2}$$

From equalities (4.1) and (4.2) we get

$$g(i) - \#D(i) = i + 1 - \nu_i$$

which leads to the desired result. □

The next lemma shows that if a numerical semigroup is non-trivial then there exists at least one value k such that $\nu_k = \nu_{k+1}$.

Lemma 4.21. *Suppose $\Lambda \neq \mathbb{N}_0$ and suppose that c and g are the conductor and the genus of Λ. Let $k = 2c - g - 2$. Then $\nu_k = \nu_{k+1}$.*

Proof. Since $\Lambda \neq \mathbb{N}_0$, $c \geqslant 2$ and so $2c - 2 \geqslant c$. This implies $k = \lambda^{-1}(2c - 2)$ and $g(k) = g$. By Lemma 4.20, $\nu_k = k - g + \#D(k) + 1$. But $D(k) = \{c - 1\}$. So, $\nu_k = k - g + 2$. On the other hand, $g(k+1) = g$ and $D(k+1) = \emptyset$. By Lemma 4.20 again, $\nu_{k+1} = k - g + 2 = \nu_k$. □

Lemma 4.22. *The trivial semigroup is the unique numerical semigroup with ν sequence equal to $1, 2, 3, 4, 5, \ldots$.*

Proof. As a consequence of Lemma 4.21 for any other numerical semigroup there is a value in the ν sequence that appears at least three times. □

The next result was proved in [5, 6].

Theorem 4.3. *The ν sequence of a numerical semigroup determines it.*

Proof. If $\Lambda = \mathbb{N}_0$ then, by Lemma 4.22, its ν sequence is unique.

Suppose that $\Lambda \neq \mathbb{N}_0$. Then we can determine the genus and the conductor from the ν sequence. Indeed, let $k = 2c - g - 2$. In the following we will show how to determine k without the knowledge of c and g. By Lemma 4.21 it holds that $\nu_k = \nu_{k+1} = k - g + 2$ and by Lemma 4.20 $\nu_i = i - g + 1$ for all $i > k$, which means that $\nu_{i+1} = \nu_i + 1$ for all $i > k$. So,

$$k = \max\{i : \nu_i = \nu_{i+1}\}.$$

We can now determine the genus as
$$g = k + 2 - \nu_k$$
and the conductor as
$$c = \frac{k+g+2}{2}.$$
At this point we know that $\{0\} \in \Lambda$ and $\{i \in \mathbb{N}_0 : i \geqslant c\} \subseteq \Lambda$ and, furthermore, $\{1, c-1\} \subseteq \mathbb{N}_0 \setminus \Lambda$. It remains to determine for all $i \in \{2, \ldots, c-2\}$ whether $i \in \Lambda$. Let us assume $i \in \{2, \ldots, c-2\}$.

On one hand, $c - 1 + i - g > c - g$ and so $\lambda_{c-1+i-g} > c$. This means that $g(c - 1 + i - g) = g$ and hence
$$\nu_{c-1+i-g} = c - 1 + i - g - g + \#D(c - 1 + i - g) + 1. \tag{4.3}$$

On the other hand, if we define $\tilde{D}(i)$ to be
$$\tilde{D}(i) = \{l \in \mathbb{N}_0 \setminus \Lambda : c - 1 + i - l \in \mathbb{N}_0 \setminus \Lambda, i < l < c - 1\}$$
then
$$D(c - 1 + i - g) = \begin{cases} \tilde{D}(i) \cup \{c-1, i\} & \text{if } i \in \mathbb{N}_0 \setminus \Lambda \\ \tilde{D}(i) & \text{otherwise.} \end{cases} \tag{4.4}$$

So, from (4.3) and (4.4),
$$i \text{ is a non-gap} \iff \nu_{c-1+i-g} = c + i - 2g + \#\tilde{D}(i).$$

This gives an inductive procedure to decide whether i belongs to Λ decreasingly from $i = c - 2$ to $i = 2$. □

Remark 4.4. From the proof of Theorem 4.3 we see that a semigroup can be determined by $k = \max\{i : \nu_i = \nu_{i+1}\}$ and the values ν_i for $i \in \{c - g + 1, \ldots, 2c - g - 3\}$.

Remark 4.5. Lemma 4.18 is a consequence of Lemma 4.17 and Theorem 4.3.

Conversely to Theorem 4.3 we next prove that any finite set of ν values is shared by an infinite number of semigroups. This was proved in [7]. Thus, the construction just given to determine a numerical semigroup from its ν sequence can only be performed if we know the behavior of the infinitely many values in the ν sequence.

Lemma 4.23. *Let k be a positive integer. Let Λ be a numerical semigroup with enumeration λ and let d be an integer with $d \geqslant 2$. Define the semigroup $\Lambda' = d\Lambda \cup \{i \in \mathbb{N} : i \geqslant d\lambda_k\}$ and let ν^Λ and $\nu^{\Lambda'}$ be the ν sequence* corresponding to Λ and Λ' respectively. Then $\nu_i^{\Lambda'} = \nu_i^{\Lambda}$ for all $i \leqslant k$ and $\Lambda' \neq \Lambda$.

Proof. It is obvious that $\Lambda' \neq \Lambda$. Let λ' be the enumeration of Λ'. For all $i \leqslant k$, $\lambda_i' = d\lambda_i$. In particular, if $j \leqslant i \leqslant k$, then $\lambda_i' - \lambda_j' = d(\lambda_i - \lambda_j) \in \Lambda' \iff \lambda_i - \lambda_j \in \Lambda$. Hence, $\nu_i^{\Lambda'} = \nu_i^{\Lambda}$. □

As a consequence of Lemma 4.23, although the sequence ν of a numerical semigroup uniquely determines it, any subset $(\nu_i)_{0 \leqslant i \leqslant k-1}$ is exactly the set of the first k values of the ν sequence of infinitely many semigroups. In fact, by varying d among the positive integers, we get an infinite set of semigroups, all of them sharing the first k values in the ν sequence.

It would be interesting to find which sequences of positive integers correspond to the sequence ν of a numerical semigroup. By now, only some necessary conditions can be stated, for instance,

- $\nu_0 = 1$,
- $\nu_1 = 2$,
- $\nu_i \leqslant i + 1$ for all $i \in \mathbb{N}_0$,
- there exists k such that $\nu_{i+1} = \nu_i + 1$ for all $i \geqslant k$,

4.2.4.5. Characterization of a numerical semigroup by τ

In this section we show that a numerical semigroup is determined by its τ sequence. This was proved in [10].

Lemma 4.24. Let Λ be a numerical semigroup with enumeration λ, conductor $c > 2$, genus g, and dominant d. Then

(1) $\tau_{(2c-g-2)+2i} = \tau_{(2c-g-2)+2i+1} = (c-g-1) + i$ for all $i \geqslant 0$
(2) At least one of the following statements holds

- $\tau_{(2c-g-2)-1} = c - g - 1$
- $\tau_{(2c-g-2)-2} = c - g - 1$.

Proof.

(1) If $i \geqslant 1$ then $\lambda_{(2c-g-2)+2i} = 2c - 2 + 2i$ and $\lambda_{(2c-g-2)+2i}/2 = c - 1 + i \in \Lambda$. So $\tau_{(2c-g-2)+2i} = \lambda^{-1}(c - 1 + i) = c - 1 + i - g$. On the other hand, $\lambda_{(2c-g-2)+2i+1} = 2c - 2 + 2i + 1 = (c - 1 + i) + (c - 1 + i + 1)$ and so $\tau_{(2c-g-2)+2i+1} = \lambda^{-1}(c - 1 + i) = c - 1 + i - g$.
If $i = 0$ then $\lambda_{(2c-g-2)+2i} = \lambda_{2c-g-2}$ and since $c > 2$ this is equal to $2c - 2$. Now $\lambda_{2c-g-2}/2 = c - 1$ and the largest non-gap which is at most

$c - 1$ is d. On the other hand, $\lambda_{2c-g-2} - d = 2c - 2 - d \geqslant c$ because $c \geqslant d + 2$. Consequently $\lambda_{2c-g-2} - d \in \Lambda$ and $\tau_{2c-g-2} = \lambda^{-1}(d) = c - g - 1$. Similarly, the largest non-gap which is at most $\lambda_{2c-g-1}/2$ is d and $\lambda_{2c-g-1} - d = 2c - 1 - d \in \Lambda$. So, $\tau_{2c-g-1} = c - g - 1$.

(2) If $c = 3$ then $g = 2$ and $\lambda_{(2c-g-2)-2} = \lambda_0$ and $\tau_{(2c-g-2)-2} = 0 = c - g - 1$. Assume $c \geqslant 4$. If $d = c - 2$ then $\lambda_{(2c-g-2)-2} = 2c - 4 = 2d$, so $\tau_{(2c-g-2)-2} = \lambda^{-1}(d) = c - g - 1$. If $d = c - 3$ then $\lambda_{(2c-g-2)-1} = 2c - 3 = d + c$, so $\tau_{(2c-g-2)-1} = \lambda^{-1}(d) = c - g - 1$. Suppose now $d \leqslant c - 4$. In this case $\lambda_{(2c-g-2)-2}/2 = c - 2$, which is between d and c, and $\lambda_{(2c-g-2)-2} - d = 2c - 4 - d \geqslant c$. So $\lambda_{(2c-g-2)-2} - d \in \Lambda$. This makes $\tau_{(2c-g-2)-2} = \lambda^{-1}(d) = c - g - 1$. □

Lemma 4.25. *The trivial semigroup is the unique numerical semigroup with τ sequence equal to $0, 0, 1, 1, 2, 2, 3, 3, 4, 4, 5, 5, \ldots$.*

Proof. It is enough to check that for any other numerical semigroup there is a value in the τ sequence that appears at least three times. If $c = 2$ then $\tau_0 = \tau_1 = \tau_2 = 0$. If $c > 2$, by Lemma 4.24, $\tau_{2c-g-2} = \tau_{2c-g-1}$ and they are equal to at least one of τ_{2c-g-3} and τ_{2c-g-4}. □

Theorem 4.4. *The τ sequence of a numerical semigroup determines it.*

Proof. Let k be the minimum integer such that $\tau_{k+2i} = \tau_{k+2i+1}$ and $\tau_{k+2i+2} = \tau_{k+2i+1} + 1$ for all $i \in \mathbb{N}_0$. If $k = 0$, by Lemma 4.25, $\Lambda = \mathbb{N}_0$. Assume $k > 0$.

By Lemma 4.24, if $c > 2$, $k = 2c - g - 2$ and $\tau_k = c - g - 1$. So,
$$\begin{cases} c = k - \tau_k + 1 \\ g = k - 2\tau_k \end{cases}$$

This result can be extended to the case $c = 2$ since in this case $c = 2$, $g = 1$, $k = 1$ and $\tau_k = 0$.

This determines $\lambda_i = i + g$ for all $i \geqslant c - g$. Now we can determine $\lambda_{c-g-1}, \lambda_{c-g-2}$, and so on using that the smallest j for which $\tau_j = i$ corresponds to $\lambda_j = 2\lambda_i$. That is, $\lambda_i = \frac{1}{2}\min\{\lambda_j : \tau_j = i\}$. □

We have just seen that any numerical semigroup is uniquely determined by its τ sequence. The next lemma shows that no finite subset of τ can determine the numerical semigroup. This result is analogous to [7, Proposition 2.2.]. In this case it refers to the ν sequence instead of the τ sequence.

Lemma 4.26. *Let r be a positive integer. Let Λ be a numerical semigroup with enumeration λ and let m be an integer with $m \geqslant 2$. Define the semigroup $\Lambda' = m\Lambda \cup \{i \in \mathbb{N}_0 : i \geqslant m\lambda_r\}$ and let τ^Λ and $\tau^{\Lambda'}$ be the τ sequence corresponding to Λ and Λ' respectively. Then $\tau_i^{\Lambda'} = \tau_i^\Lambda$ for all $i \leqslant r$ and $\Lambda' \neq \Lambda$.*

Proof. It is obvious that $\Lambda' \neq \Lambda$. Let λ' be the enumeration of Λ'. For all $i \leqslant r$, $\lambda'_i = m\lambda_i$. In particular, if $j \leqslant i \leqslant r$, then there exists k with $j \leqslant k \leqslant i$ and $\lambda_j + \lambda_k = \lambda_i$ if and only if there exists k with $j \leqslant k \leqslant i$ and $\lambda'_j + \lambda'_k = \lambda'_i$. Hence, by the definition of the τ sequence, $\tau_i^{\Lambda'} = \tau_i^\Lambda$. $\square$

As a consequence of Lemma 4.26, although the sequence τ of a numerical semigroup uniquely determines it, any subset $(\tau_i)_{0 \leqslant i \leqslant r-1}$ is exactly the set of the first r values of the τ sequence of infinitely many semigroups. In fact, by varying m among the positive integers, we get an infinite set of semigroups, all of them sharing the first r values in the τ sequence.

4.2.5. *Counting*

We are interested on the number n_g of numerical semigoups of genus g. It is obvious that $n_0 = 1$ since $\mathbb{N}_0$ is the unique numerical semigroup of genus 0. On the other hand, if 1 is in a numerical semigroup, then any non-negative integer must belong also to the numerical semigroup, because any non-negative integer is a finite sum of 1's. Thus, the unique numerical semigroup with genus 1 is $\{0\} \cup \{i \in \mathbb{N}_0 : i \geqslant 2\}$ and $n_1 = 1$. In [40] all terms of the sequence n_g are computed up to genus 37 and the terms of genus up to 50 are computed in [8]. Recently we computed $n_{51} = 164253200784$ and $n_{52} = 266815155103$. It is conjectured in [8] that the sequence given by the numbers n_g of numerical semigroups of genus g asymptotically behaves like the Fibonacci numbers and so it increases by a portion of the golden ratio. More precisely, it is conjectured that (1) $n_g \geqslant n_{g-1} + n_{g-2}$, (2) $\lim_{g \to \infty} \frac{n_{g-1}+n_{g-2}}{n_g} = 1$, (3) $\lim_{g \to \infty} \frac{n_g}{n_{g-1}} = \phi$, where $\phi = \frac{1+\sqrt{5}}{2}$ is the golden ratio. Notice that (2) and (3) are equivalent. By now, only some bounds are known for n_g which become very poor when g approaches infinity [9, 20]. Other contributions related to this sequence can be found in [3, 11, 12, 37, 40, 41, 44, 65, 74].

In Table 4.2 there are the results obtained for all numerical semigroups with genus up to 52. For each genus we wrote the number of numerical semigroups of the given genus, the Fibonacci-like-estimated value given by the sum of the number of semigroups of the two previous genus, the value of

Table 4.2. Computational results on the number of numerical semigroups up to genus 52.

g	n_g	$n_{g-1}+n_{g-2}$	$\frac{n_{g-1}+n_{g-2}}{n_g}$	$\frac{n_g}{n_{g-1}}$
0	1			
1	1			1
2	2	2	1	2
3	4	3	0.75	2
4	7	6	0.857143	1.75
5	12	11	0.916667	1.71429
6	23	19	0.826087	1.91667
7	39	35	0.897436	1.69565
8	67	62	0.925373	1.71795
9	118	106	0.898305	1.76119
10	204	185	0.906863	1.72881
11	343	322	0.938776	1.68137
12	592	547	0.923986	1.72595
13	1001	935	0.934066	1.69088
14	1693	1593	0.940933	1.69131
15	2857	2694	0.942947	1.68754
16	4806	4550	0.946733	1.68218
17	8045	7663	0.952517	1.67395
18	13467	12851	0.954259	1.67396
19	22464	21512	0.957621	1.66808
20	37396	35931	0.960825	1.66471
21	62194	59860	0.962472	1.66312
22	103246	99590	0.964589	1.66006
23	170963	165440	0.967695	1.65588
24	282828	274209	0.969526	1.65432
25	467224	453791	0.971249	1.65197
26	770832	750052	0.973042	1.64981
27	1270267	1238056	0.974642	1.64792
28	2091030	2041099	0.976121	1.64613
29	3437839	3361297	0.977735	1.64409
30	5646773	5528869	0.979120	1.64254
31	9266788	9084612	0.980341	1.64108
32	15195070	14913561	0.981474	1.63973
33	24896206	24461858	0.982554	1.63844
34	40761087	40091276	0.983567	1.63724
35	66687201	65657293	0.984556	1.63605
36	109032500	107448288	0.985470	1.63498
37	178158289	175719701	0.986312	1.63399
38	290939807	287190789	0.987114	1.63304
39	474851445	469098096	0.987884	1.63213
40	774614284	765791252	0.988610	1.63128
41	1262992840	1249465729	0.989290	1.63048
42	2058356522	2037607124	0.989919	1.62975
43	3353191846	3321349362	0.990504	1.62906
44	5460401576	5411548368	0.991053	1.62842
45	8888486816	8813593422	0.991574	1.62781
46	14463633648	14348888392	0.992067	1.62723
47	23527845502	23352120464	0.992531	1.62669
48	38260496374	37991479150	0.992969	1.62618
49	62200036752	61788341876	0.993381	1.62570
50	101090300128	100460533126	0.993770	1.62525
51	164253200784	163290336880	0.994138	1.62482
52	266815155103	265343500912	0.994484	1.62441

the quotient $\frac{n_{g-1}+n_{g-2}}{n_g}$, and the value of the quotient $\frac{n_g}{n_{g-1}}$. In Figure 4.2 and Figure 4.3 we depicted the behavior of these quotients. From these graphs one can predict that $\frac{n_{g-1}+n_{g-2}}{n_g}$ approaches 1 as g approaches infinity whereas $\frac{n_g}{n_{g-1}}$ approaches the golden ratio as g approaches infinity.

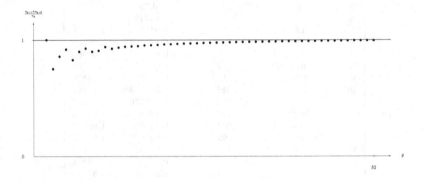

Fig. 4.2. Behavior of the quotient $\frac{n_{g-1}+n_{g-2}}{n_g}$. The values in this graph correspond to the values in Table 4.2.

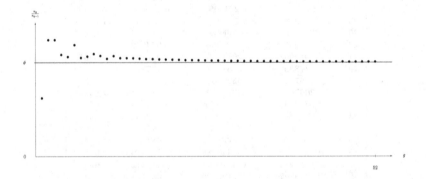

Fig. 4.3. Behavior of the quotient $\frac{n_g}{n_{g-1}}$. The values in this graph correspond to the values in Table 4.2.

The number n_g is usually studied by means of the tree rooted at the semigroup $\mathbb{N}_0$ and for which the children of a semigroup are the semigroups obtained by taking out one by one its generators larger than or equal to its conductor [8, 9, 11, 20]. This tree was previously used in [63, 64]. It is illustrated in Figure 4.4. It contains all semigroups exactly once and the semigroups at depth g have genus g. So, n_g is the number of nodes of the tree at depth g. Some alternatives for counting semigroups of a given genus without using this tree have been considered in [3, 12, 74].

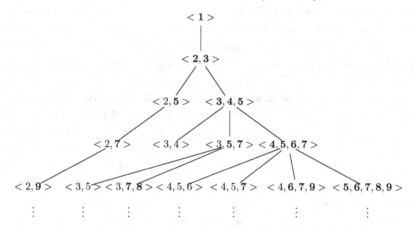

Fig. 4.4. Recursive construction of numerical semigroups of genus g from numerical semigroups of genus $g - 1$. Generators larger than the conductor are written in bold face.

4.3. Numerical semigroups and codes

4.3.1. *One-point codes and their decoding*

4.3.1.1. *One-point codes*

Linear codes A *linear code* C of length n over the alphabet $\mathbb{F}_q$ is a vector subspace of $\mathbb{F}_q^n$. Its elements are called *code words*. The *dimension* k of the code is the dimension of C as a subspace of $\mathbb{F}_q^n$. The *dual code* of C is $C^\perp = \{v \in \mathbb{F}_q^n : v \cdot c = 0 \text{ for all } c \in C\}$. It is a linear code with the same length as C and with dimension $n - k$.

The *Hamming distance* between two vectors of the same length is the number of positions in which they do not agree. The *weight* of a vector is the number of its non-zero components or, equivalently, its Hamming distance to the zero vector. The *minimum distance* d of a linear code C is the minimum Hamming distance between two code words in C. Equivalently, it is the minimum weight of all code words in C. The *correction capability* of a code is the maximum number of errors that can be added to any code word, with the code word being still uniquely identifiable. The correction capability of a linear code with minimum distance d is $\lfloor \frac{d-1}{2} \rfloor$.

One-point codes Let P be a rational point of the algebraic smooth curve $\mathcal{X}_F$ defined over $\mathbb{F}_q$ with Weierstrass semigroup Λ. Suppose that the enumeration of Λ is λ. Recall that $A = \bigcup_{m \geqslant 0} L(mP)$ is the ring of rational functions having poles only at P. We will say that the *order* of $f \in A \setminus \{0\}$ is s if $v_P(f) = -\lambda_s$. The order of 0 is considered to be either -1 [52] or $-\infty$ [35]. In the present work we will consider the order of 0 to be -1 although both would be fine. We denote the order of f by $\rho(f)$.

One can find an infinite basis $z_0, z_1, \ldots, z_i, \ldots$ of A such that $v_P(z_i) = -\lambda_i$ or, equivalently, $\rho(z_i) = i$. Consider a set of rational points $P_1, \ldots, P_n$ different from P and the map $\varphi : A \longrightarrow \mathbb{F}_q^n$ defined by $\varphi(f) = (f(P_1), \ldots, f(P_n))$. To each finite subset $W \subseteq \mathbb{N}_0$ we associate the *one-point* code

$$C_W = < \varphi(z_i) : i \in W >^\perp = < (z_i(P_1), \ldots, z_i(P_n)) : i \in W >^\perp.$$

We say that W is the set of *parity checks* of C_W. The one-point codes for which $W = \{0, 1, \ldots, m\}$ are called *classical* one-point codes. In this case we write C_m for C_W.

4.3.1.2. *Decoding one-point codes*

This section presents a sketch of a decoding algorithm for C_W. The aim is to justify the conditions guaranteeing correction of errors. Suppose that a code word $c \in C_W$ is sent and that an error e is added to it so that the received word is $u = c + e$. We will use t for the number of non-zero positions in e.

Definition 4.13. A polynomial f is an *error-locator* of an error vector e if and only if $f(P_i) = 0$ whenever $e_i \neq 0$. The *footprint* of e is the set

$$\Delta_e = \mathbb{N}_0 \setminus \{\rho(f) : f \text{ is an error-locator}\}.$$

It is well known that $\#\Delta_e = t$ and that Δ_e is $\preccurlyeq$-closed. That is, if $i \preccurlyeq j$ and $j \in \Delta_e$ then $i \in \Delta_e$. If for each minimal element in $\mathbb{N}_0 \setminus \Delta_e$ with respect to $\preccurlyeq$ we can find an error-locator with that order then localization of errors is guaranteed.

Definition 4.14. Define the *syndrome* of orders i, j as

$$s_{ij} = \sum_{l=1}^n z_i(P_l) z_j(P_l) e_l.$$

The *syndrome matrix* $S^{rr'}$ is the $(r+1) \times (r'+1)$ matrix with $S^{rr'}_{ij} = s_{ij}$ for $0 \leqslant i \leqslant r$ and $0 \leqslant j \leqslant r'$.

The matrix $S^{rr'}$ is the matrix $S^{r'r}$, transposed. By Lemma 4.1, if $i \oplus j = k$ then there exist $a_0, \ldots, a_k$ such that $z_i z_j = a_k z_k + \cdots + a_0 z_0$. Define $s_k = \sum_{l=1}^n z_k(P_l) e_l$. Then

$$s_{ij} = a_k s_k + \cdots + a_0 s_0. \tag{4.5}$$

The syndromes depend on e which is initially unknown. So, in general, s_{ij} and s_k are unknown. For a polynomial $f = z_r + a_{r-1} z_{r-1} + \cdots + a_0 z_0$, being an error locator means that $(a_0, \ldots, a_r) S^{rr'} = 0$ for all $r' > 0$.[a] Conversely, there exists M such that if $(a_0, \ldots, a_r) S^{rr'} = 0$ for all r' with $r \oplus r' \leqslant M$ then f is an error locator.[b]

Hence, we look for pairs (r, r') with $r \oplus r'$ large enough such that $(x_0, \ldots, x_{r-1}, 1) S^{rr'} = 0$ has a non-zero solution (and indeed we look for this solution). Notice that if $(x_0, \ldots, x_{r-1}, 1) S^{rr'} = 0$ has a non-zero solution then so does $(x_0, \ldots, x_{r-1}, 1) S^{rr''}$ for all $r'' < r'$.

The first difficulty is that only a few syndromes are known. This is overcome by using a majority voting procedure.

We proceed iteratively, considering the non-gaps of Λ by increasing order. Suppose that all syndromes s_{ij} are known for $i \oplus j < k$ and we want to compute the syndromes s_{ij} with $i \oplus j = k$. By Eq. (4.5) this is equivalent to finding s_k. If $k \in W$ then the computation can be done by just using the definition of C_W: $s_k = \sum_{l=1}^n z_k(P_l) e_l = \sum_{l=1}^n z_k(P_l) u_l - \sum_{l=1}^n z_k(P_l) c_l = \sum_{l=1}^n z_k(P_l) u_l$. Otherwise we establish a voting procedure to determine s_k

In the voting procedure the voters are the elements $i \preccurlyeq k$ for which $(x_0, \ldots, x_{i-1}, 1) S^{i, k \ominus i - 1} = 0$ and $(y_0, \ldots, y_{k \ominus i - 1}, 1) S^{k \ominus i, i - 1} = 0$ have non-zero solutions. We consider the value

$\tilde{s}_{i,k \ominus i} = (s_{0,(k \ominus i)}, \ldots, s_{i-1,(k \ominus i)}) \cdot (x_0, \ldots, x_{i-1}) = (s_{i,0}, \ldots, s_{i,(k \ominus i)-1}) \cdot (y_0, \ldots, y_{(k \ominus i)-1})^c$

[a] $(a_0, \ldots, a_r) S^{rr'}$ = $(\sum_{i=0}^r a_i s_{i0}, \ldots, \sum_{i=0}^r a_i s_{ir'})$ =
$(\sum_{i=0}^r a_i \sum_{k=1}^n z_i(P_k) z_0(P_k) e_k, \ldots, \sum_{i=0}^r a_i \sum_{k=1}^n z_i(P_k) z_{r'}(P_k) e_k)$ =
$(\sum_{k=1}^n z_0(P_k) e_k \sum_{i=0}^r a_i z_i(P_k), \ldots, \sum_{k=1}^n z_{r'}(P_k) e_k \sum_{i=0}^r a_i z_i(P_k))$ =
$(\sum_{k=1}^n z_0(P_k) e_k f(P_k), \ldots, \sum_{k=1}^n z_{r'}(P_k) e_k f(P_k)) = (0, \ldots, 0)$
[b] An explanation for this can be found in [14, 35, 56].
[c] If $(x_0, \ldots, x_{i-1}, 1) S^{i,j-1} = 0$ and $(y_0, \ldots, y_{j-1}, 1) S^{j,i-1} = 0$ then $(s_{0,j}, \ldots, s_{i-1,j}) \cdot (x_0, \ldots, x_{i-1}) = (s_{i,0}, \ldots, s_{i,j-1}) \cdot (y_0, \ldots, y_{j-1})$. Indeed, $(x_0, \ldots, x_{i-1}) S^{i,j-1} = 0$ implies that $(s_{i,0}, \ldots, s_{i,j-1})$ = $-(x_0, \ldots, x_{i-1}) S^{i-1,j-1}$ and similarly, $(y_0, \ldots, y_{j-1}, 1) S^{j,i-1} = 0$ implies that $(s_{0,j}, \ldots, s_{i-1,j}) = -(y_0, \ldots, y_{j-1}) S^{j-1,i-1}$. Now, $(s_{0,j}, \ldots, s_{i-1,j}) \cdot (x_0, \ldots, x_{i-1}) = -(y_0, \ldots, y_{j-1}) S^{j-1,i-1} \cdot (x_0, \ldots, x_{i-1}) = -(y_0, \ldots, y_{j-1}) S^{j-1,i-1} (x_0, \ldots, x_{i-1})^T = -(x_0, \ldots, x_{i-1}) S^{i-1,j-1} (y_0, \ldots, y_{j-1})^T = -(x_0, \ldots, x_{i-1}) S^{i-1,j-1} \cdot (y_0, \ldots, y_{j-1}) = (s_{i,0}, \ldots, s_{i,j-1}) \cdot (y_0, \ldots, y_{j-1})$.

as a candidate for $s_{i,k\ominus i}$. Notice that if $s_{i,k\ominus i} = \tilde{s}_{i,k\ominus i}$ then $(x_0, \ldots, x_{i-1}, 1)S^{i,k\ominus i} = 0$ and $(y_0, \ldots, y_{k\ominus i-1}, 1)S^{k\ominus i, i} = 0$. Otherwise, if $s_{i,k\ominus i} \neq \tilde{s}_{i,k\ominus i}$ then there exist no error-locators of order i and no error-locators of order $k \ominus i$. Since $\tilde{s}_{i,k\ominus i}$ is a candidate for $s_{i,k\ominus i}$, the associated candidate $\tilde{s}_k$ for s_k will be derived from the equation $\tilde{s}_{i,k\ominus i} = a_k \tilde{s}_k + a_{k-1} s_{k-1} + \cdots + a_0 s_0$, where $a_0, \ldots, a_k$ are such that $z_i z_{k\ominus i} = a_k z_k + \cdots + a_0 z_0$. That is, $\tilde{s}_k = \frac{\tilde{s}_{i,k\ominus i} - a_{k-1} s_{k-1} - \cdots - a_0 s_0}{a_k}$.

Lemma 4.27.

- If $i \in N_k$ and $i, k \ominus i \notin \Delta_e$ then i is a voter and its vote coincides with s_k.
- If a voter i votes for a wrong cadidate for s_k then $i, k \ominus i \in \Delta_e$.
- If $\nu_k > 2\#(N_k \cap \Delta_e)$ then a majority of voters vote for the right value s_k.

Proof. The first two items are deduced from what has been said before. Consider the sets $A = \{i \in N_k : i, k \ominus i \in \Delta_e\}$, $B = \{i \in N_k : i \in \Delta_e, k \ominus i \notin \Delta_e\}$, $C = \{i \in N_k : i \notin \Delta_e, k \ominus i \in \Delta_e\}$, $D = \{i \in N_k : i, k \ominus i \notin \Delta_e\}$. By the previous items, the wrong votes are at most $\#A$ while the right votes are at least $\#D$. Obviously, $\nu_k = \#A + \#B + \#C + \#D$, $\#(N_k \cap \Delta_e) = \#A + \#B = \#A + \#C$. So, the difference between the right and the wrong votes is at least $\#D - \#A = \nu_k - 2\#A - \#B - \#C = \nu_k - 2\#(N_k \cap \Delta_e) > 0$. □

The conclusion of this section is the next theorem.

Theorem 4.5. *If $\nu_i > 2\#(N_i \cap \Delta_e)$ for all $i \notin W$ then e is correctable by C_W.*

4.3.2. The ν sequence, classical codes, and Feng-Rao improved codes

From the equality $\#\Delta_e = t$ we deduce the next lemma.

Lemma 4.28. *If the number t of errors in e satisfies $t \leqslant \lfloor \frac{\nu_i - 1}{2} \rfloor$, then $\nu_i > 2\#(N_i \cap \Delta_e)$.*

4.3.2.1. The ν sequence and the minimum distance of classical codes

Theorem 4.5 and Lemma 4.28 can be used in order to get an estimate of the minimum distance of a one-point code. The next definition arises from [23, 35, 39].

Definition 4.15. The *order bound* on the minimum distance of the classical code C_W, with $W = \{0, \ldots, m\}$ is

$$d_{ORD}(C_m) = \min\{\nu_i : i > m\}.$$

The order bound is also referred to as the *Feng-Rao bound*. The order bound is proved to be a lower bound on the minimum distance for classical codes [23, 35, 39].

Lemma 4.29. $d(C_m) \geqslant d_{ORD}(C_m)$.

From Lemma 4.20 we deduce that $\nu_{i+1} \leqslant \nu_{i+2}$ and so $d_{ORD}(C_i) = \nu_{i+1}$ for all $i \geqslant 2c - g - 2$.

A refined version of the order bound is

$$d_{ORD}^{P_1,\ldots,P_n}(C_m) = \min\{\nu_i : i > m, C_i \neq C_{i+1}\}.$$

While d_{ORD} only depends on the Weierstrass semigroup, $d_{ORD}^{P_1,\ldots,P_n}$ depends also on the points $P_1, \ldots, P_n$. Since our point of view is that of numerical semigroups we will concentrate on d_{ORD}.

Generalized Hamming weights are a generalization of the minimum distance of a code with many applications to coding theory but also to other fields such as cryptography. For the generalized Hamming weights of one-point codes there is a generalization of the order bound based also on the associated Weierstrass semigroups. We will not discuss this topic here but the reader interested in it can see [22, 34].

4.3.2.2. *On the order bound on the minimum distance*

In this section we will find a formula for the smallest m for which $d_{ORD}(C_i) = \nu_{i+1}$ for all $i \geqslant m$, for the case of acute semigroups. At the end we will use Munuera-Torres and Oneto-Tamone's results to generalize this formula.

Remark 4.6. Let Λ be a non-ordinary numerical semigroup with conductor c, subconductor c' and dominant d. Then, $c' + d \geqslant c$. Indeed, $c' + d \in \Lambda$ and by Remark 4.2 it is strictly larger than d. So, it must be larger than or equal to c.

Theorem 4.6. *Let Λ be a non-ordinary acute numerical semigroup with enumeration λ, conductor c, subconductor c' and dominant d. Let*

$$m = \min\{\lambda^{-1}(c + c' - 2), \lambda^{-1}(2d)\}. \tag{4.6}$$

Then,

(1) $\nu_m > \nu_{m+1}$
(2) $\nu_i \leqslant \nu_{i+1}$ for all $i > m$.

Proof. Following the notations in Lemma 4.20, for $i \geqslant \lambda^{-1}(c)$, $g(i) = g$. Thus, for $i \geqslant \lambda^{-1}(c)$ we have

$$\nu_i \leqslant \nu_{i+1} \text{ if and only if } \#D(i+1) \geqslant \#D(i) - 1. \tag{4.7}$$

Let $l = c - d - 1$. Notice that l is the number of gaps between the conductor and the dominant. Since Λ is acute, the l integers before c' are also gaps. Let us call $k = \lambda^{-1}(c' + d)$. For all $1 \leqslant i \leqslant l$, both $(c' - i)$ and $(d + i)$ are in $D(k)$ because they are gaps and

$$(c' - i) + (d + i) = c' + d.$$

Moreover, there are no more gaps in $D(k)$ because, if $j \leqslant c' - l - 1$ then $c' + d - j \geqslant d + l + 1 = c$ and so $c' + d - j \in \Lambda$. Therefore,

$$D(k) = \{c' - i : 1 \leqslant i \leqslant l\} \cup \{d + i : 1 \leqslant i \leqslant l\}.$$

Now suppose that $j \geqslant k$. By Remark 4.6, $\lambda_k \geqslant c$ and so $\lambda_j = \lambda_k + j - k = c' + d + j - k$. Then,

$$D(j) = A(j) \cup B(j),$$

where

$$A(j) = \begin{cases} \{c' - i : 1 \leqslant i \leqslant l - j + k\} \\ \cup \{d + i : j - k + 1 \leqslant i \leqslant l\} & \text{if } \lambda_k \leqslant \lambda_j \leqslant c + c' - 2, \\ \emptyset & \text{otherwise.} \end{cases}$$

$$B(j) = \begin{cases} \emptyset & \text{if } \lambda_k \leqslant \lambda_j \leqslant 2d + 1 \\ \{d + i : 1 \leqslant i \leqslant \lambda_j - 2d - 1\} & \text{if } 2d + 2 \leqslant \lambda_j \leqslant c + d, \\ \{d + i : \lambda_j - d - c + 1 \leqslant i \leqslant l\} & \text{if } c + d \leqslant \lambda_j \leqslant 2c - 2, \\ \emptyset & \text{if } \lambda_j \geqslant 2c - 1. \end{cases}$$

Notice that $A(j) \cap B(j) = \emptyset$ and hence

$$\#D(j) = \#A(j) + \#B(j).$$

We have

$$\#A(j) = \begin{cases} 2(l - j + k) & \text{if } \lambda_k \leqslant \lambda_j \leqslant c + c' - 2, \\ 0 & \text{otherwise.} \end{cases}$$

$$\#B(j) = \begin{cases} 0 & \text{if } \lambda_k \leqslant \lambda_j \leqslant 2d + 1, \\ \lambda_j - 2d - 1 & \text{if } 2d + 2 \leqslant \lambda_j \leqslant c + d, \\ 2c - 1 - \lambda_j & \text{if } c + d \leqslant \lambda_j \leqslant 2c - 2, \\ 0 & \text{if } \lambda_j \geqslant 2c - 1. \end{cases}$$

So,

$$\#A(j+1) = \begin{cases} \#A(j) - 2 & \text{if } \lambda_k \leqslant \lambda_j \leqslant c + c' - 2, \\ \#A(j) & \text{otherwise.} \end{cases}$$

$$\#B(j+1) = \begin{cases} \#B(j) & \text{if } \lambda_k \leqslant \lambda_j \leqslant 2d \\ \#B(j) + 1 & \text{if } 2d + 1 \leqslant \lambda_j \leqslant c + d - 1 \\ \#B(j) - 1 & \text{if } c + d \leqslant \lambda_j \leqslant 2c - 2 \\ \#B(j) & \text{if } \lambda_j \geqslant 2c - 1. \end{cases}$$

Notice that $c + c' - 2 < c + d$. Thus, for $\lambda_j \geqslant c + d$,

$$\#D(j+1) = \begin{cases} \#D(j) - 1 & \text{if } c + d \leqslant \lambda_j \leqslant 2c - 2, \\ \#D(j) & \text{if } \lambda_j \geqslant 2c - 1. \end{cases}$$

Hence, by (4.7), $\nu_i \leqslant \nu_{i+1}$ for all $i \geqslant \lambda^{-1}(c+d)$ because $\lambda^{-1}(c+d) \geqslant \lambda^{-1}(c)$. Now, let us analyze what happens if $\lambda_j < c + d$.

If $c + c' - 2 \leqslant 2d$ then

$$\#D(j+1) = \begin{cases} \#D(j) - 2 & \text{if } \lambda_k \leqslant \lambda_j \leqslant c + c' - 2, \\ \#D(j) & \text{if } c + c' - 1 \leqslant \lambda_j \leqslant 2d, \\ \#D(j) + 1 & \text{if } 2d + 1 \leqslant \lambda_j \leqslant c + d - 1 \end{cases}$$

and if $2d + 1 \leqslant c + c' - 2$ then

$$\#D(j+1) = \begin{cases} \#D(j) - 2 & \text{if } \lambda_k \leqslant \lambda_j \leqslant 2d, \\ \#D(j) - 1 & \text{if } 2d + 1 \leqslant \lambda_j \leqslant c + c' - 2, \\ \#D(j) + 1 & \text{if } c + c' - 1 \leqslant \lambda_j \leqslant c + d - 1. \end{cases}$$

So, by (4.7) and since both $c + c' - 2$ and $2d$ are larger than or equal to c, the result follows. □

Corollary 4.1. *Let Λ be a non-ordinary acute numerical semigroup with enumeration λ, conductor c and subconductor c'. Let*

$$m = \min\{\lambda^{-1}(c + c' - 2), \lambda^{-1}(2d)\}.$$

Then, m is the smallest integer for which

$$d_{ORD}(C_i) = \nu_{i+1}$$

for all $i \geqslant m$.

Example 4.15. Recall the Weierstrass semigroup at the point P_0 on the Klein quartic that we presented in Example 4.2. Its conductor is 5, its dominant is 3 and its subconductor is 3. In Table 4.3 we have, for each integer from 0 to $\lambda^{-1}(2c - 2)$, the values λ_i, ν_i and $d_{ORD}(C_i)$.

Table 4.3. Klein quartic.

i	λ_i	ν_i	$d_{ORD}(C_i)$
0	0	1	2
1	3	2	2
2	5	2	2
3	6	3	2
4	7	2	4
5	8	4	4

For this example, $\lambda^{-1}(c+c'-2) = \lambda^{-1}(2d) = 3$ and so, $m = \min\{\lambda^{-1}(c+c'-2), \lambda^{-1}(2d)\} = 3$. We can check that, as stated in Theorem 4.6, $\nu_3 > \nu_4$ and $\nu_i \leqslant \nu_{i+1}$ for all $i > 3$. Moreover, as stated in Corollary 4.1, $d_{ORD}(C_i) = \nu_{i+1}$ for all $i \geqslant 3$ while $d_{ORD}(C_2) \neq \nu_3$.

Lemma 4.30. *Let Λ be a non-ordinary numerical semigroup with conductor c, subconductor c' and dominant d.*

(1) If Λ is symmetric then $\min\{c + c' - 2, 2d\} = c + c' - 2 = 2c - 2 - \lambda_1$.
(2) If Λ is pseudo-symmetric then $\min\{c + c' - 2, 2d\} = c + c' - 2$.
(3) If Λ is Arf then $\min\{c + c' - 2, 2d\} = 2d$.
(4) If Λ is generated by an interval then $\min\{c + c' - 2, 2d\} = c + c' - 2$.

Proof.

(1) We already saw in the proof of Lemma 4.16 that if Λ is symmetric then $d = c - 2$. So, $c + c' - 2 = d + c' \leqslant 2d$ because $c' \leqslant d$. Moreover, by Lemma 4.4, any non-negative integer i is a gap if and only if $c-1-i \in \Lambda$. This implies that $c' - 1 = c - 1 - \lambda_1$ and so $c' = c - \lambda_1$. Therefore, $c + c' - 2 = 2c - 2 - \lambda_1$.

(2) If Λ is pseudo-symmetric and non-ordinary then $d = c - 2$ because 1 is a gap different from $(c-1)/2$. So, $c + c' - 2 = d + c' \leqslant 2d$.

(3) If Λ is Arf then $c' = d$. Indeed, if $c' < d$ then $d - 1 \in \Lambda$ and, by Lemma 4.6, $d - 1 \geqslant c$, a contradiction. Since $d \leqslant c - 2$, we have $2d \leqslant c + c' - 2$.

(4) Suppose Λ is generated by the interval $\{i, i+1, \ldots, j\}$. By Lemma 4.12, there exists k such that $c = ki$ and $d = (k-1)j$. We have that $c-d \leqslant j-i$, because otherwise $(k+1)i - kj = c - d - (j-i) > 1$, and hence $kj + 1$ would be a gap greater than c. On the other hand $d - c' \geqslant j - i$, and hence $2d - (c + c' - 2) = d - c + d - c' + 2 \geqslant i - j + j - i + 2 = 2$. □

Table 4.4. Hermitian curve.

i	λ_i	ν_i	$d_{ORD}(C_i)$
0	0	1	2
1	4	2	2
2	5	2	3
3	8	3	3
4	9	4	3
5	10	3	4
6	12	4	4
7	13	6	4
8	14	6	4
9	15	4	5
10	16	5	8
11	17	8	8
12	18	9	8
13	19	8	9
14	20	9	10
15	21	10	12
16	22	12	12

Example 4.16. Consider the Hermitian curve over $\mathbb{F}_{16}$. Its numerical semigroup is generated by 4 and 5. So, this is a symmetric numerical semigroup because it is generated by two coprime integers, and it is also a semigroup generated by the interval $\{4,5\}$.

In Table 4.4 we include, for each integer from 0 to 16, the values λ_i, ν_i and $d_{ORD}(C_i)$. Notice that in this case the conductor is 12, the dominant is 10 and the subconductor is 8. We do not give the values in the table for $i > \lambda^{-1}(2c - 1) - 1 = 16$ because $d_{ORD}(C_i) = \nu_{i+1}$ for all $i \geqslant \lambda^{-1}(2c - 1) - 1$. We can check that, as follows from Theorem 4.6 and Lemma 4.30, $\lambda^{-1}(c + c' - 2) = 12$ is the largest integer m with $\nu_m > \nu_{m+1}$ and so the smallest integer for which $d_{ORD}(C_i) = \nu_{i+1}$ for all $i \geqslant m$. Notice also that, as pointed out in Lemma 4.30, $c + c' - 2 = 2c - 2 - \lambda_1$.

Furthermore, in this example there are 64 rational points on the curve different from P_∞ and the map φ evaluating the functions of A at these 64 points satisfies that the words $\varphi(f_0), \ldots, \varphi(f_{57})$ are linearly independent whereas $\varphi(f_{58})$ is linearly dependent to the previous ones. So, $d_{ORD}^{P_1,\ldots,P_n}(C_i) = d_{ORD}(C_i)$ for all $i \leqslant 56$.

Table 4.5. Garcia-Stichtenoth tower.

i	λ_i	ν_i	$d_{ORD}(C_i)$
0	0	1	2
1	16	2	2
2	20	2	2
3	24	2	2
4	25	2	2
5	26	2	2
6	27	2	2
7	28	2	2
8	29	2	2
9	30	2	2
10	31	2	2
11	32	3	2
12	33	2	2
13	34	2	2
14	35	2	2
15	36	4	2
16	37	2	2
17	38	2	2
18	39	2	4
19	40	5	4
20	41	4	4
21	42	4	4
22	43	4	6
23	44	6	6
24	45	6	6
25	46	6	6

Example 4.17. Let us consider now the semigroup of the fifth code associated to the second tower of Garcia and Stichtenoth over $\mathbb{F}_4$. As noticed in Example 4.6, this is an Arf numerical semigroup. We set in Table 4.5 the values λ_i, ν_i and $d_{ORD}(C_i)$ for each integer from 0 to 25. In this case the conductor is 24, the dominant is 20 and the subconductor is 20. As before, we do not give the values for $i > \lambda^{-1}(2c-1) - 1 = 25$. We can check that, as follows from Theorem 4.6 and Lemma 4.30, $\lambda^{-1}(2d) = 19$ is the largest integer m with $\nu_m > \nu_{m+1}$ and so, the smallest integer for which $d_{ORD}(C_i) = \nu_{i+1}$ for all $i \geqslant m$.

Munuera and Torres in [45] and Oneto and Tamone in [47] proved that for *any* numerical semigroup $m \leqslant \min\{c + c' - 2 - g, 2d - g\}$. Notice that for acute semigroups this inequality is an equality.

Munuera and Torres in [45] introduced the definition of near-acute semigroups. They proved that the formula $m = \min\{c + c' - 2 - g, 2d - g\}$ not only applies for acute semigroups but also for near-acute semigroups. Next we give the definition of near-acute semigroups.

Definition 4.16. A numerical semigroup with conductor c, dominant d and subdominant d' is said to be a *near-acute semigroup* if either $c - d \leqslant d - d'$ or $2d - c + 1 \notin \Lambda$.

Oneto and Tamone in [47] proved that $m = \min\{c + c' - 2 - g, 2d - g\}$ if and *only if* $c + c' - 2 \leqslant 2d$ or $2d - c + 1 \notin \Lambda$. Let us see next that these conditions in Oneto and Tamone's result are equivalent to having a near-acute semigroup.

Lemma 4.31. *For a numerical semigroup the following are equivalent*

(1) $c - d \leqslant d - d'$ *or* $2d - c + 1 \notin \Lambda$,
(2) $c + c' - 2 \leqslant 2d$ *or* $2d - c + 1 \notin \Lambda$.

Proof. Let us see first that (1) implies (2). If $2d - c + 1 \notin \Lambda$ then it is obvious. Otherwise the condition $c - d \leqslant d - d'$ is equivalent to $d' \leqslant 2d - c$ which, together with $2d - c + 1 \in \Lambda$ implies $c' \leqslant 2d - c + 1$ by definition of c'. This in turn implies that $c + c' - 2 < c + c' - 1 \leqslant 2d$.

To see that (1) is a consequence of (2) notice that by definition, $d' \leqslant c' - 2$. Then, if $c + c' - 2 \leqslant 2d$, we have $d - d' \geqslant d - c' + 2 \geqslant c - d$. □

From all these results one concludes the next theorem.

Theorem 4.7.

(1) For any *numerical semigroup* $m \leqslant \min\{c + c' - 2 - g, 2d - g\}$.
(2) $m = \min\{c + c' - 2 - g, 2d - g\}$ *if and only if the corresponding numerical semigroup is near-acute.*

In [48] Oneto and Tamone give further results on m and in [49] the same authors conjecture that for any numerical semigroup,

$$\lambda_m \geqslant c + d - \lambda_1.$$

4.3.2.3. The ν sequence and Feng-Rao improved codes

The one-point codes whose set W of parity checks is selected so that the orders outside W satisfy the hypothesis of Lemma 4.28 and W is minimal with this property are called *Feng-Rao improved codes*. They were defined in [24, 35].

Definition 4.17. Given a rational point P of an algebraic smooth curve $\mathcal{X}_F$ defined over $\mathbb{F}_q$ with Weierstrass semigroup Λ and sequence ν with associated basis $z_0, z_1, \ldots$ and given n other different points $P_1, \ldots, P_n$ of $\mathcal{X}_F$, the associated *Feng-Rao improved code* guaranteeing correction of t errors is defined as

$$C_{\tilde{R}(t)} = < (z_i(P_1), \ldots, z_i(P_n)) : i \in \tilde{R}(t) >^{\perp},$$

where

$$\tilde{R}(t) = \{i \in \mathbb{N}_0 : \nu_i < 2t + 1\}.$$

4.3.2.4. On the improvement of the Feng-Rao improved codes

Feng-Rao improved codes will actually give an improvement with respect to classical codes only if ν_i is decreasing at some i. We next study this condition.

Lemma 4.32. *If Λ is an ordinary numerical semigroup with enumeration λ then*

$$\nu_i = \begin{cases} 1 & \text{if } i = 0, \\ 2 & \text{if } 1 \leqslant i \leqslant \lambda_1, \\ i - \lambda_1 + 2 & \text{if } i > \lambda_1. \end{cases}$$

Proof. It is obvious that $\nu_0 = 1$ and that $\nu_i = 2$ whenever $0 < \lambda_i < 2\lambda_1$. So, since $2\lambda_1 = \lambda_{\lambda_1+1}$, we have that $\nu_i = 2$ for all $1 \leqslant i \leqslant \lambda_1$. Finally, if $\lambda_i \geqslant 2\lambda_1$ then all non-gaps up to $\lambda_i - \lambda_1$ are in N_i as well as λ_i, and none of the remaining non-gaps are in N_i. Now, if the genus of Λ is g, then $\nu_i = \lambda_i - \lambda_1 + 2 - g$ and $\lambda_i = i + g$. So, $\nu_i = i - \lambda_1 + 2$. □

As a consequence of Lemma 4.32, the ν sequence is non-decreasing if Λ is an ordinary numerical semigroup. We will see in this section that ordinary numerical semigroups are in fact the only semigroups for which the ν sequence is non-decreasing.

Lemma 4.33. *Suppose that for the semigroup* Λ *the* ν *sequence is non-decreasing. Then* Λ *is Arf.*

Proof. Let λ be the enumeration of Λ. Let us see by induction that, for any non-negative integer i,

(i) $N_{\lambda^{-1}(2\lambda_i)} = \{j \in \mathbb{N}_0 : j \leqslant i\} \sqcup \{\lambda^{-1}(2\lambda_i - \lambda_j) : 0 \leqslant j < i\}$, where $\sqcup$ means the union of disjoint sets.

(ii) $N_{\lambda^{-1}(\lambda_i + \lambda_{i+1})} = \{j \in \mathbb{N}_0 : j \leqslant i\} \sqcup \{\lambda^{-1}(\lambda_i + \lambda_{i+1} - \lambda_j) : 0 \leqslant j \leqslant i\}$.

Notice that if (i) is satisfied for all i, then $\{j \in \mathbb{N}_0 : j \leqslant i\} \subseteq N_{\lambda^{-1}(2\lambda_i)}$ for all i, and hence by Lemma 4.10 Λ is Arf.

It is obvious that both (i) and (ii) are satisfied for the case $i = 0$.

Suppose $i > 0$. By the induction hypothesis, $\nu_{\lambda^{-1}(\lambda_{i-1} + \lambda_i)} = 2i$. Now, since (ν_i) is not decreasing and $2\lambda_i > \lambda_{i-1} + \lambda_i$, we have $\nu_{\lambda^{-1}(2\lambda_i)} \geqslant 2i$. On the other hand, if $j, k \in \mathbb{N}_0$ are such that $j \leqslant k$ and $\lambda_j + \lambda_k = 2\lambda_i$ then $\lambda_j \leqslant \lambda_i$ and $\lambda_k \geqslant \lambda_i$. So, $\lambda(N_{\lambda^{-1}(2\lambda_i)}) \subseteq \{\lambda_j : 0 \leqslant j \leqslant i\} \sqcup \{2\lambda_i - \lambda_j : 0 \leqslant j < i\}$ and hence $\nu_{\lambda^{-1}(2\lambda_i)} \geqslant 2i$ if and only if $N_{\lambda^{-1}(2\lambda_i)} = \{j \in \mathbb{N}_0 : j \leqslant i\} \sqcup \{\lambda^{-1}(2\lambda_i - \lambda_j) : 0 \leqslant j < i\}$. This proves (i).

Finally, (i) implies $\nu_{\lambda^{-1}(2\lambda_i)} = 2i + 1$ and (ii) follows by an analogous argumentation. $\square$

Theorem 4.8. *The only numerical semigroups for which the* ν *sequence is non-decreasing are ordinary numerical semigroups.*

Proof. It is a consequence of Lemma 4.33, Lemma 4.16, Theorem 4.6 and Lemma 4.32. $\square$

Corollary 4.2. *The only numerical semigroup for which the* ν *sequence is strictly increasing is the trivial numerical semigroup.*

Proof. It is a consequence of Theorem 4.8 and Lemma 4.32. $\square$

As a consequence of Theorem 4.8 we can show that the only numerical semigroups for which the associated classical codes are not improved by the Feng-Rao improved codes, at least for one value of t, are ordinary semigroups.

Corollary 4.3. *Given a numerical semigroup* Λ *define* $m(\delta) = \max\{i \in \mathbb{N}_0 : \nu_i < \delta\}$. *There exists at least one value of* δ *for which* $\{i \in \mathbb{N}_0 : \nu_i < \delta\} \subsetneq \{i \in \mathbb{N}_0 : i \leqslant m(\delta)\}$ *if and only if* Λ *is non-ordinary.*

4.3.3. *Generic errors and the τ sequence*

All the results in these sections are based on [10, 14, 15]. Correction of generic errors has already been considered in [36, 50, 53].

4.3.3.1. *Generic errors*

Definition 4.18. The points $P_{i_1}, \ldots, P_{i_t}$ ($P_{i_j} \neq P$) are *generically distributed* if no non-zero function generated by $z_0, \ldots, z_{t-1}$ vanishes in all of them. In the context of one-point codes, *generic errors* are those errors whose non-zero positions correspond to generically distributed points. Equivalently, e is generic if and only if $\Delta_e = \Delta_t := \{0, \ldots, t-1\}$.

Generic errors of weight t can be a very large proportion of all possible errors of weight t [33]. Thus, by restricting the errors to be corrected to generic errors the decoding requirements become weaker and we are still able to correct almost all errors. In some of these references generic errors are called *independent errors*.

Example 4.18 (Generic sets of points in $\mathcal{H}_q$). Recall that the Hermitian curve from Example 4.1. It is defined over $\mathbb{F}_{q^2}$ and its affine equation is $x^{q+1} = y^q + y$.

The unique point at infinity is $P_\infty = (0:1:0)$. If $b \in \mathbb{F}_q$ then $b^q + b = Tr(b) = 0$ and the unique affine point with $y = b$ is $(0, b)$. There are a total of q such points. If $b \in \mathbb{F}_{q^2} \setminus \mathbb{F}_q$ then $b^q + b = Tr(b) \in \mathbb{F}_q \setminus \{0\}$ and there are $q+1$ solutions of $x^{q+1} = b^q + b$, so there are $q+1$ different affine points with $y = b$. There are a total of $(q^2 - q)(q+1)$ such points. The total number of affine points is then $q + (q^2 - q)(q+1) = q^3$.

If we distinguish the point P_∞, we can take $z_0 = 1, z_1 = x, z_2 = y$, $z_3 = x^2, z_4 = xy, z_5 = y^2 \ldots$

Non-generic sets of two points are pairs of points satisfying $x^{q+1} = y^q + y$ and simultaneously vanishing at $f = z_1 + az_0 = x + a$ for some $a \in \mathbb{F}_{q^2}$. The expression $x + a$ represents a line with q points. There are q^2 such lines. There are a total of $q^2 \binom{q}{2}$ pairs of colinear points over lines of the form $x + a$ and so $q^2 \binom{q}{2}$ non-generic errors.

Consequently, the portion of non-generic errors of weight 2 is $\dfrac{q^2 \binom{q}{2}}{\binom{q^3}{2}} = \dfrac{1}{q^2 + q + 1}$.

A set of three points is non-generic if the points satisfy $x^{q+1} = y^q + y$ and simultaneously vanish at $f = z_1 + az_0 = x + a$ for some $a \in \mathbb{F}_{q^2}$ or at $f = z_2 + az_1 + bz_0 = y + ax + b$ for some $a, b \in \mathbb{F}_{q^2}$.

The expression $x + a$ represents a line (which we call of type 1) with q points. There are q^2 lines of type 1.

The line $y + ax + b$ is called of type 2 if $a^{q+1} = b^q + b$ and of type 3 otherwise. There are q^3 lines of type 2 and $q^4 - q^3$ lines of type 3.

Lines of type 2 have only one point. Indeed, a point on $\mathcal{H}_q$ and on the line $y + ax + b$ must satisfy $x^{q+1} = (-ax - b)^q + (-ax - b) = -(ax)^q - ax - a^{q+1}$. Notice that $(x + a^q)^{q+1} = (x + a^q)^q(x + a^q) = (x^q + a)(x + a^q) = x^{q+1} + x^q a^q + ax + a^{q+1}$. So, $x = -a^q$ is the unique solution to $x^{q+1} = -(ax)^q - ax - a^{q+1}$ and so the unique point of $\mathcal{H}_q$ on the line $y + ax + b$ is $(-a^q, a^{q+1} - b)$.

Lines of type 3 have $q + 1$ points. This follows by a counting argument. On one hand, as seen before, a point on $\mathcal{H}_q$ and on the line $y + ax + b$ must satisfy $x^{q+1} = -(ax)^q - ax - b^q - b$. There are at most $q + 1$ different values of x satisfying this equation and so at most $q + 1$ different points of $\mathcal{H}_q$ on the line $y + ax + b$. On the other hand there are a total of $\binom{q^3}{2}$ pairs of affine points. Each pair meets only in one line. The number of pairs sharing lines of type 1 is $q^2\binom{q}{2}$, the number of pairs sharing lines of type 2 is 0 and the number of pairs sharing lines of type 3 is at most $q^3(q-1)\binom{q+1}{2}$, with equality only if all lines of type 3 have $q+1$ points. Since $q^2\binom{q}{2} + q^3(q-1)\binom{q+1}{2} = \binom{q^3}{2}$, we deduce that all the lines of type 3 must have $q + 1$ points.

In total there are $q^2\binom{q}{3}$ sets of three points sharing a line of type 1 and $(q^4 - q^3)\binom{q+1}{3}$ sets of three points sharing a line of type 3.

The portion of non-generic errors of weight 3 is then $\frac{q^2\binom{q}{3} + q^3(q-1)\binom{q+1}{3}}{\binom{q^3}{3}} = \frac{1}{q^2+q+1}$.

4.3.3.2. Conditions for correcting generic errors

In the next lemma we find conditions guaranteeing the majority voting step for generic errors. It is a reformulation of results that appeared in [10, 14, 51].

Lemma 4.34. Let $\Sigma_t = \mathbb{N}_0 \setminus \Delta_t = t + \mathbb{N}_0$. The following conditions are equivalent.

(1) $\nu_k > 2\#(N_k \cap \Delta_t)$,
(2) $k \in \Sigma_t \oplus \Sigma_t$,
(3) $\tau_k \geqslant t$.

Proof. Let $A = \{i \in N_k : i, k \ominus i \in \Delta_t\}$, $D = \{i \in N_k : i, k \ominus i \in \Sigma_t\}$. By an argument analogous to that in the proof of Lemma 4.27, $\nu_k > 2\#(N_k \cap \Delta_e)$ is equivalent to $\#D > \#A$. If this inequality is satisfied then $\#D > 0$ and so $k \in \Sigma_t \oplus \Sigma_t$. On the other hand, $\min \Sigma_t \oplus \Sigma_t = t \oplus t > (t-1) \oplus (t-1) = \max \Delta_t \oplus \Delta_t$. So, $\Sigma_t \oplus \Sigma_t \cap \Delta_t \oplus \Delta_t = \emptyset$ and, if $k \in \Sigma_t \oplus \Sigma_t$ then $k \notin \Delta_t \oplus \Delta_t$ and so $\#A = 0$ implying $\#D > \#A$.

The equivalence of $k \in \Sigma_t \oplus \Sigma_t$ and $\tau_k \geqslant t$ is straightforward. □

The one-point codes whose set W of parity checks is selected so that the orders outside W satisfy the hypothesis of Lemma 4.34 and W is minimal with this property are called improved codes correcting generic errors. They were defined in [4, 14].

Definition 4.19. Given a rational point P of an algebraic smooth curve $\mathcal{X}_F$ defined over $\mathbb{F}_q$ with Weierstrass semigroup Λ and sequence ν with associated basis $z_0, z_1, \ldots$ and given n other different points $P_1, \ldots, P_n$ of $\mathcal{X}_F$, the associated *improved code* guaranteeing correction of t generic errors is defined as

$$C_{\tilde{R}^*(t)} = <(z_i(P_1), \ldots, z_i(P_n)) : i \in \tilde{R}^*(t) >^\perp,$$

where

$$\tilde{R}^*(t) = \{i \in \mathbb{N}_0 : \tau_i < t\}.$$

4.3.3.3. Comparison of improved codes and classical codes correcting generic errors

Classical evaluation codes are those codes for which the set of parity checks corresponds to all the elements up to a given order. Thus, the classical evaluation code with maximum dimension correcting t generic errors is defined by the set of checks $R^*(t) = \{i \in \mathbb{N}_0 : i \leqslant m(t)\}$ where $m(t) = \max\{i \in \mathbb{N}_0 : \tau_i < t\}$. Then, by studying the monotonicity of the τ sequence we can compare $\tilde{R}^*(t)$ and $R^*(t)$ and the associated codes.

It is easy to check that for the trivial numerical semigroup one has $\tau_{2i} = \tau_{2i+1} = i$ for all $i \in \mathbb{N}_0$. That is, the τ sequence is

$$0, 0, 1, 1, 2, 2, 3, 3, 4, 4, 5, 5, \ldots$$

The next lemma determines the τ sequence of all non-trivial ordinary semigroups.

Lemma 4.35. *The non-trivial ordinary numerical semigroup with conductor c has τ sequence given by*

$$\tau_i = \begin{cases} 0 & \text{if } i \leqslant c \\ \lfloor \frac{i-c+1}{2} \rfloor & \text{if } i > c. \end{cases}$$

Proof. Suppose that the numerical semigroup has enumeration λ. On one hand, $\lambda_1, \ldots, \lambda_c$ are all generators and thus $\tau_i = 0$ for $i \leqslant c$. For $i > c$, $\lambda_i = c + i - 1 \geqslant 2c$. So, if λ_i is even (which is equivalent to $c + i$ being odd) then $\tau_i = \lambda^{-1}(\frac{\lambda_i}{2}) = \frac{c+i-1}{2} - c + 1 = \frac{i-c+1}{2} = \lfloor \frac{i-c+1}{2} \rfloor$. If λ_i is odd (which is equivalent to either both c and i being even or being odd) then $\tau_i = \lambda^{-1}(\frac{\lambda_i - 1}{2}) = \frac{c+i-2}{2} - c + 1 = \frac{i-c}{2} = \lfloor \frac{i-c+1}{2} \rfloor$. □

Remark 4.7. The formula in Lemma 4.35 can be reformulated as $\tau_j = 0$ for all $j \leqslant c$ and, for all $i \geqslant 0$, $\tau_{c+2i+1} = \tau_{c+2i+2} = i + 1$.

The next lemma gives, for non-ordinary semigroups, the smallest index m for which τ is non-decreasing from τ_m on. We will use the notation $\lfloor a \rfloor_\Lambda$ to denote the semigroup floor of a non-negative integer a, that is, the largest non-gap of Λ which is at most a.

Lemma 4.36. *Let Λ be a non-ordinary semigroup with dominant d and let $m = \lambda^{-1}(2d)$, then*

(1) $\tau_m = c - g - 1 > \tau_{m+1}$,
(2) $\tau_i < c - g - 1$ for all $i < m$,
(3) $\tau_i \leqslant \tau_{i+1}$ for all $i > m$.

Proof. For statement 1 notice that both $2d$ and $2d+1$ belong to Λ because they must be larger than the conductor. Furthermore, $\tau_{\lambda^{-1}(2d)} = \lambda^{-1}(d) = c - g - 1$ while $\tau_{\lambda^{-1}(2d+1)} = \tau_{\lambda^{-1}(2d)+1} < \lambda^{-1}(d)$ because $d + 1 \notin \Lambda$.

Statement 2 follows from the fact that if $\lambda_i < 2d$ then $\tau_i < \lambda^{-1}(d) = c - g - 1$.

For statement 3 suppose that $i > m$. Notice that $2d$ is the largest non-gap that can be written as a sum of two non-gaps both of them smaller than the conductor c. Then if $j \leqslant k \leqslant i$ and $\lambda_j + \lambda_k = \lambda_i$ it must be $\lambda_k \geqslant c$ and so $\tau_i = \lambda^{-1}(\lfloor \lambda_i - c \rfloor_\Lambda)$. Since both λ^{-1} and $\lfloor \cdot \rfloor_\Lambda$ are non-decreasing, so is τ_i for $i > m$. □

Corollary 4.4. *The only numerical semigroups for which the τ sequence is non-decreasing are ordinary semigroups.*

A direct consequence of Corollary 4.4 is that the classical code determined by $R^*(t)$ is always worse than the improved code determined by $\widetilde{R}^*(t)$ at least for one value of t unless the corresponding numerical semigroup is ordinary. From Lemma 4.36 we can derive that $\widetilde{R}^*(t)$ and $R^*(t)$ coincide from a certain point and we can find this point. We summarize the results of this section in the next Corollary.

Corollary 4.5.

(1) $\widetilde{R}^*(t) \subseteq R^*(t)$ for all $t \in \mathbb{N}_0$.
(2) $\widetilde{R}^*(t) = R^*(t)$ for all $t \geqslant c - g$.
(3) $\widetilde{R}^*(t) = R^*(t)$ for all $t \in \mathbb{N}_0$ if and only if the associated numerical semigroup is ordinary.

Proof. Statement 1 is a consequence of the definition of $R^*(t)$. Statement 2 is clear if the associated semigroup is ordinary. Otherwise it follows from the fact proved in Lemma 4.36 that the largest value of τ_i before it starts being non-decreasing is precisely $c - g - 1$ and that before that all values of τ_i are smaller than $c - g - 1$. Statement 3 is a consequence of Corollary 4.4. □

4.3.3.4. Comparison of improved codes correcting generic errors and Feng–Rao improved codes

In the next theorem we compare τ_i with $\lfloor \frac{\nu_i - 1}{2} \rfloor$ and this will give a new characterization of Arf semigroups. Recall that $t \leqslant \lfloor \frac{\nu_i - 1}{2} \rfloor$ guarantees the computation of syndromes of order i when performing majority voting (Lemma 4.27).

Theorem 4.9. *Let Λ be a numerical semigroup with conductor c, genus g, and associated sequences τ and ν. Then*

(1) $\tau_i \geqslant \lfloor \frac{\nu_i - 1}{2} \rfloor$ for all $i \in \mathbb{N}_0$,
(2) $\tau_i = \lfloor \frac{\nu_i - 1}{2} \rfloor$ for all $i \geqslant 2c - g - 1$,
(3) $\tau_i = \lfloor \frac{\nu_i - 1}{2} \rfloor$ for all $i \in \mathbb{N}_0$ if and only if Λ is Arf.

Proof. Let λ be the enumeration of Λ.

(1) Suppose that the elements in N_i are ordered $N_{i,0} < N_{i,1} < N_{i,2} < \cdots < N_{i,\nu_i - 1}$. On one hand $\tau_i = N_{i, \lfloor \frac{\nu_i - 1}{2} \rfloor}$. On the other hand $N_{i,j} \geqslant j$ and this finishes the proof of the first statement.

(2) The result is obvious for the trivial semigroup. Thus we can assume that $c \geqslant g + 1$. Notice that $\tau_i = N_{i, \lfloor \frac{\nu_i - 1}{2} \rfloor} = \lfloor \frac{\nu_i - 1}{2} \rfloor$ if and only if all

integers less than or equal to $\lfloor \frac{\nu_i-1}{2} \rfloor$ belong to N_i. Now let us prove that if $i \geqslant 2c-g-1$ then all integers less than or equal to $\lfloor \frac{\nu_i-1}{2} \rfloor$ belong to N_i. Indeed, if $j \leqslant \lfloor \frac{\nu_i-1}{2} \rfloor$ then $\lambda_j \leqslant \lambda_i/2$ and $\lambda_i - \lambda_j \geqslant \lambda_i - \lambda_i/2 = \lambda_i/2 \geqslant c - 1/2$. Since $\lambda_i - \lambda_j \in \mathbb{N}_0$ this means that $\lambda_i - \lambda_j \geqslant c$ and so $\lambda_i - \lambda_j \in \Lambda$.

(3) Suppose that Λ is Arf. We want to show that for any non-negative integer i, all non-negative integers less than or equal to $\lfloor \frac{\nu_i-1}{2} \rfloor$ belong to N_i. By definition of τ_i there exists k with $\tau_i \leqslant k \leqslant i$ and $\lambda_{\tau_i} + \lambda_k = \lambda_i$. Now, if j is a non-negative integer with $j \leqslant \lfloor \frac{\nu_i-1}{2} \rfloor$, by statement 1 it also satisfies $j \leqslant \tau_i$. Then $\lambda_i - \lambda_j = \lambda_{\tau_i} + \lambda_k - \lambda_j \in \Lambda$ by the Arf property, and so $j \in N_i$.

On the other hand, suppose that $\tau_i = \lfloor \frac{\nu_i-1}{2} \rfloor$ for all non-negative integer i. This means that all integers less than or equal to τ_r belong to N_r for any non-negative integer r. If $i \geqslant j \geqslant k$ then $\tau_{\lambda^{-1}(\lambda_i+\lambda_j)} \geqslant j \geqslant k$ and by hypothesis $k \in N_{\lambda^{-1}(\lambda_i+\lambda_j)}$, which means that $\lambda_i + \lambda_j - \lambda_k \in \Lambda$. This implies that Λ is Arf. □

Statement 1) of Lemma 4.24 for the case when $i > 0$ is a direct consequence of Theorem 4.9 and Lemma 4.20.

Finally, Theorem 4.9 together with Lemma 4.24 has the next corollary. Different versions of this result appeared in [4, 10, 14]. The importance of the result is that it shows that the improved codes correcting generic errors do always require at most as many checks as the Feng–Rao improved codes correcting any kind of errors. It also states conditions under which their redundancies are equal and characterizes Arf semigroups as the unique semigroups for which there is no improvement.

Corollary 4.6.

(1) $\widetilde{R}^*(t) \subseteq \widetilde{R}(t)$ for all $t \in \mathbb{N}_0$.
(2) $\widetilde{R}^*(t) = \widetilde{R}(t)$ for all $t \geqslant c - g$.
(3) $\widetilde{R}^*(t) = \widetilde{R}(t)$ for all $t \in \mathbb{N}_0$ if and only if the associated numerical semigroup is Arf.

Proof. Statement 1. and 3. follow immediately from Theorem 4.9 and the fact that $\widetilde{R}(t) = \{i \in \mathbb{N}_0 : \lfloor \frac{\nu_i-1}{2} \rfloor < t\}$ and $\widetilde{R}^*(t) = \{i \in \mathbb{N}_0 : \tau_i < t\}$. For statement 2., we can use that for $i \geqslant 2c - g - 1$, $\tau_i = \lfloor \frac{\nu_i-1}{2} \rfloor$ (Theorem 4.9) and that for $i \geqslant 2c - g - 1$, $\tau_i \geqslant c - g - 1$ (Lemma 4.24), being $c - g - 1$ the largest value of τ_j before it starts being non-decreasing (Lemma 4.36). □

Further reading

We tried to cite the specific bibliography related to each section within the text. Next we mention some more general references: The book [59] has many results on numerical semigroups, including some of the problems presented in the first section of this chapter but also many others. The book [57] is also devoted to numerical semigroups from the perspective of the Frobenius' coin exchange problem. Algebraic geometry codes have been widely explained in different books such as [56, 67, 72]. For one-point codes and also their relation with Weierstrass semigroups an important reference is the chapter [35].

Acknowledgments

The author would like to thank Michael E. O'Sullivan, Ruud Pellikaan and Pedro A. García-Sánchez for many helpful discussions. She would also like to thank all the coauthors of the papers involved in this chapter. They are, by order of appearance of the differet papers: Michael E. O'Sullivan, Pedro A. García-Sánchez, Anna de Mier, and Stanislav Bulygin.

This work was partly supported by the Spanish Government through projects TIN2009-11689 "RIPUP" and CONSOLIDER INGENIO 2010 CSD2007-00004 "ARES", and by the Government of Catalonia under grant 2009 SGR 1135.

References

[1] Roger Apéry. Sur les branches superlinéaires des courbes algébriques. *C. R. Acad. Sci. Paris*, 222:1198–1200, 1946.

[2] Valentina Barucci, David E. Dobbs, and Marco Fontana. Maximality properties in numerical semigroups and applications to one-dimensional analytically irreducible local domains. *Mem. Amer. Math. Soc.*, 125(598):x+78, 1997.

[3] Victor Blanco, Pedro A. García Sánchez, and Justo Puerto. Counting numerical semigroups with short generating functions. *International Journal of Algebra and Computation*, 21:1217–1235, 2011.

[4] Maria Bras-Amorós. Improvements to evaluation codes and new characterizations of Arf semigroups. In *Applied algebra, algebraic algorithms and error-correcting codes (Toulouse, 2003)*, volume 2643 of *Lecture Notes in Comput. Sci.*, pages 204–215. Springer, Berlin, 2003.

[5] Maria Bras-Amorós. Acute semigroups, the order bound on the minimum distance, and the Feng-Rao improvements. *IEEE Trans. Inform. Theory*, 50(6):1282–1289, 2004.

[6] Maria Bras-Amorós. Addition behavior of a numerical semigroup. In *Arithmetic, geometry and coding theory (AGCT 2003)*, volume 11 of *Sémin. Congr.*, pages 21–28. Soc. Math. France, Paris, 2005.
[7] Maria Bras-Amorós. A note on numerical semigroups. *IEEE Trans. Inform. Theory*, 53(2):821–823, 2007.
[8] Maria Bras-Amorós. Fibonacci-like behavior of the number of numerical semigroups of a given genus. *Semigroup Forum*, 76(2):379–384, 2008.
[9] Maria Bras-Amorós. Bounds on the number of numerical semigroups of a given genus. *J. Pure Appl. Algebra*, 213(6):997–1001, 2009.
[10] Maria Bras-Amorós. On numerical semigroups and the redundancy of improved codes correcting generic errors. *Des. Codes Cryptogr.*, 53(2):111–118, 2009.
[11] Maria Bras-Amorós and Stanislav Bulygin. Towards a better understanding of the semigroup tree. *Semigroup Forum*, 79(3):561–574, 2009.
[12] Maria Bras-Amorós and Anna de Mier. Representation of numerical semigroups by Dyck paths. *Semigroup Forum*, 75(3):677–682, 2007.
[13] Maria Bras-Amorós and Pedro A. García-Sánchez. Patterns on numerical semigroups. *Linear Algebra Appl.*, 414(2-3):652–669, 2006.
[14] Maria Bras-Amorós and Michael E. O'Sullivan. The correction capability of the Berlekamp-Massey-Sakata algorithm with majority voting. *Appl. Algebra Engrg. Comm. Comput.*, 17(5):315–335, 2006.
[15] Maria Bras-Amorós and Michael E. O'Sullivan. Duality for some families of correction capability optimized evaluation codes. *Adv. Math. Commun.*, 2(1):15–33, 2008.
[16] Antonio Campillo and José Ignacio Farrán. Computing Weierstrass semigroups and the Feng-Rao distance from singular plane models. *Finite Fields Appl.*, 6(1):71–92, 2000.
[17] Antonio Campillo, José Ignacio Farrán, and Carlos Munuera. On the parameters of algebraic-geometry codes related to Arf semigroups. *IEEE Trans. Inform. Theory*, 46(7):2634–2638, 2000.
[18] Frank Curtis. On formulas for the Frobenius number of a numerical semigroup. *Math. Scand.*, 67(2):190–192, 1990.
[19] David E. Dobbs and Gretchen L. Matthews. On a question of Wilf concerning numerical semigroups. *International Journal of Commutative Rings*, 3(2), 2003.
[20] Sergi Elizalde. Improved bounds on the number of numerical semigroups of a given genus. *Journal of Pure and Applied Algebra*, 214:1404–1409, 2010.
[21] Hershel M. Farkas and Irwin Kra. *Riemann surfaces*, volume 71 of *Graduate Texts in Mathematics*. Springer-Verlag, New York, second edition, 1992.
[22] José Ignacio Farrán and Carlos Munuera. Goppa-like bounds for the generalized Feng-Rao distances. *Discrete Appl. Math.*, 128(1):145–156, 2003. International Workshop on Coding and Cryptography (WCC 2001) (Paris).
[23] Gui-Liang Feng and T. R. N. Rao. A simple approach for construction of algebraic-geometric codes from affine plane curves. *IEEE Trans. Inform. Theory*, 40(4):1003–1012, 1994.
[24] Gui-Liang Feng and T. R. N. Rao. Improved geometric Goppa codes. I. Basic

theory. *IEEE Trans. Inform. Theory*, 41(6, part 1):1678–1693, 1995. Special issue on algebraic geometry codes.
[25] Ralf Fröberg, Christian Gottlieb, and Roland Häggkvist. On numerical semigroups. *Semigroup Forum*, 35(1):63–83, 1987.
[26] William Fulton. *Algebraic curves*. Advanced Book Classics. Addison-Wesley Publishing Company Advanced Book Program, Redwood City, CA, 1989. An introduction to algebraic geometry, Notes written with the collaboration of Richard Weiss, Reprint of 1969 original.
[27] Arnaldo Garcia and Henning Stichtenoth. On the asymptotic behaviour of some towers of function fields over finite fields. *J. Number Theory*, 61(2):248–273, 1996.
[28] Pedro A. García-Sánchez and José C. Rosales. Numerical semigroups generated by intervals. *Pacific J. Math.*, 191(1):75–83, 1999.
[29] Olav Geil. On codes from norm-trace curves. *Finite Fields Appl.*, 9(3):351–371, 2003.
[30] Olav Geil and Ryutaroh Matsumoto. Bounding the number of $\mathbb{F}_q$-rational places in algebraic function fields using Weierstrass semigroups. *J. Pure Appl. Algebra*, 213(6):1152–1156, 2009.
[31] Massimo Giulietti. Notes on algebraic-geometric codes. www.math.kth.se/math/forskningsrapporter/Giulietti.pdf.
[32] David M. Goldschmidt. *Algebraic functions and projective curves*, volume 215 of *Graduate Texts in Mathematics*. Springer-Verlag, New York, 2003.
[33] Johan P. Hansen. Dependent rational points on curves over finite fields—Lefschetz theorems and exponential sums. In *International Workshop on Coding and Cryptography (Paris, 2001)*, volume 6 of *Electron. Notes Discrete Math.*, page 13 pp. (electronic). Elsevier, Amsterdam, 2001.
[34] Petra Heijnen and Ruud Pellikaan. Generalized Hamming weights of q-ary Reed-Muller codes. *IEEE Trans. Inform. Theory*, 44(1):181–196, 1998.
[35] Tom Høholdt, Jacobus H. van Lint, and Ruud Pellikaan. *Algebraic Geometry codes*, pages 871–961. North-Holland, Amsterdam, 1998.
[36] Helge Elbrønd Jensen, Rasmus Refslund Nielsen, and Tom Høholdt. Performance analysis of a decoding algorithm for algebraic-geometry codes. *IEEE Trans. Inform. Theory*, 45(5):1712–1717, 1999.
[37] Nathan Kaplan. Counting numerical semigroups by genus and some cases of a question of Wilf. Preprint.
[38] Seon Jeong Kim. Semigroups which are not Weierstrass semigroups. *Bull. Korean Math. Soc.*, 33(2):187–191, 1996.
[39] Christoph Kirfel and Ruud Pellikaan. The minimum distance of codes in an array coming from telescopic semigroups. *IEEE Trans. Inform. Theory*, 41(6, part 1):1720–1732, 1995. Special issue on algebraic geometry codes.
[40] Jiryo Komeda. On non-Weierstrass gap sequqences. Technical report, Kanagawa Institute of Technology B-13, 1989. Research Reports of Kanagawa Institute of Technology B-13.
[41] Jiryo Komeda. Non-Weierstrass numerical semigroups. *Semigroup Forum*, 57(2):157–185, 1998.
[42] Joseph Lewittes. Places of degree one in function fields over finite fields. *J.*

Pure Appl. Algebra, 69(2):177–183, 1990.
[43] Monica Madero-Craven and Kurt Herzinger. Apery sets of numerical semigroups. Comm. Algebra, 33(10):3831–3838, 2005.
[44] Nivaldo Medeiros. Listing of numerical semigroups. http://w3.impa.br/~nivaldo/algebra/semigroups/.
[45] Carlos Munuera and Fernando Torres. A note on the order bound on the minimum distance of AG codes and acute semigroups. Adv. Math. Commun., 2(2):175–181, 2008.
[46] Carlos Munuera, Fernando Torres, and Juan E. Villanueva. Sparse numerical semigroups. In *Applied algebra, algebraic algorithms and error-correcting codes (Tarragona, 2009)*, volume 5527 of *Lecture Notes in Comput. Sci.*, pages 23–31, Berlin, 2009. Springer.
[47] Anna Oneto and Grazia Tamone. On numerical semigroups and the order bound. J. Pure Appl. Algebra, 212(10):2271–2283, 2008.
[48] Anna Oneto and Grazia Tamone. On the order bound of one-point algebraic geometry codes. J. Pure Appl. Algebra, 213(6):1179–1191, 2009.
[49] Anna Oneto and Grazia Tamone. On some invariants in numerical semigroups and estimations of the order bound. Semigroup Forum, 81(3):483–509, 2010.
[50] Michael E. O'Sullivan. Decoding Hermitian codes beyond $(d_{min} - 1)/2$. In *IEEE Int. Symp. Information Theory*, 1997.
[51] Michael E. O'Sullivan. A generalization of the Berlekamp-Massey Sakata algorithm. Preprint, 2001.
[52] Michael E. O'Sullivan. New codes for the Berlekamp-Massey-Sakata algorithm. Finite Fields Appl., 7(2):293–317, 2001.
[53] Ruud Pellikaan. On decoding by error location and dependent sets of error positions. Discrete Math., 106/107:369–381, 1992. A collection of contributions in honour of Jack van Lint.
[54] Ruud Pellikaan, Henning Stichtenoth, and Fernando Torres. Weierstrass semigroups in an asymptotically good tower of function fields. Finite Fields Appl., 4(4):381–392, 1998.
[55] Ruud Pellikaan and Fernando Torres. On Weierstrass semigroups and the redundancy of improved geometric Goppa codes. IEEE Trans. Inform. Theory, 45(7):2512–2519, 1999.
[56] Oliver Pretzel. *Codes and algebraic curves*, volume 8 of *Oxford Lecture Series in Mathematics and its Applications*. The Clarendon Press Oxford University Press, New York, 1998.
[57] Jorge L. Ramírez Alfonsín. *The Diophantine Frobenius problem*, volume 30 of *Oxford Lecture Series in Mathematics and its Applications*. Oxford University Press, Oxford, 2005.
[58] José Carlos Rosales and Manuel B. Branco. Irreducible numerical semigroups. Pacific J. Math., 209(1):131–143, 2003.
[59] José Carlos Rosales and Pedro A. García-Sánchez. *Numerical semigroups*, volume 20 of *Developments in Mathematics*. Springer, New York, 2009.
[60] José Carlos Rosales, Pedro A. García-Sánchez, Juan Ignacio García-García, and Manuel B. Branco. Systems of inequalities and numerical semigroups.

[61] J. London Math. Soc. (2), 65(3):611–623, 2002.
[61] José Carlos Rosales, Pedro A. García-Sánchez, Juan Ignacio García-García, and Manuel B. Branco. Arf numerical semigroups. *J. Algebra*, 276(1):3–12, 2004.
[62] José Carlos Rosales, Pedro A. García-Sánchez, Juan Ignacio García-García, and Manuel B. Branco. Numerical semigroups with a monotonic Apéry set. *Czechoslovak Math. J.*, 55(130)(3):755–772, 2005.
[63] José Carlos Rosales, Pedro A. García-Sánchez, Juan Ignacio García-García, and José Antonio Jiménez Madrid. The oversemigroups of a numerical semigroup. *Semigroup Forum*, 67(1):145–158, 2003.
[64] José Carlos Rosales, Pedro A. García-Sánchez, Juan Ignacio García-García, and José Antonio Jiménez Madrid. Fundamental gaps in numerical semigroups. *J. Pure Appl. Algebra*, 189(1-3):301–313, 2004.
[65] Neil J. A. Sloane. The On-Line Encyclopedia of Integer Sequences, A007323. http://www.research.att.com/ njas/sequences/A007323.
[66] Henning Stichtenoth. A note on Hermitian codes over GF(q^2). *IEEE Trans. Inform. Theory*, 34(5, part 2):1345–1348, 1988. Coding techniques and coding theory.
[67] Henning Stichtenoth. *Algebraic function fields and codes*. Universitext. Springer-Verlag, Berlin, 1993.
[68] Karl-Otto Stöhr and José Felipe Voloch. Weierstrass points and curves over finite fields. *Proc. London Math. Soc. (3)*, 52(1):1–19, 1986.
[69] James J. Sylvester. Mathematical questions with their solutions. *Educational Times*, 41:21, 1884.
[70] Fernando Torres. Notes on Goppa codes. www.ime.unicamp.br/~ftorres/RESEARCH/ARTS_PDF/codes.pdf.
[71] Fernando Torres. On certain N-sheeted coverings of curves and numerical semigroups which cannot be realized as Weierstrass semigroups. *Comm. Algebra*, 23(11):4211–4228, 1995.
[72] Jacobus H. van Lint and Gerard van der Geer. *Introduction to coding theory and algebraic geometry*, volume 12 of *DMV Seminar*. Birkhäuser Verlag, Basel, 1988.
[73] Herbert S. Wilf. A circle-of-lights algorithm for the "money-changing problem". *Amer. Math. Monthly*, 85(7):562–565, 1978.
[74] Yufei Zhao. Constructing numerical semigroups of a given genus. *Semigroup Forum*, 80(2):242–254, 2010.

Chapter 5

Codes, Arrangements and Matroids

Relinde Jurrius[*] and Ruud Pellikaan[†]

Eindhoven University of Technology
Department of Mathematics and Computer Science, Coding and Crypto
P.O. Box 513, NL-5600 MB Eindhoven, The Netherlands
[*]r.p.m.j.jurrius@tue.nl   [†]g.r.pellikaan@tue.nl

This chapter treats error-correcting codes and their weight enumerator as the center of several closely related topics such as arrangements of hyperplanes, graph theory, matroids, posets and geometric lattices and their characteristic, chromatic, Tutte, Möbius and coboundary polynomial, respectively. Their interrelations and many examples and counterexamples are given. It is concluded with a section with references to the literature for further reading and open questions.

AMS classification: 05B35, 05C31, 06A07, 14N20, 94B27, 94B70, 94C15

Contents

5.1	Introduction	220
5.2	Error-correcting codes	221
	5.2.1 Codes and Hamming distance	221
	5.2.2 Linear codes	227
	5.2.3 Generator matrix	229
	5.2.4 Parity check matrix	230
	5.2.5 Inner product and dual codes	232
	5.2.6 The Hamming and simplex codes	232
	5.2.7 Singleton bound and MDS codes	234
5.3	Weight enumerators and error probability	236
	5.3.1 Weight spectrum	236
	5.3.2 The decoding problem	239
	5.3.3 The q-ary symmetric channel	241
	5.3.4 Error probability	243
5.4	Codes, projective systems and arrangements	245
5.5	The extended and generalized weight enumerator	246
	5.5.1 Generalized weight enumerators	248

    5.5.2 Extended weight enumerator 252
    5.5.3 Puncturing and shortening of codes 256
    5.5.4 Connections 260
    5.5.5 MDS-codes 263
5.6 Matroids and codes 267
    5.6.1 Matroids 267
    5.6.2 Graphs, codes and matroids 269
    5.6.3 The weight enumerator and the Tutte polynomial 276
    5.6.4 Deletion and contraction of matroids 279
    5.6.5 MacWilliams type property for duality 279
5.7 Posets and lattices 282
    5.7.1 Posets, the Möbius function and lattices 282
    5.7.2 Geometric lattices 289
    5.7.3 Geometric lattices and matroids 293
5.8 The characteristic polynomial 295
    5.8.1 The characteristic and coboundary polynomial 295
    5.8.2 The Möbius polynomial and Whitney numbers 298
    5.8.3 Minimal codewords and subcodes 300
    5.8.4 The characteristic polynomial of an arrangement 301
    5.8.5 The characteristic polynomial of a code 304
    5.8.6 Examples and counterexamples 307
5.9 Overview of polynomial relations 312
5.10 Further reading and open problems 314
    5.10.1 Multivariate and other polynomials 314
    5.10.2 The coset leader weight enumerator 315
    5.10.3 Graph codes 316
    5.10.4 The reconstruction problem 316
    5.10.5 Questions concerning the Möbius polynomial 316
    5.10.6 Monomial conjectures 317
    5.10.7 Complexity issues 318
    5.10.8 The zeta function 318
References 319

5.1. Introduction

A lot of mathematical objects are closely related to each other. While studying certain aspects of a mathematical object, one tries to find a way to "view" the object in a way that is most suitable for a specific problem. Or in other words, one tries to find the best way to model the problem. Many related fields of mathematics have evolved from one another this way. In practice, it is very useful to be able to transform your problem into other terminology: it gives a lot more available knowledge that can be helpful to solve a problem.

In this chapter we give a broad overview of closely related fields, starting from the weight enumerator of an error-correcting code. We explain the importance of this polynomial in coding theory. From various methods of determining the weight enumerator, we naturally run into other ways to view an error-correcting code. We will introduce and link the following mathematical objects:

- linear codes and their weight enumerator (§5.2, §5.3, §5.5);
- arrangements and their characteristic polynomial (§5.4, §5.8);
- graphs and their chromatic polynomial (§5.6.2);
- matroids and their Tutte polynomial (§5.6);
- posets and their Möbius function (§5.7);
- geometric lattices and their coboundary polynomial (§5.7, §5.8).

A nice example to show the power of these connections is the MacWilliams identities, that relate the polynomials associated to an object and its dual. This will be treated in Section 5.6.5. Several examples and counterexamples are given in Section 5.8.6 and an overview is given to show which polynomials determine each other in Section 5.9.

These notes are based on the Master's thesis [1], ongoing research [2, 3] and the lecture notes [4] of the Soria Summer School in 2009. The whole chapter is self-contained, and various references to further reading, background knowledge and open problems are given in Section 5.10.

5.2. Error-correcting codes

The basics of the theory of error-correcting codes can be found in [5–8].

5.2.1. *Codes and Hamming distance*

The idea of *redundant* information is a well-known phenomenon in reading a newspaper. Misspellings usually go unnoticed for a casual reader, while the meaning is still grasped. In Semitic languages such as Hebrew, and even older in the hieroglyphics in the tombs of the pharaohs of Egypt, only the consonants are written while the vowels are left out, so that we do not know for sure how to pronounce these words nowadays. The letter "e" is the most frequent occurring symbol in the English language, and leaving out all these letters would still give in almost all cases an understandable text to the expense of greater attention of the reader. The art and science of deleting redundant information in a clever way such that it can be stored in less memory or space and still can be expanded to the original message, is called *data compression* or *source coding*. It is not the topic of this chapter. So we can compress data but an error made in a compressed text would give a different message that is most of the time completely meaningless. The idea in *error-correcting codes* is the converse. One adds redundant information in such a way that it is possible to detect or even correct errors after transmission.

Legend goes that Hamming was so frustrated the computer halted every time it detected an error after he handed in a stack of punch cards, he thought about a way the computer would be able not only to detect the error but also to correct it automatically. He came with the nowadays famous code named after him. Whereas the theory of Hamming [9] is about the actual construction, the encoding and decoding of codes and uses tools from *combinatorics* and *algebra*, the approach of Shannon [10] leads to *information theory* and his theorems tell us what is and what is not possible in a *probabilistic* sense.

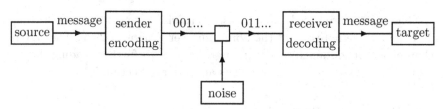

Fig. 5.1. Block diagram of a communication system

According to Shannon we have a message **m** in a certain alphabet and of a certain length. We encode **m** to **c** by expanding the length of the message and adding redundant information. One can define the *information rate* R that measures the slowing down of the transmission of the data. The encoded message **c** is sent over a noisy channel such that the symbols are changed, according to certain probabilities that are characteristic of the channel. The received word **r** is decoded to **m**′. Now given the characteristics of the channel one can define the *capacity* C of the channel and it has the property that for every $R < C$ it is possible to find an encoding and decoding scheme such that the *error probability* that $\mathbf{m}' \neq \mathbf{m}$ is arbitrarily small. For $R > C$ such a scheme is not possible. The capacity is explicitly known as a function of the characteristic probability for quite a number of channels.

The notion of a channel must be taken in a broad sense. Not only the transmission of data via satellite or telephone but also the storage of information on a hard disk of a computer or a compact disc for music and film can be modeled by a channel.

The theorem of Shannon tells us the existence of certain encoding and decoding schemes, and even tells that they exist in abundance and that almost all schemes satisfy the required conditions, but it does not tell us how to construct a specific efficient scheme.

Example 5.1. Replacing every symbol by a threefold repetition gives the possibility of correcting one error in every 3-tuple of symbols in a received word by a majority vote. We call this a *repetition code*. The price one has to pay is that the transmission is three times slower. We see here the two conflicting demands of error-correction: to correct as many errors as possible and to transmit as fast a possible. Notice furthermore that in case two errors are introduced by transmission the majority decoding rule will introduce a *decoding error*.

Example 5.2. An improvement of the repetition code of rate 1/3 is given by Hamming. Suppose we have a message (m_1, m_2, m_3, m_4) of 4 bits. Put them in the middle of the *Venn-diagram* of three intersecting circles as given in Figure 5.2. Complete the three empty areas of the circles according to the rule that the number of ones in every circle is even. In this way we get 3 redundant bits (r_1, r_2, r_3) that we add to the message and which we transmit over the channel.

Fig. 5.2. Venn diagram of the Hamming code

In every block of 7 bits the receiver can correct one error, since the parity in every circle should be even. So if the parity is even we declare the circle correct, if the parity is odd we declare the circle incorrect. The error is in the incorrect circles and in the complement of the correct circles. We see that every pattern of at most one error can be corrected in this way. For instance, if $\mathbf{m} = (1, 1, 0, 1)$ is the message, then $\mathbf{r} = (0, 0, 1)$ is the redundant information added and $\mathbf{c} = (1,1,0,1,0,0,1)$ the codeword sent. Suppose that after transmission one symbol is flipped and $\mathbf{y} = (1,0,0,1,0,0,1)$ is the received word as given in Figure 5.3.

Fig. 5.3. Venn diagram of a received word for the Hamming code

Then we conclude that the error is in the left and upper circle, but not in the right one. And we conclude that the error is at m_2. But in case of 2 errors, if for instance the word $\mathbf{y}' = (1,0,0,1,1,0,1)$ is received, then the receiver would assume that the error occurred in the upper circle and not in the two lower circles, and would therefore conclude that the transmitted codeword was $(1,0,0,1,1,0,0)$. Hence the decoding scheme creates an extra error.

The redundant information $\mathbf{r}$ can be obtained from the message $\mathbf{m}$ by means of three linear equations or parity checks modulo 2:
$$\begin{cases} r_1 = m_2 + m_3 + m_4 \\ r_2 = m_1 + m_3 + m_4 \\ r_3 = m_1 + m_2 + m_4 \end{cases}$$
Let $\mathbf{c} = (\mathbf{m}, \mathbf{r})$ be the codeword. Then $\mathbf{c}$ is a codeword if and only if $H\mathbf{c}^T = 0$, where
$$H = \begin{pmatrix} 0 & 1 & 1 & 1 & 1 & 0 & 0 \\ 1 & 0 & 1 & 1 & 0 & 1 & 0 \\ 1 & 1 & 0 & 1 & 0 & 0 & 1 \end{pmatrix}.$$

The information rate is improved from 1/3 for the repetition code to 4/7 for the Hamming code.

In general the alphabets of the message word and the encoded word might be distinct. Furthermore the length of both the message word and the encoded word might vary such as in a *convolutional code*. We restrict ourselves to $[n, k]$ *block codes*: that is, the message words have a fixed length of k symbols and the encoded words have a fixed length of n symbols both from the same *alphabet* Q. For the purpose of error control, before transmission, we add redundant symbols to the message in a clever way.

Let Q be a set of q symbols called the *alphabet*. Let Q^n be the set of all n-tuples $\mathbf{x} = (x_1, \ldots, x_n)$, with entries $x_i \in Q$. A *block code* C of *length* n over Q is a non-empty subset of Q^n. The elements of C are called *codewords*. If C contains M codewords, then M is called the *size* of the code. We call a code with length n and size M an (n, M) code. If $M = q^k$, then C is called an $[n, k]$ code. For an (n, M) code defined over Q, the value $n - \log_q(M)$ is called the *redundancy*. The *information rate* is defined as $R = \log_q(M)/n$.

Example 5.3. The repetition code has length 3 and 2 codewords, so its information rate is $1/3$. The Hamming code has length 7 and 2^4 codewords, therefore its rate is $4/7$. These are the same values as we found in Examples 5.1 and 5.2.

Example 5.4. Let C be the binary block code of length n consisting of all words with exactly two ones. This is an $(n, n(n-1)/2)$ code. In this example the number of codewords is not a power of the size of the alphabet.

Let C be an $[n, k]$ block code over Q. An *encoder* of C is a one-to-one map

$$\mathcal{E} : Q^k \longrightarrow Q^n$$

such that $C = \mathcal{E}(Q^k)$. Let $\mathbf{c} \in C$ be a codeword. Then there exists a unique $\mathbf{m} \in Q^k$ with $\mathbf{c} = \mathcal{E}(\mathbf{m})$. This $\mathbf{m}$ is called the *message* or *source word* of $\mathbf{c}$.

In order to measure the difference between two distinct words and to evaluate the error-correcting capability of the code, we need to introduce an appropriate metric to Q^n. A natural metric used in coding theory is the *Hamming distance*. For $\mathbf{x} = (x_1, \ldots, x_n)$, $\mathbf{y} = (y_1, \ldots, y_n) \in Q^n$, the Hamming distance $d(\mathbf{x}, \mathbf{y})$ is defined as the number of places where they differ, that is

$$d(\mathbf{x}, \mathbf{y}) = |\{i : x_i \neq y_i\}|.$$

Proposition 5.1. *The Hamming distance is a metric on Q^n, that means that it has the following properties for all $\mathbf{x}, \mathbf{y}, \mathbf{z} \in Q^n$:*

(1) $d(\mathbf{x}, \mathbf{y}) \geq 0$ and equality hods if and only if $\mathbf{x} = \mathbf{y}$,
(2) $d(\mathbf{x}, \mathbf{y}) = d(\mathbf{y}, \mathbf{x})$ (symmetry),
(3) $d(\mathbf{x}, \mathbf{z}) \leq d(\mathbf{x}, \mathbf{y}) + d(\mathbf{y}, \mathbf{z})$ (triangle inequality).

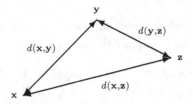

Fig. 5.4. Triangle inequality

Proof. Properties (1) and (2) are trivial from the definition. We leave (3) to the reader as an exercise. □

The *minimum (Hamming) distance* of a code C of length n is defined as

$$d = d(C) = \min\{d(\mathbf{x},\mathbf{y}) : \mathbf{x},\mathbf{y} \in C, \mathbf{x} \neq \mathbf{y}\}$$

if C consists of more than one element, and is by definition $n+1$ if C consists of one word. We denote by (n, M, d) a code C with length n, size M and minimum distance d.

The main problem of error-correcting codes from "Hamming's point of view" is to construct for a given length and number of codewords a code with the largest possible minimum distance, and to find efficient encoding and decoding algorithms for such a code.

Example 5.5. The triple repetition code consists of two codewords: $(0,0,0)$ and $(1,1,1)$, so its minimum distance is 3. The Hamming code corrects one error. So the minimum distance is at least 3, by the triangle inequality. The Hamming code has minimum distance 3. Notice that both codes have the property that $\mathbf{x}+\mathbf{y}$ is again a codeword if $\mathbf{x}$ and $\mathbf{y}$ are codewords.

Let $\mathbf{x} \in Q^n$. The *ball of radius r around* $\mathbf{x}$, denoted by $B_r(\mathbf{x})$, is defined by $B_r(\mathbf{x}) = \{\mathbf{y} \in Q^n : d(\mathbf{x},\mathbf{y}) \leq r\}$. The *sphere of radius r around* $\mathbf{x}$ is denoted by $S_r(\mathbf{x})$ and defined by $S_r(\mathbf{x}) = \{\mathbf{y} \in Q^n : d(\mathbf{x},\mathbf{y}) = r\}$.

Figure 5.5 shows the ball in the Euclidean plane. This is misleading in some respects, but gives an indication what we should have in mind.

Figure 5.6 shows Q^2, where the alphabet Q consists of 5 elements. The ball $B_0(\mathbf{x})$ consists of the point in the circle, $B_1(\mathbf{x})$ is depicted by the points inside the cross, and $B_2(\mathbf{x})$ consists of all 25 dots.

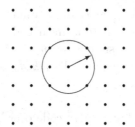

Fig. 5.5. Ball of radius $\sqrt{2}$ in the Euclidean plane

Proposition 5.2. *Let Q be an alphabet of q elements and $\mathbf{x} \in Q^n$. Then*

$$|S_i(\mathbf{x})| = \binom{n}{i}(q-1)^i \quad \text{and} \quad |B_r(\mathbf{x})| = \sum_{i=0}^{r} \binom{n}{i}(q-1)^i.$$

Proof. Let $\mathbf{y} \in S_i(\mathbf{x})$. Let I be the subset of $\{1,\ldots,n\}$ consisting of all positions j such that $y_j \neq x_j$. Then the number of elements of I is equal to i, and $(q-1)^i$ is the number of words $\mathbf{y} \in S_i(\mathbf{x})$ that have the same fixed I. The number of possibilities to choose the subset I with a fixed number of elements i is equal to $\binom{n}{i}$. This shows the formula for the number of elements of $S_i(\mathbf{x})$.

Furthermore $B_r(\mathbf{x})$ is the disjoint union of the subsets $S_i(\mathbf{x})$ for $i = 0,\ldots,r$. This proves the statement about the number of elements of $B_r(\mathbf{x})$. □

5.2.2. Linear codes

If the alphabet Q is a finite field, which is the case for instance when $Q = \{0,1\} = \mathbb{F}_2$, then Q^n is a vector space. Therefore it is natural to look at codes in Q^n that have more structure, in particular that are linear subspaces.

A *linear code* C is a linear subspace of $\mathbb{F}_q^n$, where $\mathbb{F}_q$ stands for the finite

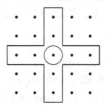

Fig. 5.6. Balls of radius 0 and 1 in the Hamming metric

field with q elements. The *dimension* of a linear code is its dimension as a linear space over $\mathbb{F}_q$. We denote a linear code C over $\mathbb{F}_q$ of length n and dimension k by $[n, k]_q$, or simply by $[n, k]$. If furthermore the minimum distance of the code is d, then we call $[n, k, d]_q$ or $[n, k, d]$ the *parameters* of the code.

It is clear that for a linear $[n, k]$ code over $\mathbb{F}_q$, its size is $M = q^k$. The information rate is $R = k/n$ and the redundancy is $n - k$.

Let C and D be linear codes in $\mathbb{F}_q^n$. Then C is called *permutation equivalent* to D, if there exists a permutation matrix Π such that $\Pi(C) = D$. If moreover $C = D$, then Π is called a *permutation automorphism* of C. The code C is called *generalized equivalent* or *monomial equivalent* to D, if there exists a monomial matrix M such that $M(C) = D$. If moreover $C = D$, then M is called a *monomial automorphism* of C.

For a word $\mathbf{x} \in \mathbb{F}_q^n$, its *support*, $\mathrm{supp}(\mathbf{x})$, is defined as the set of nonzero coordinate positions, so $\mathrm{supp}(\mathbf{x}) = \{i : x_i \neq 0\}$. The *weight* of $\mathbf{x}$ is defined as the number of elements of its support, which is denoted by $\mathrm{wt}(\mathbf{x})$. The *minimum weight* of a code C is defined as the minimal value of the weights of the nonzero codewords in case there is a nonzero codeword, and $n + 1$ otherwise.

Proposition 5.3. *The minimum distance of a linear code C of dimension $k > 0$ is equal to its minimum weight.*

Proof. Since C is a linear code, we have that $0 \in C$ and for any $\mathbf{c}_1, \mathbf{c}_2 \in C$, $\mathbf{c}_1 - \mathbf{c}_2 \in C$. Then the conclusion follows from the fact that $\mathrm{wt}(\mathbf{c}) = d(0, \mathbf{c})$ and $d(\mathbf{c}_1, \mathbf{c}_2) = \mathrm{wt}(\mathbf{c}_1 - \mathbf{c}_2)$. □

Now let us see some examples of linear codes.

Example 5.6. The repetition code over $\mathbb{F}_q$ of length n consists of all words $\mathbf{c} = (c, c, \ldots, c)$ with $c \in \mathbb{F}_q$. This is a linear code of dimension 1 and minimum distance n.

Example 5.7. Let n be an integer with $n \geq 2$. The *even weight* code C of length n over $\mathbb{F}_q$ consists of all words in $\mathbb{F}_q^n$ of even weight. The minimum weight of C is 2 by definition, the minimum distance of C is 2 if $q = 2$ and 1 otherwise. The code C is linear if and only if $q = 2$.

Example 5.8. The *Hamming code* C of Example 5.2 consists of all the words $\mathbf{c} \in \mathbb{F}_2^7$ satisfying $H\mathbf{c}^T = \mathbf{0}$, where

$$H = \begin{pmatrix} 0 & 1 & 1 & 1 & 1 & 0 & 0 \\ 1 & 0 & 1 & 1 & 0 & 1 & 0 \\ 1 & 1 & 0 & 1 & 0 & 0 & 1 \end{pmatrix}.$$

This code is linear of dimension 4, since it is given by the solutions of three independent homogeneous linear equations in 7 variables. The minimum weight is 3 as shown in Example 5.5. So it is a $[7, 4, 3]$ code.

5.2.3. *Generator matrix*

Let C be a linear $[n, k]$ code over $\mathbb{F}_q$. Since C is a k-dimensional linear subspace of $\mathbb{F}_q^n$, there exists a *basis* that consists of k linearly independent codewords, say $\mathbf{g}_1, \ldots, \mathbf{g}_k$. Suppose $\mathbf{g}_i = (g_{i1}, \ldots, g_{in})$ for $i = 1, \ldots, k$. Denote

$$G = \begin{pmatrix} \mathbf{g}_1 \\ \mathbf{g}_2 \\ \vdots \\ \mathbf{g}_k \end{pmatrix} = \begin{pmatrix} g_{11} & g_{12} & \cdots & g_{1n} \\ g_{21} & g_{22} & \cdots & g_{2n} \\ \vdots & \vdots & \vdots & \vdots \\ g_{k1} & g_{k2} & \cdots & g_{kn} \end{pmatrix}.$$

Every codeword $\mathbf{c}$ can be written uniquely as a linear combination of the basis elements, so $\mathbf{c} = m_1\mathbf{g}_1 + \cdots + m_k\mathbf{g}_k$ where $m_1, \ldots, m_k \in \mathbb{F}_q$. Let $\mathbf{m} = (m_1, \ldots, m_k) \in \mathbb{F}_q^k$. Then $\mathbf{c} = \mathbf{m}G$. The *encoding*

$$\mathcal{E} : \mathbb{F}_q^k \longrightarrow \mathbb{F}_q^n,$$

from the message word $\mathbf{m} \in \mathbb{F}_q^k$ to the codeword $\mathbf{c} \in \mathbb{F}_q^n$ can be done efficiently by a matrix multiplication

$$\mathbf{c} = \mathcal{E}(\mathbf{m}) := \mathbf{m}G.$$

A $k \times n$ matrix G with entries in $\mathbb{F}_q$ is called a *generator matrix* of a $\mathbb{F}_q$-linear code C if the rows of G are a basis of C.

A given $[n, k]$ code C can have more than one generator matrix, however every generator matrix of C is a $k \times n$ matrix of rank k. Conversely every $k \times n$ matrix of rank k is the generator matrix of a $\mathbb{F}_q$-linear $[n, k]$ code.

Example 5.9. The linear codes with parameters $[n, 0, n + 1]$ and $[n, n, 1]$ are the *trivial codes* $\{\mathbf{0}\}$ and $\mathbb{F}_q^n$, and they have the empty matrix and the $n \times n$ identity matrix I_n as generator matrix, respectively.

Example 5.10. The repetition code of length n has generator matrix

$$G = (\begin{matrix} 1 & 1 & \cdots & 1 \end{matrix}).$$

Example 5.11. The Hamming code C of Example 5.2 is a $[7,4]$ code. The message symbols m_i for $i = 1, \ldots, 4$ are free to choose. If we take $m_i = 1$ and the remaining $m_j = 0$ for $j \neq i$ we get the codeword $\mathbf{g}_i$. In this way we get the basis $\mathbf{g}_1, \mathbf{g}_2, \mathbf{g}_3, \mathbf{g}_4$. Therefore, C has the following generator matrix

$$G = \begin{pmatrix} 1 & 0 & 0 & 0 & 0 & 1 & 1 \\ 0 & 1 & 0 & 0 & 1 & 0 & 1 \\ 0 & 0 & 1 & 0 & 1 & 1 & 0 \\ 0 & 0 & 0 & 1 & 1 & 1 & 1 \end{pmatrix}.$$

Let C be an $[n,k]$ code. The code is called *systematic at the positions* $(j_1, \ldots, j_k)$ if for all $\mathbf{m} \in \mathbb{F}_q^k$ there exists a unique codeword $\mathbf{c}$ such that $c_{j_i} = m_i$ for all $i = 1, \ldots, k$. In that case, the set $(j_1, \ldots, j_k)$ is called an *information set*. A generator matrix G of C is called systematic at the positions $(j_1, \ldots, j_k)$ if the $k \times k$ submatrix G' consisting of the k columns of G at the positions $(j_1, \ldots, j_k)$ is the identity matrix. For such a matrix G the mapping $\mathbf{m} \mapsto \mathbf{m}G$ is called *systematic encoding*.

5.2.4. *Parity check matrix*

There are two standard ways to describe a subspace, *explicitly* by giving a basis, or *implicitly* by the solution space of a set of homogeneous linear equations. Therefore there are two ways to describe a linear code. That is explicitly as we have seen by a generator matrix, or implicitly by a set of homogeneous linear equations, that is, by the null space of a matrix.

Let C be a $\mathbb{F}_q$-linear $[n,k]$ code. Suppose that H is an $m \times n$ matrix with entries in $\mathbb{F}_q$. Let C be the null space of H. So C is the set of all $\mathbf{c} \in \mathbb{F}_q^n$ such that $H\mathbf{c}^T = 0$. These m homogeneous linear equations are called *parity check equations*, or simply *parity checks*. The dimension k of C is at least $n - m$. If there are dependent rows in the matrix H, that is if $k > n - m$, then we can delete a few rows until we obtain an $(n-k) \times n$ matrix H' with independent rows and with the same null space as H. So H' has rank $n - k$. An $(n-k) \times n$ matrix of rank $n - k$ is called a *parity check matrix* of an $[n,k]$ code C if C is the null space of this matrix.

Remark 5.1. The parity check matrix of a code can be used for *error detection*. This is useful in a communication channel where one asks for *retransmission* in case more than a certain number of errors occurred. Suppose that C is a linear code of minimum distance d and H is a parity check matrix of C. Suppose that the codeword $\mathbf{c}$ is transmitted and $\mathbf{r} = \mathbf{c} + \mathbf{e}$ is received. Then $\mathbf{e}$ is called the *error vector* and wt($\mathbf{e}$) the *number of errors*. Now $H\mathbf{r}^T = 0$ if there is no error and $H\mathbf{r}^T \neq 0$ for all $\mathbf{e}$ such that $0 < \text{wt}(\mathbf{e}) < d$. Therefore we can detect any pattern of t errors with $t < d$. But not more, since if the error vector is equal to a nonzero codeword of minimal weight d, then the receiver would assume that no errors have been made. The vector $H\mathbf{r}^T$ is called the *syndrome* of the received word.

We show that every linear code has a parity check matrix and we give a method to obtain such a matrix in case we have a generator matrix G of the code.

Proposition 5.4. *Suppose C is an $[n,k]$ code. Let I_k be the $k \times k$ identity matrix. Let P be a $k \times (n-k)$ matrix. Then, $(I_k|P)$ is a generator matrix of C if and only if $(-P^T|I_{n-k})$ is a parity check matrix of C.*

Proof. Every codeword $\mathbf{c}$ is of the form $\mathbf{m}G$ with $\mathbf{m} \in \mathbb{F}_q^k$. Suppose that the generator matrix G is systematic at the first k positions. So $\mathbf{c} = (\mathbf{m}, \mathbf{r})$ with $\mathbf{r} \in \mathbb{F}_q^{n-k}$ and $\mathbf{r} = \mathbf{m}P$. Hence for a word of the form $\mathbf{c} = (\mathbf{m}, \mathbf{r})$ with $\mathbf{m} \in \mathbb{F}_q^k$ and $\mathbf{r} \in \mathbb{F}_q^{n-k}$ the following statements are equivalent:

$\mathbf{c}$ is a codeword,
$\iff -\mathbf{m}P + \mathbf{r} = 0,$
$\iff -P^T\mathbf{m}^T + \mathbf{r}^T = 0,$
$\iff \left(-P^T|I_{n-k}\right)(\mathbf{m},\mathbf{r})^T = 0,$
$\iff \left(-P^T|I_{n-k}\right)\mathbf{c}^T = 0.$

Hence $\left(-P^T|I_{n-k}\right)$ is a parity check matrix of C. The converse is proved similarly. □

Example 5.12. The trivial codes $\{0\}$ and $\mathbb{F}_q^n$ have I_n and the empty matrix as parity check matrix, respectively.

Example 5.13. As a consequence of Proposition 5.4 we see that a parity check matrix of the binary even weight code is equal to the generator matrix (1 1 $\cdots$ 1) of the repetition code.

Example 5.14. The generator matrix G of the Hamming code C in Example 5.11 is of the form $(I_4|P)$ and in Example 5.8 we see that the parity check matrix is equal to $(P^T|I_3)$.

5.2.5. Inner product and dual codes

The *inner product* on $\mathbb{F}_q^n$ is defined by

$$\mathbf{x} \cdot \mathbf{y} = x_1 y_1 + \cdots + x_n y_n$$

for $\mathbf{x}, \mathbf{y} \in \mathbb{F}_q^n$. This inner product is *bilinear*, *symmetric* and *nondegenerate*, but the notion of "positive definite" makes no sense over a finite field as it does over the real numbers. For instance for a binary word $\mathbf{x} \in \mathbb{F}_2^n$ we have that $\mathbf{x} \cdot \mathbf{x} = 0$ if and only if the weight of $\mathbf{x}$ is even.

For an $[n, k]$ code C we define the *dual* or *orthogonal code* $C^\perp$ as

$$C^\perp = \{ \mathbf{x} \in \mathbb{F}_q^n : \mathbf{c} \cdot \mathbf{x} = 0 \text{ for all } \mathbf{c} \in C \}.$$

Proposition 5.5. *Let C be an $[n, k]$ code with generator matrix G. Then $C^\perp$ is an $[n, n-k]$ code with parity check matrix G.*

Proof. From the definition of dual codes, the following statements are equivalent:

$$\mathbf{x} \in C^\perp,$$
$$\iff \mathbf{c} \cdot \mathbf{x} = 0 \text{ for all } \mathbf{c} \in C,$$
$$\iff \mathbf{m} G \mathbf{x}^T = 0 \text{ for all } \mathbf{m} \in \mathbb{F}_q^k,$$
$$\iff G \mathbf{x}^T = 0.$$

This means that $C^\perp$ is the null space of G. Because G is a $k \times n$ matrix of rank k, the linear space $C^\perp$ has dimension $n - k$ and G is a parity check matrix of $C^\perp$. □

Example 5.15. The trivial codes $\{0\}$ and $\mathbb{F}_q^n$ are dual codes.

Example 5.16. The binary even weight code and the repetition code of the same length are dual codes.

5.2.6. The Hamming and simplex codes

The following proposition gives a method to determine the minimum distance of a code in terms of the number of dependent columns of the parity check matrix.

Proposition 5.6. *Let H be a parity check matrix of a code C. Then the minimum distance d of C is the smallest integer d such that d columns of H are linearly dependent.*

Proof. Let $\mathbf{h}_1, \ldots, \mathbf{h}_n$ be the columns of H. Let $\mathbf{c}$ be a nonzero codeword of weight w. Let $\mathrm{supp}(\mathbf{c}) = \{j_1, \ldots, j_w\}$ with $1 \leq j_1 < \cdots < j_w \leq n$. Then $H\mathbf{c}^T = 0$, so $c_{j_1}\mathbf{h}_{j_1} + \cdots + c_{j_w}\mathbf{h}_{j_w} = 0$ with $c_{j_i} \neq 0$ for all $i = 1, \ldots, w$. Therefore the columns $\mathbf{h}_{j_1}, \ldots, \mathbf{h}_{j_w}$ are dependent. Conversely if $\mathbf{h}_{j_1}, \ldots, \mathbf{h}_{j_w}$ are dependent, then there exist constants $a_1, \ldots, a_w$, not all zero, such that $a_1 \mathbf{h}_{j_1} + \cdots + a_w \mathbf{h}_{j_w} = 0$. Let $\mathbf{c}$ be the word defined by $c_j = 0$ if $j \neq j_i$ for all i, and $c_j = a_i$ if $j = j_i$ for some i. Then $H\mathbf{c}^T = 0$. Hence $\mathbf{c}$ is a nonzero codeword of weight at most w. □

Let H be a parity check matrix of a code C. As a consequence of Proposition 5.6 we have the following special cases. The minimum distance of code is 1 if and only if H has a zero column. Now suppose that H has no zero column, then the minimum distance of C is at least 2. The minimum distance is equal to 2 if and only if H has two columns say $\mathbf{h}_{j_1}, \mathbf{h}_{j_2}$ that are dependent. In the binary case that means $\mathbf{h}_{j_1} = \mathbf{h}_{j_2}$. In other words the minimum distance of a binary code is at least 3 if and only if H has no zero columns and all columns are mutually distinct. This is the case for the Hamming code of Example 5.8. For a given redundancy r the length of a binary linear code C of minimum distance 3 is at most $2^r - 1$, the number of all nonzero binary columns of length r. For arbitrary $\mathbb{F}_q$, the number of nonzero columns with entries in $\mathbb{F}_q$ is $q^r - 1$. Two such columns are dependent if and only if one is a nonzero multiple of the other. Hence the length of a $\mathbb{F}_q$-linear code C with $d(C) \geq 3$ and redundancy r is at most $(q^r - 1)/(q - 1)$.

Let $n = (q^r - 1)/(q - 1)$. Let $H_r(q)$ be a $r \times n$ matrix over $\mathbb{F}_q$ with nonzero columns, such that no two columns are dependent. The code $\mathcal{H}_r(q)$ with $H_r(q)$ as parity check matrix is called a q-ary *Hamming code*. The code with $H_r(q)$ as generator matrix is called a q-ary *simplex code* and is denoted by $\mathcal{S}_r(q)$. The simplex code $\mathcal{S}_r(q)$ and the Hamming code $\mathcal{H}_r(q)$ are dual codes.

Proposition 5.7. *Let $r \geq 2$. Then the q-ary Hamming code $\mathcal{H}_r(q)$ has parameters $[(q^r - 1)/(q - 1), (q^r - 1)/(q - 1) - r, 3]$.*

Proof. The rank of the matrix $H_r(q)$ is r, since the r standard basis vectors of weight 1 are among the columns of the matrix. So indeed $H_r(q)$ is a parity check matrix of a code with redundancy r. Any 2 columns are independent by construction. And a column of weight 2 is a linear combination of two columns of weight 1, and such a triple of columns exists, since $r \geq 2$. Hence the minimum distance is 3 by Proposition 5.6. □

Example 5.17. Consider the following ternary Hamming code $\mathcal{H}_3(3)$ of redundancy 3 of length 13 with parity check matrix

$$H_3(3) = \begin{pmatrix} 1 & 1 & 1 & 1 & 1 & 1 & 1 & 1 & 1 & 0 & 0 & 0 & 0 \\ 2 & 2 & 2 & 1 & 1 & 1 & 0 & 0 & 0 & 1 & 1 & 1 & 0 \\ 2 & 1 & 0 & 2 & 1 & 0 & 2 & 1 & 0 & 2 & 1 & 0 & 1 \end{pmatrix}.$$

By Proposition 5.7 the code $\mathcal{H}_3(3)$ has parameters $[13, 10, 3]$. Notice that all rows of $H_3(3)$ have weight 9. In fact every linear combination $\mathbf{x}H_3(3)$ with $\mathbf{x} \in \mathbb{F}_3^3$ and $\mathbf{x} \neq 0$ has weight 9. So all nonzero codewords of the ternary simplex code of dimension 3 have weight 9. Hence $\mathcal{S}_3(3)$ is a *constant weight* code. This is a general fact of simplex codes as is stated in the following proposition.

Proposition 5.8. *The q-ary simplex code $\mathcal{S}_r(q)$ is a constant weight code with parameters $[(q^r - 1)/(q - 1), r, q^{r-1}]$.*

Proof. We have already seen in Proposition 5.7 that $H_r(q)$ has rank r, so it is indeed a generator matrix of a code of dimension r. Let $\mathbf{c}$ be a nonzero codeword of the simplex code. Then $\mathbf{c} = \mathbf{m}H_r(q)$ for some nonzero $\mathbf{m} \in \mathbb{F}_q^r$. Let $\mathbf{h}_j^T$ be the j-th column of $H_r(q)$. Then $c_j = 0$ if and only if $\mathbf{m} \cdot \mathbf{h}_j = 0$. Now $\mathbf{m} \cdot \mathbf{x} = 0$ is a nontrivial homogeneous linear equation. This equation has q^{r-1} solutions $\mathbf{x} \in \mathbb{F}_q^r$, it has $q^{r-1} - 1$ nonzero solutions. It has $(q^{r-1} - 1)/(q - 1)$ solutions $\mathbf{x}$ such that $\mathbf{x}^T$ is a column of $H_r(q)$, since for every nonzero $\mathbf{x} \in \mathbb{F}_q^r$ there is exactly one column in $H_r(q)$ that is a nonzero multiple of $\mathbf{x}^T$. So the number of zeros of $\mathbf{c}$ is $(q^{r-1} - 1)/(q - 1)$. Hence the weight of $\mathbf{c}$ is the number of nonzero coordinates which is q^{r-1}. □

5.2.7. *Singleton bound and MDS codes*

The following bound gives us the maximal minimum distance of a code with a given length and dimension. This bound is called the *Singleton bound*.

Theorem 5.1. (The Singleton Bound) *If C is an $[n, k, d]$ code, then*

$$d \leq n - k + 1.$$

Proof. Let H be a parity check matrix of C. This is an $(n-k) \times n$ matrix of row rank $n-k$. The minimum distance of C is the smallest integer d such that H has d linearly dependent columns, by Proposition 5.6. This means that every $d-1$ columns of H are linearly independent. Hence, the column rank of H is at least $d-1$. By the fact that the column rank of a matrix is equal to the row rank, we have $n - k \geq d - 1$. This implies the Singleton bound. □

Let C be an $[n, k, d]$ code. If $d = n - k + 1$, then C is called a *maximum distance separable code* or a *MDS code*, for short. From the Singleton bound, a maximum distance separable code achieves the maximum possible value for the minimum distance given the code length and dimension.

Example 5.18. The minimum distance of the zero code of length n is $n + 1$, by definition. Hence the zero code has parameters $[n, 0, n + 1]$ and is MDS. Its dual is the whole space $\mathbb{F}_q^n$ with parameters $[n, n, 1]$ and is also MDS. The n-fold repetition code has parameters $[n, 1, n]$ and its dual is an $[n, n - 1, 2]$ code and both are MDS.

Proposition 5.9. *For an $[n, k, d]$ code over $\mathbb{F}_q$, the following statements are equivalent:*

(1) C is an MDS code,
(2) every $n - k$ column of a parity check matrix H of C is linearly independent,
(3) every k column of a generator matrix G of C is linearly independent.

Proof. Let H be a parity check matrix of an $[n, k, d]$ code C. As the minimum distance of C is d, any $d-1$ columns of H are linearly independent by Proposition 5.6. Now $d \leq n - k + 1$ by the Singleton bound. So $d = n - k + 1$ if and only if every $n - k$ column of H is independent. Hence (1) and (2) are equivalent.

Now let us assume (3). Let $\mathbf{c}$ be an element of C that is zero at k given coordinates. Let $\mathbf{c} = \mathbf{x}G$ for some $\mathbf{x} \in \mathbb{F}_q^k$. Let G' be the square matrix consisting of the k columns of G corresponding to the k given zero coordinates of $\mathbf{c}$. Then $\mathbf{x}G' = 0$. Hence $\mathbf{x} = 0$, since the k columns of G' are independent by assumption. So $\mathbf{c} = 0$. This implies that the minimum distance of C is at least $n-(k-1) = n-k+1$. Therefore C is an $[n, k, n-k+1]$ MDS code, by the Singleton bound.

Assume that C is MDS. Let G be a generator matrix of C. Let G' be the square matrix consisting of k chosen columns of G. Let $\mathbf{x} \in \mathbb{F}_q^k$ such that $\mathbf{x}G' = 0$. Then $\mathbf{c} = \mathbf{x}G$ is a codeword and its weight is at most $n - k$. So $\mathbf{c} = 0$, since the minimum distance is $n - k + 1$. Hence $\mathbf{x} = 0$, since the rank of G is k. Therefore the k columns are independent. □

Proposition 5.10. *Let $n \leq q$. Let $\mathbf{a} = (a_1, \ldots, a_n)$ be an n-tuple of mutually distinct elements of $\mathbb{F}_q$. Let k be an integer such that $0 \leq k \leq n$. Define* the matrices $G(\mathbf{a})$ and $G'(\mathbf{a})$ by

$$G_k(\mathbf{a}) = \begin{pmatrix} 1 & \cdots & 1 \\ a_1 & \cdots & a_n \\ \vdots & \ddots & \vdots \\ a_1^{k-1} & \cdots & a_n^{k-1} \end{pmatrix} \quad \text{and} \quad G'_k(\mathbf{a}) = \begin{pmatrix} 1 & \cdots & 1 & 0 \\ a_1 & \cdots & a_n & 0 \\ \vdots & \ddots & \vdots & \vdots \\ a_1^{k-1} & \cdots & a_n^{k-1} & 1 \end{pmatrix}.$$

The codes with generator matrix $G_k(\mathbf{a})$ and $G'_k(\mathbf{a})$ are MDS.

Proof. All $k \times k$ submatrices are Vandermonde matrices, and their determinant is not zero, since the a_i are mutually distinct. □

5.3. Weight enumerators and error probability

5.3.1. Weight spectrum

The weight spectrum of a code is an important invariant, that provides useful information for both the code structure and practical applications of the code.

Let C be a code of length n. The *weight spectrum* or *weight distribution* is the following set

$$\{(w, A_w) : w = 0, 1, \ldots, n\}$$

where A_w denotes the number of codewords in C of weight w.

The so-called *weight enumerator* of a code C is a convenient representation of the weight spectrum. It is defined as the following polynomial:

$$W_C(Z) = \sum_{w=0}^{n} A_w Z^w.$$

The *homogeneous weight enumerator* of C is defined as

$$W_C(X, Y) = \sum_{w=0}^{n} A_w X^{n-w} Y^w.$$

Note that $W_C(Z)$ and $W_C(X, Y)$ are equivalent in representing the weight spectrum. They determine each other uniquely by the following equations:

$$W_C(Z) = W_C(1, Z) \quad \text{and} \quad W_C(X, Y) = X^n W_C(X^{-1} Y).$$

Given the weight enumerator or the homogeneous weight enumerator, the weight spectrum is determined completely by the coefficients.

Clearly, the weight enumerator and homogeneous weight enumerator can be written in another form, that is

$$W_C(Z) = \sum_{\mathbf{c} \in C} Z^{\text{wt}(\mathbf{c})}$$

and

$$W_C(X, Y) = \sum_{\mathbf{c} \in C} X^{n-\text{wt}(\mathbf{c})} Y^{\text{wt}(\mathbf{c})}.$$

Example 5.19. The zero code has one codeword, and its weight is zero. Hence the homogeneous weight enumerator of this code is $W_{\{0\}}(X, Y) = X^n$. The number of words of weight w in the trivial code $\mathbb{F}_q^n$ is $A_w = \binom{n}{w}(q-1)^w$. So

$$W_{\mathbb{F}_q^n}(X, Y) = \sum_{w=0}^{n} \binom{n}{w}(q-1)^w X^{n-w} Y^w = (X + (q-1)Y)^n.$$

Example 5.20. The n-fold repetition code C has homogeneous weight enumerator

$$W_C(X, Y) = X^n + (q-1)Y^n.$$

In the binary case its dual is the even weight code. Hence it has homogeneous weight enumerator

$$W_{C^\perp}(X, Y) = \sum_{t=0}^{\lfloor n/2 \rfloor} \binom{n}{2t} X^{n-2t} Y^{2t} = \frac{1}{2}\left((X+Y)^n + (X-Y)^n\right).$$

Example 5.21. The nonzero entries of the weight distribution of the [7,4,3] binary Hamming code are given by $A_0 = 1$, $A_3 = 7$, $A_4 = 7$, $A_7 = 1$, as is seen by inspecting the weights of all 16 codewords. Hence its homogeneous weight enumerator is

$$X^7 + 7X^4Y^3 + 7X^3Y^4 + Y^7.$$

Example 5.22. The simplex code $\mathcal{S}_r(q)$ is a constant weight code by Proposition 5.8 with parameters $[(q^r - 1)/(q - 1), r, q^{r-1}]$. Hence its homogeneous weight enumerator is

$$W_{\mathcal{S}_r(q)}(X, Y) = X^n + (q^r - 1)X^{n-q^{r-1}} Y^{q^{r-1}}.$$

Let C be a linear code. Then $A_0 = 1$ and the minimum distance $d(C)$, which is equal to the minimum weight, is determined by the weight enumerator as follows:

$$d(C) = \min\{i : A_i \neq 0, i > 0\}.$$

It also determines the dimension k of C, since

$$W_C(1,1) = \sum_{w=0}^{n} A_w = q^k.$$

Although there is no apparent relation between the minimum distances of a code and its dual, the weight enumerators satisfy the *MacWilliams identity*.

Theorem 5.2 (MacWilliams). *Let C be an $[n,k]$ code over $\mathbb{F}_q$. Then*

$$W_{C^\perp}(X,Y) = q^{-k} W_C(X + (q-1)Y, X - Y).$$

Proof. See [8, Ch.5. §2. Theorem 1] for a proof for binary codes. A general proof will be given via matroids in Theorem 5.13. □

The computation of the minimum distance and the weight enumerator of a code is NP-hard [11–13].

Example 5.23. The zero code C has homogeneous weight enumerator X^n and its dual $\mathbb{F}_q^n$ has homogeneous weight enumerator $(X + (q-1)Y)^n$, by Example 5.19, which is indeed equal to $q^0 W_C(X + (q-1)Y, X - Y)$ and confirms MacWilliams identity.

Example 5.24. The n-fold repetition code C has homogeneous weight enumerator $X^n + (q-1)Y^n$ and the homogeneous weight enumerator of its dual code in the binary case is $\frac{1}{2}((X+Y)^n + (X-Y)^n)$, by Example 5.20, which is equal to $2^{-1} W_C(X+Y, X-Y)$, confirming the MacWilliams identity for $q = 2$. For arbitrary q we have

$$\begin{aligned}
W_{C^\perp}(X,Y) &= q^{-1} W_C(X + (q-1)Y, X - Y) \\
&= q^{-1}((X + (q-1)Y)^n + (q-1)(X-Y)^n) \\
&= \sum_{w=0}^{n} \binom{n}{w} \frac{(q-1)^w + (q-1)(-1)^w}{q} X^{n-w} Y^w.
\end{aligned}$$

5.3.2. The decoding problem

Let C be a linear code in $\mathbb{F}_q^n$ of minimum distance d. If $\mathbf{c}$ is a transmitted codeword and $\mathbf{r}$ is the received word, then $\{i : r_i \neq c_i\}$ is the set of *error positions* and the number of error positions is called the *number of errors* of the received word. Let $\mathbf{e} = \mathbf{r} - \mathbf{c}$. Then $\mathbf{e}$ is called the *error vector* and $\mathbf{r} = \mathbf{c} + \mathbf{e}$. Hence $\mathrm{supp}(\mathbf{e})$ is the set of error positions and $\mathrm{wt}(\mathbf{e})$ the number of errors. The e_i's are called the *error values*.

If $t' = d(C, \mathbf{r})$ is the distance of $\mathbf{r}$ to the code C, then there exists a *nearest codeword* $\mathbf{c}'$ such that $t' = d(\mathbf{c}', \mathbf{r})$. So there exists an error vector $\mathbf{e}'$ such that $\mathbf{r} = \mathbf{c}' + \mathbf{e}'$ and $\mathrm{wt}(\mathbf{e}') = t'$. If the number of errors t is at most $(d-1)/2$, then we are sure that $\mathbf{c} = \mathbf{c}'$ and $\mathbf{e} = \mathbf{e}'$. In other words, the nearest codeword to $\mathbf{r}$ is unique when $\mathbf{r}$ has distance at most $(d-1)/2$ to C. The number $\lfloor (d(C)-1)/2 \rfloor$ is called the *error-correcting capacity* of the code C and is denoted by $e(C)$.

A *decoder* $\mathcal{D}$ for the code C is a map

$$\mathcal{D} : \mathbb{F}_q^n \longrightarrow \mathbb{F}_q^n \cup \{*\}$$

such that $\mathcal{D}(\mathbf{c}) = \mathbf{c}$ for all $\mathbf{c} \in C$.

If $\mathcal{E} : \mathbb{F}_q^k \to \mathbb{F}_q^n$ is an encoder of C and $\mathcal{D} : \mathbb{F}_q^n \to \mathbb{F}_q^k \cup \{*\}$ is a map such that $\mathcal{D}(\mathcal{E}(\mathbf{m})) = \mathbf{m}$ for all $\mathbf{m} \in \mathbb{F}_q^k$, then $\mathcal{D}$ is called a *decoder with respect to the encoder* $\mathcal{E}$. Then $\mathcal{E} \circ \mathcal{D}$ is a decoder of C.

It is allowed that the decoder gives as outcome the symbol $*$ in case it fails to find a codeword. This is called a *decoding failure*. If $\mathbf{c}$ is the codeword sent and $\mathbf{r}$ is the received word and $\mathcal{D}(\mathbf{r}) = \mathbf{c}' \neq \mathbf{c}$, then this is called a *decoding error*. If $\mathcal{D}(\mathbf{r}) = \mathbf{c}$, then $\mathbf{r}$ is *decoded correctly*. Notice that a decoding failure is noted on the receiving end, whereas there is no way that the decoder can detect a decoding error.

A *complete decoder* is a decoder that always gives a codeword in C as outcome. A *nearest neighbor decoder*, also called a *minimum distance decoder*, is a complete decoder with the property that $\mathcal{D}(\mathbf{r})$ is a nearest codeword. A decoder $\mathcal{D}$ for a code C is called a *t-bounded distance* decoder or a decoder that *corrects t errors* if $\mathcal{D}(\mathbf{r})$ is a nearest codeword for all received words $\mathbf{r}$ with $d(C, \mathbf{r}) \leq t$ errors. A decoder for a code C with error-correcting capacity $e(C)$ *decodes up to half the minimum distance* if it is an $e(C)$-bounded distance decoder, where $e(C) = \lfloor (d(C)-1)/2 \rfloor$ is the error-correcting capacity of C.

If $\mathcal{D}$ is a t-bounded distance decoder, then it is not required that $\mathcal{D}$ gives a decoding failure as outcome for a received word $\mathbf{r}$ if the distance of $\mathbf{r}$ to the code is strictly larger than t. In other words, $\mathcal{D}$ is also a t'-bounded distance decoder for all $t' \leq t$.

The *covering radius* $\rho(C)$ of a code C is the smallest ρ such that $d(C, \mathbf{y}) \leq \rho$ for all $\mathbf{y}$. A nearest neighbor decoder is a t-bounded distance decoder for all $t \leq \rho(C)$. A $\rho(C)$-bounded distance decoder is a nearest neighbor decoder, since $d(C, \mathbf{r}) \leq \rho(C)$ for all received words $\mathbf{r}$.

Let $\mathbf{r}$ be a received word with respect to a code C. We call the set $\mathbf{r} + C = \{\mathbf{r} + \mathbf{c} : \mathbf{c} \in C\}$ the *coset* of $\mathbf{r}$ in C. If $\mathbf{r}$ is a codeword, the coset is equal to the code itself. If $\mathbf{r}$ is not a codeword, the coset is not a linear subspace. A *coset leader* of $\mathbf{r} + C$ is a choice of an element of minimal weight in the coset $\mathbf{r} + C$.

The choice of a coset leader of the coset $\mathbf{r} + C$ is unique if $d(C, \mathbf{r}) \leq (d-1)/2$. Let $\rho(C)$ be the covering radius of the code, then there is at least one codeword $\mathbf{c}$ such that $d(\mathbf{c}, \mathbf{r}) \leq \rho(C)$. Hence the weight of a coset leader is at most $\rho(C)$.

Let $\mathbf{r}$ be a received word. Let $\mathbf{e}$ be the chosen coset leader of the coset $\mathbf{r} + C$. The *coset leader decoder* gives $\mathbf{r} - \mathbf{e}$ as output. The coset leader decoder is a nearest neighbor decoder. A *list decoder* gives as output the collection of all nearest codewords.

It is nice to know the existence of a decoder from a theoretical point of view, in practice the problem is to find an *efficient algorithm* that computes the outcome of the decoder. Whereas finding the closest vector of a given vector to a linear subspace in Euclidean n-space can be computed efficiently by an orthogonal projection to the subspace, the corresponding problem for linear codes is in general not such an easy task. In fact it is an NP-hard problem [11].

5.3.3. The q-ary symmetric channel

The *q-ary symmetric channel* (qSC) is a channel where q-ary words are sent with independent errors with the same *cross-over probability* p at each coordinate, with $0 \leq p \leq \frac{q-1}{q}$, such that all the $q-1$ wrong symbols occur with the same probability $p/(q-1)$. So a symbol is transmitted correctly with probability $1-p$. The special case $q = 2$ is called the *binary symmetric channel* (BSC).

Remark 5.2. Let $P(\mathbf{x})$ be the probability that the codeword $\mathbf{x}$ is sent. Then this probability is assumed to be the same for all codewords. Hence $P(\mathbf{x}) = \frac{1}{|C|}$ for all $\mathbf{x} \in C$. Let $P(\mathbf{y}|\mathbf{x})$ be the probability that $\mathbf{y}$ is received given that $\mathbf{x}$ is sent. Then

$$P(\mathbf{y}|\mathbf{x}) = \left(\frac{p}{q-1}\right)^{d(\mathbf{x},\mathbf{y})} (1-p)^{n-d(\mathbf{x},\mathbf{y})}$$

for a q-ary symmetric channel.

Let C be a code of minimum distance d. Consider the decoder that corrects up to t errors with $2t + 1 \leq d$. Let $\mathbf{c}$ be the codeword that is sent. Let $\mathbf{r}$ be the received word. In case the distance of $\mathbf{r}$ to the code is at most t, then the decoder will produce a unique closest codeword $\mathbf{c}'$. If $\mathbf{c} = \mathbf{c}'$, then this is called *correct decoding* which is the case if $d(\mathbf{r}, \mathbf{c}) \leq t$. If $\mathbf{c} \neq \mathbf{c}'$ then it is called a *decoding error*. If $d(\mathbf{r}, C) > t$ the decoding algorithm fails to produce a codeword and such an instance is called a *decoding failure*.

For every decoding scheme and channel one defines three probabilities $P_{cd}(p)$, $P_{de}(p)$ and $P_{df}(p)$, that is the *probability of correct decoding, decoding error* and *decoding failure*, respectively. Then

$$P_{cd}(p) + P_{de}(p) + P_{df}(p) = 1 \text{ for all } 0 \leq p \leq \frac{q-1}{q}.$$

So it suffices to find formulas for two of these three probabilities. The *error probability*, also called the *error rate* is defined by $P_{err}(p) = 1 - P_{cd}(p)$. Hence

$$P_{err}(p) = P_{de}(p) + P_{df}(p).$$

Proposition 5.11. *The probability of correct decoding of a decoder that corrects up to t errors with $2t + 1 \leq d$ of a code C of minimum distance d*

on a *q-ary symmetric channel with cross-over probability p is given by*

$$P_{cd}(p) = \sum_{w=0}^{t} \binom{n}{w} p^w (1-p)^{n-w}.$$

Proof. Every codeword has the same probability of transmission. So

$$P_{cd}(p) = \sum_{\mathbf{x} \in C} P(\mathbf{x}) \sum_{d(\mathbf{x},\mathbf{y}) \leq t} P(\mathbf{y}|\mathbf{x})$$

$$= \frac{1}{|C|} \sum_{\mathbf{x} \in C} \sum_{d(\mathbf{x},\mathbf{y}) \leq t} P(\mathbf{y}|\mathbf{x})$$

$$= \sum_{w=0}^{t} \binom{n}{w} (q-1)^w \left(\frac{p}{q-1}\right)^w (1-p)^{n-w}$$

by Proposition 5.2 and Remark 5.2. Clearing the factor $(q-1)^w$ in the numerator and the denominator give the desired result. □

In Proposition 5.14 a formula will be derived for the probability of decoding error for a decoding algorithm that corrects errors up to half the minimum distance.

Example 5.25. Consider the binary triple repetition code. Assume that $(0,0,0)$ is transmitted. In case the received word has weight 0 or 1, then it is correctly decoded to $(0,0,0)$. If the received word has weight 2 or 3, then it is decoded to $(1,1,1)$ which is a decoding error. Hence there are no decoding failures and

$$P_{cd}(p) = (1-p)^3 + 3p(1-p)^2 = 1 - 3p^2 + 2p^3 \text{ and } P_{err}(p) = P_{de}(p) = 3p^2 - 2p^3.$$

If the Hamming code is used, then there are no decoding failures and

$$P_{cd}(p) = (1-p)^7 + 7p(1-p)^6 \text{ and}$$

$$P_{err}(p) = P_{de}(p) = 21p^2 - 70p^3 + 105p^4 - 84p^5 + 35p^6 - 6p^7.$$

This shows that the error probabilities of the repetition code is smaller than the one for the Hamming code. This comparison is not fair, since only one bit of information is transmitted with the repetition code and 4 bits with the Hamming code. One could transmit 4 bits of information by using the repetition code four times. This would give the error probability

$$1 - (1 - 3p^2 + 2p^3)^4 = 12p^2 - 8p^3 - 54p^4 + 72p^5 + 84p^6 - 216p^7 + \cdots$$

Suppose that four bits of information are transmitted uncoded, by the Hamming code and the triple repetition code, respectively. Then the error probabilities are 0.04, 0.002 and 0.001, respectively, if the cross-over probability is 0.01. The error probability for the repetition code is in fact smaller than that of the Hamming code for all $p \leq \frac{1}{2}$, but the transmission by the Hamming code is almost twice as fast as the repetition code.

Example 5.26. Consider the binary n-fold repetition code. Let $t = (n-1)/2$. Use the decoding algorithm correcting all patterns of t errors. Then by Proposition 5.11 we have

$$P_{err}(p) = \sum_{i=t+1}^{n} \binom{n}{i} p^i (1-p)^{n-i}.$$

Hence the error probability becomes arbitrarily small for increasing n. The price one has to pay is that the information rate $R = 1/n$ tends to 0. The remarkable result of Shannon [10] states that for a fixed rate $R < C(p)$, where

$$C(p) = 1 + p \log_2(p) + (1-p) \log_2(1-p)$$

is the *capacity* of the binary symmetric channel, one can devise encoding and decoding schemes such that $P_{err}(p)$ becomes arbitrarily small.

The main problem of error-correcting codes from "Shannon's point of view" is to construct efficient encoding and decoding algorithms of codes with the smallest error probability for a given information rate and cross-over probability.

5.3.4. *Error probability*

Consider the q-ary symmetric channel where the receiver checks whether the received word **r** is a codeword or not, for instance by computing whether $H\mathbf{r}^T$ is zero or not for a chosen parity check matrix H, and asks for *retransmission* in case **r** is not a codeword, as explained in Remark 5.1. Now it may occur that **r** is again a codeword but not equal to the codeword that was sent. This is called an *undetected error*. See [14].

Proposition 5.12. *Let $W_C(X,Y)$ be the weight enumerator of C. Then the probability of undetected error on a q-ary symmetric channel with cross-over probability p is given by*

$$P_{ue}(p) = W_C\left(1-p, \frac{p}{q-1}\right) - (1-p)^n.$$

Proof. Every codeword has the same probability of transmission and the code is linear. So without loss of generality we may assume that the zero word is sent. Hence

$$P_{ue}(p) = \frac{1}{|C|} \sum_{\mathbf{x} \in C} \sum_{\mathbf{x} \neq \mathbf{y} \in C} P(\mathbf{y}|\mathbf{x}) = \sum_{0 \neq \mathbf{y} \in C} P(\mathbf{y}|0).$$

If the received codeword $\mathbf{y}$ has weight w, then w symbols are changed and the remaining $n - w$ symbols remained the same. So

$$P(\mathbf{y}|0) = (1-p)^{n-w} \left(\frac{p}{q-1}\right)^w$$

by Remark 5.2. Hence

$$P_{ue}(p) = \sum_{w=1}^{n} A_w (1-p)^{n-w} \left(\frac{p}{q-1}\right)^w.$$

Substituting $X = 1 - p$ and $Y = p/(q-1)$ in $W_C(X,Y)$ gives the desired result, since $A_0 = 1$. □

Now $P_{retr}(p) = 1 - P_{ue}(p)$ is the probability of *retransmission*.

Example 5.27. Let C be the binary triple repetition code. Then $P_{ue}(p) = p^3$, since $W_C(X,Y) = X^3 + Y^3$ by Example 5.20.

Example 5.28. Let C be the [7,4,3] Hamming code. Then

$$P_{ue}(p) = 7(1-p)^4 p^3 + 7(1-p)^3 p^4 + p^7 = 7p^3 - 21p^4 + 21p^5 - 7p^6 + p^7$$

by Example 5.21.

Proposition 5.13. *Let $N(v,w,s)$ be the number of error patterns in $\mathbb{F}_q^n$ of weight w that are at distance s from a given word of weight v. Then*

$$N(v,w,s) = \sum_{\substack{0 \leq i,j \leq n \\ i+2j+w=s+v}} \binom{n-v}{j+w-v} \binom{v}{i} \binom{v-i}{j} (q-1)^{j+w-v} (q-2)^i.$$

Proof. See [6]. Consider a given word $\mathbf{x}$ of weight v. Let $\mathbf{y}$ be a word of weight w and distance s to $\mathbf{x}$. Suppose that $\mathbf{y}$ has k nonzero coordinates in the complement of the support of $\mathbf{x}$, j zero coordinates in the support of $\mathbf{x}$, and i nonzero coordinates in the support of $\mathbf{x}$ that are distinct from the coordinates of $\mathbf{x}$. Then $s = d(\mathbf{x}, \mathbf{y}) = i + j + k$ and $\text{wt}(\mathbf{y}) = w = v + k - j$. There are $\binom{n-v}{k}$ possible subsets of k elements in the complement of the support of $\mathbf{x}$ and there are $(q-1)^k$ possible choices for the nonzero symbols at the corresponding coordinates. There are $\binom{v}{i}$ possible subsets of i elements in the support of $\mathbf{x}$ and there are $(q-2)^i$ possible choices of the symbols at those positions that are distinct from the coordinates of $\mathbf{x}$. There are $\binom{v-i}{j}$ possible subsets of j elements in the support of $\mathbf{x}$ that are zero at those positions. Hence

$$N(v,w,s) = \sum_{\substack{i+j+k=s \\ v+k-j=w}} \left[\binom{n-v}{k}(q-1)^k\right]\left[\binom{v}{i}(q-2)^i\right]\binom{v-i}{j}.$$

Rewriting this formula using $k = j + w - v$ gives the desired result. □

Proposition 5.14. *The probability of decoding error of a decoder that corrects up to t errors with $2t+1 \leq d$ of a code C of minimum distance d on a q-ary symmetric channel with cross-over probability p is given by*

$$P_{de}(p) = \sum_{w=0}^{n} \left(\frac{p}{q-1}\right)^w (1-p)^{n-w} \sum_{s=0}^{t} \sum_{v=1}^{n} A_v N(v,w,s).$$

Proof. This is left as an exercise. □

5.4. Codes, projective systems and arrangements

Let $\mathbb{F}$ be a field. A *projective system* $\mathcal{P} = (P_1, \ldots, P_n)$ in $\mathbb{P}^r(\mathbb{F})$, the projective space over $\mathbb{F}$ of dimension r, is an n-tuple of points P_j in this projective space, such that not all these points lie in a hyperplane. See [15–17].

Let P_j be given by the homogeneous coordinates $(p_{0j} : p_{1j} : \ldots : p_{rj})$. Let $G_\mathcal{P}$ be the $(r+1) \times n$ matrix with $(p_{0j}, p_{1j}, \ldots, p_{rj})^T$ as j-th column. Then $G_\mathcal{P}$ has rank $r+1$, since not all points lie in a hyperplane. If $\mathbb{F}$ is a finite field, then $G_\mathcal{P}$ is the generator matrix of a nondegenerate code over $\mathbb{F}$ of length n and dimension $r+1$. Conversely, let G be a generator matrix of a nondegenerate code C of dimension k over $\mathbb{F}_q$. Then G has no zero columns. Take the columns of G as homogeneous coordinates of points in $\mathbb{P}^{k-1}(\mathbb{F}_q)$. This gives the projective system $\mathcal{P}_G$ over $\mathbb{F}_q$ of G.

Proposition 5.15. *Let C be a nondegenerate code over $\mathbb{F}_q$ of length n and dimension k with generator matrix G. Let $\mathcal{P}_G$ be the projective system of G. The code has minimum distance d if and only if $n - d$ is the maximal number of points of $\mathcal{P}_G$ in a hyperplane of $\mathbb{P}^{k-1}(\mathbb{F})$.*

Proof. See [15–17]. □

An n-tuple $(H_1, \ldots, H_n)$ of hyperplanes in $\mathbb{F}^k$ is called an *arrangement* in $\mathbb{F}^k$. The arrangement is called *simple* if all the n hyperplanes are mutually distinct. The arrangement is called *central* if all the hyperplanes are linear subspaces. A central arrangement is called *essential* if the intersection of all its hyperplanes is equal to $\{0\}$.

Let $G = (g_{ij})$ be a generator matrix of a nondegenerate code C of dimension k. So G has no zero column. Let H_j be the linear hyperplane in $\mathbb{F}_q^k$ with equation

$$g_{1j}X_1 + \cdots + g_{kj}X_k = 0.$$

The arrangement $(H_1, \ldots, H_n)$ associated with G will be denoted by $\mathcal{A}_G$.

In case of a central arrangement one considers the hyperplanes in $\mathbb{P}^{k-1}(\mathbb{F})$. Note that projective systems and essential arrangements are dual notions and that there is a one-to-one correspondence between generalized equivalence classes of nondegenerate $[n, k]$ codes over $\mathbb{F}_q$, equivalence classes of projective systems over $\mathbb{F}_q$ of n points in $\mathbb{P}^{k-1}(\mathbb{F}_q)$ and equivalence classes of essential arrangements of n hyperplanes in $\mathbb{P}^{k-1}(\mathbb{F}_q)$.

We can translate Proposition 5.15 for an arrangement.

Proposition 5.16. *Let C be a nondegenerate code over $\mathbb{F}_q$ with generator matrix G. Let $\mathbf{c}$ be a codeword $\mathbf{c} = \mathbf{x}G$ for some $\mathbf{x} \in \mathbb{F}_q^k$. Then $n - \mathrm{wt}(\mathbf{c})$ is equal to the number of hyperplanes in $\mathcal{A}_G$ through $\mathbf{x}$.*

Proof. See [15–17]. □

A code C is called *projective* if $d(C^\perp) \geq 3$. Let G be a generator matrix of C. Then C is projective if and only if C is nondegenerate and any two columns of G are independent. So C is projective if and only if C is nondegenerate and the hyperplanes of $\mathcal{A}_G$ are mutually distinct.

5.5. The extended and generalized weight enumerator

The number A_w of codewords of weight w equals the number of points that are on exactly $n - w$ of the hyperplanes in $\mathcal{A}_G$, by Proposition 5.16. In particular A_n is equal to the number of points that is in the complement of the union of these hyperplanes in $\mathbb{F}_q^k$. This number can be computed by the *principle of inclusion/exclusion*:

$$A_n = q^k - |H_1 \cup \cdots \cup H_n|$$
$$= q^k + \sum_{w=1}^{n} (-1)^w \sum_{\substack{i_1 < \cdots < i_w \\ i_j \in [n]}} |H_{i_1} \cap \cdots \cap H_{i_w}|.$$

The following notations are introduced to find a formalism as above for the computation of the weight enumerator. This method is based on Katsman and Tsfasman [15]. Later we will encounter two more methods: by matroids and the Tutte polynomial in Section 5.6.3 and by geometric lattices and the characteristic polynomial in Section 5.7.

Definition 5.1. For a subset J of $[n] := \{1, 2, \ldots, n\}$ define

$$C(J) = \{\mathbf{c} \in C : c_j = 0 \text{ for all } j \in J\}$$
$$l(J) = \dim C(J)$$
$$B_J = q^{l(J)} - 1$$
$$B_t = \sum_{|J|=t} B_J.$$

Remark 5.3. The encoding map $\mathbf{x} \mapsto \mathbf{x}G = \mathbf{c}$ from vectors $\mathbf{x} \in \mathbb{F}_q^k$ to codewords gives the following isomorphism of vector spaces

$$\bigcap_{j \in J} H_j \cong C(J)$$

by Proposition 5.16. Furthermore B_J is equal to the number of nonzero codewords $\mathbf{c}$ that are zero at all j in J, and this is equal to the number of nonzero elements of the intersection $\bigcap_{j \in J} H_j$.

Proposition 5.17. *We have the following connection between the B_t and the weight distribution of a code:*

$$B_t = \sum_{w=d}^{n-t} \binom{n-w}{t} A_w.$$

Proof. Count in two ways the number of elements of the set

$$\{(J, \mathbf{c}) : J \subseteq [n], |J| = t, \mathbf{c} \in C(J), \mathbf{c} \neq 0\}. \qquad \square$$

We will generalize this idea to determine the generalized weight enumerators.

5.5.1. Generalized weight enumerators

The notion of the generalized weight enumerator was first introduced by Helleseth, Kløve and Mykkeltveit [18, 19] and later studied by Wei [20]. See also [21]. This notion has applications in the wire-tap channel II [22] and trellis complexity [23].

Instead of looking at words of C, we consider all the subcodes of C of a certain dimension r. We say that the *weight of a subcode* (also called the *effective length* or *support weight*) is equal to n minus the number of coordinates that are zero for every word in the subcode. The smallest weight for which a subcode of dimension r exists, is called the *r-th generalized Hamming weight* of C. To summarize:

$$\mathrm{supp}(D) = \{i \in [n] : \text{there is an } \mathbf{x} \in D : x_i \neq 0\},$$
$$\mathrm{wt}(D) = |\mathrm{supp}(D)|$$
$$d_r = \min\{\mathrm{wt}(D) : D \subseteq C \text{ subcode}, \dim D = r\}.$$

Note that $d_0 = 0$ and $d_1 = d$, the minimum distance of the code. The number of subcodes with a given weight w and dimension r is denoted by $A_w^{(r)}$. Together they form the *r-th generalized weight distribution* of the code. Just as with the ordinary weight distribution, we can make a polynomial with the distribution as coefficients: the *generalized weight enumerator*.

The r-th generalized weight enumerator is given by

$$W_C^{(r)}(X,Y) = \sum_{w=0}^{n} A_w^{(r)} X^{n-w} Y^w,$$

where $A_w^{(r)} = |\{D \subseteq C : \dim D = r, \mathrm{wt}(D) = w\}|$.

We can see from this definition that $A_0^{(0)} = 1$ and $A_0^{(r)} = 0$ for all $0 < r \leq k$. Furthermore, every 1-dimensional subspace of C contains $q-1$ nonzero codewords, so $(q-1)A_w^{(1)} = A_w$ for $0 < w \leq n$. This means we can find the original weight enumerator by using

$$W_C(X,Y) = W_C^{(0)}(X,Y) + (q-1)W_C^{(1)}(X,Y).$$

We will give a way to determine the generalized weight enumerator of a linear $[n,k]$ code C over $\mathbb{F}_q$. We give two lemmas about the determination of $l(J)$, which will become useful later.

Lemma 5.1. *Let C be a linear code with generator matrix G. Let $J \subseteq [n]$ and $|J| = t$. Let G_J be the existence of $k \times t$ submatrix of G of the columns of G indexed by J, and let $r(J)$ be the rank of G_J. Then the dimension $l(J)$ is equal to $k - r(J)$.*

Proof. Let C_J be the code generated by G_J. Consider the projection map $\pi : C \to \mathbb{F}_q^t$ given by deleting the coordinates that are not indexed by J. Then π is a linear map, the image of C under π is C_J and the kernel is $C(J)$ by definition. It follows that $\dim C_J + \dim C(J) = \dim C$. So $l(J) = k - r(J)$. $\square$

Lemma 5.2. *Let d and $d^\perp$ be the minimum distance of C and $C^\perp$, respectively. Let $J \subseteq [n]$ and $|J| = t$. Then we have*

$$l(J) = \begin{cases} k - t & \text{for all } t < d^\perp \\ 0 & \text{for all } t > n - d. \end{cases}$$

Proof. Let $t > n - d$ and let $\mathbf{c} \in C(J)$. Then J is contained in the complement of $\mathrm{supp}(\mathbf{c})$, so $t \leq n - \mathrm{wt}(\mathbf{c})$. It follows that $\mathrm{wt}(\mathbf{c}) \leq n - t < d$, so $\mathbf{c}$ is the zero word and therefore $l(J) = 0$.

Let G be a generator matrix for C, then G is also a parity check matrix for $C^\perp$. We saw in Lemma 5.1 that $l(J) = k - r(J)$, where $r(J)$ is the rank of the matrix formed by the columns of G indexed by J. Let $t < d^\perp$, then every t-tuple of columns of G is linearly independent by Proposition 5.6, so $r(J) = t$ and $l(J) = k - t$. $\square$

Note that by the Singleton bound, we have $d^\perp \leq n - (n-k) + 1 = k + 1$ and $n - d \geq k - 1$, so for $t = k$ both of the above cases apply. This is no problem, because if $t = k$ then $k - t = 0$.

We introduce the following notations:

$$[m, r]_q = \prod_{i=0}^{r-1}(q^m - q^i)$$
$$\langle r \rangle_q = [r, r]_q$$
$$\begin{bmatrix} k \\ r \end{bmatrix}_q = \frac{[k, r]_q}{\langle r \rangle_q}.$$

Remark 5.4. The first number is equal to the number of $m \times r$ matrices of rank r over $\mathbb{F}_q$. The second is the number of bases of $\mathbb{F}_q^r$. The third number is the Gaussian binomial, and it represents the number of r-dimensional subspaces of $\mathbb{F}_q^k$.

For $J \subseteq [n]$ and $r \geq 0$ an integer we define:

$$B_J^{(r)} = |\{D \subseteq C(J) : D \text{ subspace of dimension } r\}|$$
$$B_t^{(r)} = \sum_{|J|=t} B_J^{(r)}.$$

Note that $B_J^{(r)} = \begin{bmatrix} l(J) \\ r \end{bmatrix}_q$. For $r = 0$ this gives $B_t^{(0)} = \binom{n}{t}$. So we see that in general $l(J) = 0$ does not imply $B_J^{(r)} = 0$, because $\begin{bmatrix} 0 \\ 0 \end{bmatrix}_q = 1$. But if $r \neq 0$, we do have that $l(J) = 0$ implies $B_J^{(r)} = 0$ and $B_t^{(r)} = 0$.

Proposition 5.18. *Let r be a positive integer. Let d_r be the r-th generalized Hamming weight of C, and $d^\perp$ the minimum distance of the dual code $C^\perp$. Then we have*

$$B_t^{(r)} = \begin{cases} \binom{n}{t} \begin{bmatrix} k-t \\ r \end{bmatrix}_q & \text{for all } t < d^\perp \\ 0 & \text{for all } t > n - d_r. \end{cases}$$

Proof. The first case is a direct corollary of Lemma 5.2, since there are $\binom{n}{t}$ subsets $J \subseteq [n]$ with $|J| = t$. The proof of the second case goes analogous to the proof of the same lemma: let $|J| = t$, $t > n - d_r$ and suppose there is a subspace $D \subseteq C(J)$ of dimension r. Then J is contained in the complement of supp(D), so $t \leq n - \text{wt}(D)$. It follows that wt$(D) \leq n - t < d_r$, which is impossible, so such a D does not exist. So $B_J^{(r)} = 0$ for all J with $|J| = t$ and $t > n - d_r$, and therefore $B_t^{(r)} = 0$ for $t > n - d_r$. □

We can check that the formula is well-defined: if $t < d^\perp$ then $l(J) = k-t$. If also $t > n - d_r$, we have $t > n - d_r \geq k - r$ by the generalized Singleton bound. This implies $r > k - t = l(J)$, so $\begin{bmatrix} k-t \\ r \end{bmatrix}_q = 0$.

The relation between $B_t^{(r)}$ and $A_w^{(r)}$ becomes clear in the next proposition.

Proposition 5.19. *The following formula holds:*

$$B_t^{(r)} = \sum_{w=0}^{n} \binom{n-w}{t} A_w^{(r)}.$$

Proof. We will count the elements of the set

$$\mathcal{B}_t^{(r)} = \{(D, J) : J \subseteq [n], |J| = t, D \subseteq C(J) \text{ subspace of dimension } r\}$$

in two different ways. For each J with $|J| = t$ there are $B_J^{(r)}$ pairs (D, J) in $\mathcal{B}_t^{(r)}$, so the total number of elements in this set is $\sum_{|J|=t} B_J^{(r)} = B_t^{(r)}$. On the other hand, let D be an r-dimensional subcode of C with $\mathrm{wt}(D) = w$. There are $A_w^{(r)}$ possibilities for such a D. If we want to find a J such that $D \subseteq C(J)$, we have to pick t coordinates from the $n-w$ all-zero coordinates of D. Summation over all w proves the given formula. □

Note that because $A_w^{(r)} = 0$ for all $w < d_r$, we can start summation at $w = d_r$. We can end summation at $w = n-t$ because for $t > n-w$ we have $\binom{n-w}{t} = 0$. So the formula can be rewritten as

$$B_t^{(r)} = \sum_{w=d_r}^{n-t} \binom{n-w}{t} A_w^{(r)}.$$

In practice, we will often prefer the summation given in the proposition.

Theorem 5.3. *The generalized weight enumerator is given by the following formula:*

$$W_C^{(r)}(X,Y) = \sum_{t=0}^{n} B_t^{(r)} (X-Y)^t Y^{n-t}.$$

Proof. By using the previous proposition, changing the order of summation and using the binomial expansion of $X^{n-w} = ((X-Y)+Y)^{n-w}$ we have

$$\sum_{t=0}^{n} B_t^{(r)}(X-Y)^t Y^{n-t} = \sum_{t=0}^{n} \sum_{w=0}^{n} \binom{n-w}{t} A_w^{(r)} (X-Y)^t Y^{n-t}$$

$$= \sum_{w=0}^{n} A_w^{(r)} \left(\sum_{t=0}^{n-w} \binom{n-w}{t} (X-Y)^t Y^{n-w-t} \right) Y^w$$

$$= \sum_{w=0}^{n} A_w^{(r)} X^{n-w} Y^w$$

$$= W_C^{(r)}(X,Y).$$

In the second step, we can let the summation over t run to $n-w$ instead of n because $\binom{n-w}{t} = 0$ for $t > n-w$. □

It is possible to determine the $A_w^{(r)}$ directly from the $B_t^{(r)}$, by using the next proposition.

Proposition 5.20. *The following formula holds:*

$$A_w^{(r)} = \sum_{t=n-w}^{n} (-1)^{n+w+t} \binom{t}{n-w} B_t^{(r)}.$$

There are several ways to prove this proposition. One is to reverse the argument from Theorem 5.3; this method is left as an exercise. Instead, we first prove the following general lemma:

Lemma 5.3. *Let V be a vector space of dimension $n+1$ and let $\mathbf{a} = (a_0, \ldots, a_n)$ and $\mathbf{b} = (b_0, \ldots, b_n)$ be vectors in V. Then the following formulas are equivalent:*

$$a_j = \sum_{i=0}^{n} \binom{i}{j} b_i, \qquad b_j = \sum_{i=j}^{n} (-1)^{i+j} \binom{i}{j} a_i.$$

Proof. We can view the relations between $\mathbf{a}$ and $\mathbf{b}$ as linear transformations, given by the matrices with entries $\binom{i}{j}$ and $(-1)^{i+j}\binom{i}{j}$, respectively. So it is sufficient to prove that these matrices are each other's inverse. We calculate the entry on the i-th row and j-th column. Note that we can start the summation at $l = j$, because for $l < j$ we have $\binom{l}{j} = 0$.

$$\sum_{l=j}^{i} (-1)^{j+l} \binom{i}{l}\binom{l}{j} = \sum_{l=j}^{i} (-1)^{l-j} \binom{i}{j}\binom{i-j}{l-j}$$
$$= \sum_{l=0}^{i-j} (-1)^l \binom{i}{j}\binom{i-j}{l}$$
$$= \binom{i}{j}(1-1)^{i-j}$$
$$= \delta_{ij}.$$

Here δ_{ij} is the Kronecker-delta. So the product matrix is exactly the identity matrix of size $n+1$, and therefore the matrices are each other's inverse. □

Proof. (**Proposition 5.20**) The proposition is now a direct consequence of Proposition 5.19 and Lemma 5.3. □

5.5.2. Extended weight enumerator

Let G be the generator matrix of a linear $[n,k]$ code C over $\mathbb{F}_q$. Then we can form the $[n,k]$ code $C \otimes \mathbb{F}_{q^m}$ over $\mathbb{F}_{q^m}$ by taking all $\mathbb{F}_{q^m}$-linear combinations of the codewords in C. We call this the *extension code* of C over $\mathbb{F}_{q^m}$. We denote the number of codewords in $C \otimes \mathbb{F}_{q^m}$ of weight w by $A_{C \otimes \mathbb{F}_{q^m}, w}$ and the number of subspaces in $C \otimes \mathbb{F}_{q^m}$ of dimension r and weight w by $A^{(r)}_{C \otimes \mathbb{F}_{q^m}, w}$. We can determine the weight enumerator of such an extension code by using only the code C.

By embedding its entries in $\mathbb{F}_{q^m}$, we find that G is also a generator matrix for the extension code $C \otimes \mathbb{F}_{q^m}$. In Lemma 5.1 we saw that $l(J) = k - r(J)$. Because $r(J)$ is independent of the extension field $\mathbb{F}_{q^m}$, we have $\dim_{\mathbb{F}_q} C(J) = \dim_{\mathbb{F}_{q^m}} (C \otimes \mathbb{F}_{q^m})(J)$. This motivates the usage of T as a variable for q^m in the next definition, that is an extension of Definition 5.1.

Definition 5.2. Let C be a linear code over $\mathbb{F}_q$. Then we define

$$B_J(T) = T^{l(J)} - 1$$
$$B_t(T) = \sum_{|J|=t} B_J(T).$$

The *extended weight enumerator* is given by

$$W_C(X, Y, T) = X^n + \sum_{t=0}^{n} B_t(T)(X - Y)^t Y^{n-t}.$$

Note that $B_J(q^m)$ is the number of nonzero codewords in $(C \otimes \mathbb{F}_{q^m})(J)$.

Proposition 5.21. *Let d and $d^\perp$ be the minimum distance of C and $C^\perp$ respectively. Then we have*

$$B_t(T) = \begin{cases} \binom{n}{t}(T^{k-t} - 1) & \text{for all } t < d^\perp \\ 0 & \text{for all } t > n - d. \end{cases}$$

Proof. This proposition and its proof are generalizations of Proposition 5.17 and its proof. The proof is a direct consequence of Lemma 5.2. For $t < d^\perp$ we have $l(J) = k-t$, so $B_J(T) = T^{k-t} - 1$ and $B_t(T) = \binom{n}{t}(T^{k-t}-1)$. For $t > n - d$ we have $l(J) = 0$, so $B_J(T) = 0$ and $B_t(T) = 0$. □

Theorem 5.4. *The following holds:*

$$W_C(X, Y, T) = \sum_{w=0}^{n} A_w(T) X^{n-w} Y^w$$

with $A_w(T) \in \mathbb{Z}[T]$ given by $A_0(T) = 1$ and

$$A_w(T) = \sum_{t=n-w}^{n} (-1)^{n+w+t} \binom{t}{n-w} B_t(T)$$

for $0 < w \leq n$.

Proof. Note that $A_w(T) = 0$ for $0 < w < d$ because the summation is empty. By substituting $w = n - t + j$ and reversing the order of summation, we have

$$W_C(X,Y,T) = X^n + \sum_{t=0}^{n} B_t(T)(X-Y)^t Y^{n-t}$$

$$= X^n + \sum_{t=0}^{n} B_t(T) \left(\sum_{j=0}^{t} \binom{t}{j} (-1)^j X^{t-j} Y^j \right) Y^{n-t}$$

$$= X^n + \sum_{t=0}^{n} \sum_{j=0}^{t} (-1)^j \binom{t}{j} B_t(T) X^{t-j} Y^{n-t+j}$$

$$= X^n + \sum_{t=0}^{n} \sum_{w=n-t}^{n} (-1)^{t-n+w} \binom{t}{t-n+w} B_t(T) X^{n-w} Y^w$$

$$= X^n + \sum_{w=0}^{n} \sum_{t=n-w}^{n} (-1)^{n+w+t} \binom{t}{n-w} B_t(T) X^{n-w} Y^w.$$

Hence $W_C(X,Y,T)$ is of the form $\sum_{w=0}^{n} A_w(T) X^{n-w} Y^w$ with $A_w(T)$ of the form given in the theorem. $\square$

Note that in the definition of $A_w(T)$ we can let the summation over t run to $n-d$ instead of n, because $B_t(T) = 0$ for $t > n-d$.

Proposition 5.22. *The following formula holds:*

$$B_t(T) = \sum_{w=d}^{n-t} \binom{n-w}{t} A_w(T).$$

Proof. The statement is a direct consequence of Lemma 5.3 and Theorem 5.4. $\square$

As we said before, the motivation for looking at the extended weight enumerator comes from the extension codes. In the next proposition we show that the extended weight enumerator for $T = q^m$ is indeed the weight enumerator of the extension code $C \otimes \mathbb{F}_{q^m}$.

Proposition 5.23. *Let C be a linear $[n,k]$ code over $\mathbb{F}_q$. Then we have*

$$W_C(X,Y,q^m) = W_{C \otimes \mathbb{F}_{q^m}}(X,Y).$$

Proof. For $w = 0$ it is clear that $A_0(q^m) = A_{C \otimes \mathbb{F}_{q^m}, 0} = 1$, so assume $w \neq 0$. It is enough to show that $A_w(q^m) = (q^m - 1) A^{(1)}_{C \otimes \mathbb{F}_{q^m}, w}$. First we have

$$B_t(q^m) = \sum_{|J|=t} B_J(q^m)$$

$$= \sum_{|J|=t} |\{\mathbf{c} \in (C \otimes \mathbb{F}_{q^m})(J) : \mathbf{c} \neq 0\}|$$

$$= (q^m - 1) \sum_{|J|=t} |\{D \subseteq (C \otimes \mathbb{F}_{q^m})(J) : \dim D = 1\}$$

$$= (q^m - 1) B_t^{(1)}(C \otimes \mathbb{F}_{q^m}).$$

We also know that $A_w(T)$ and $B_t(T)$ are related in the same way as $A_w^{(1)}$ and $B_t^{(1)}$. Combining this proves the statement. □

Because of Proposition 5.23 we interpret $W_C(X, Y, T)$ as the weight enumerator of the extension code over the algebraic closure of $\mathbb{F}_q$. This means we can find a relation with the two variable zeta-functions of a code, see Duursma [24].

For further applications, the next way of writing the extended weight enumerator will be useful:

Proposition 5.24. *The extended weight enumerator of a linear code C can be written as*

$$W_C(X, Y, T) = \sum_{t=0}^{n} \sum_{|J|=t} T^{l(J)} (X - Y)^t Y^{n-t}.$$

Proof. By rewriting and using the binomial expansion of $((X-Y)+Y)^n$, we get

$$\sum_{t=0}^{n} \sum_{|J|=t} T^{l(J)}(X-Y)^t Y^{n-t}$$

$$= \sum_{t=0}^{n} (X-Y)^t Y^{n-t} \sum_{|J|=t} \left((T^{l(J)} - 1) + 1 \right)$$

$$= \sum_{t=0}^{n} (X-Y)^t Y^{n-t} \left(\sum_{|J|=t} (T^{l(J)} - 1) + \binom{n}{t} \right)$$

$$= \sum_{t=0}^{n} B_t(T)(X-Y)^t Y^{n-t} + \sum_{t=0}^{n} \binom{n}{t}(X-Y)^t Y^{n-t}$$

$$= \sum_{t=0}^{n} B_t(T)(X-Y)^t Y^{n-t} + X^n$$

$$= W_C(X, Y, T).$$

□

5.5.3. *Puncturing and shortening of codes*

There are several ways to get new codes from existing ones. In this section, we will focus on *puncturing* and *shortening* of codes and show how they are used in an alternative algorithm for finding the extended weight enumerator. The algorithm is based on the Tutte-Grothendieck decomposition of matrices introduced by Brylawski [25]. Greene [26] used this decomposition for the determination of the weight enumerator.

Let C be a linear $[n, k]$ code and let $J \subseteq [n]$. Then the code C *punctured by* J is obtained by deleting all the coordinates indexed by J from the codewords of C. The length of this punctured code is $n - |J|$ and the dimension is at most k. Let C be a linear $[n, k]$ code and let $J \subseteq [n]$. If we puncture the code $C(J)$ by J, we get the code C *shortened by* J. The length of this shortened code is $n - |J|$ and the dimension is $l(J)$.

The operations of puncturing and shortening a code are each other's dual: puncturing a code C by J and then taking the dual, gives the same code as shortening $C^\perp$ by J.

We have seen that we can determine the extended weight enumerator of an $[n, k]$ code C with the use of a $k \times n$ generator matrix of C. This concept can be generalized for arbitrarily matrices, not necessarily of full rank.

Let $\mathbb{F}$ be a field. Let G be a $k \times n$ matrix over $\mathbb{F}$, possibly of rank smaller than k and with zero columns. Then for each $J \subseteq [n]$ we define

$$l(J) = l(J, G) = k - r(G_J)$$

as in Lemma 5.1. Define the extended weight enumerator $W_G(X, Y, T)$ as in Definition 5.2. We can now make the following remarks about $W_G(X, Y, T)$.

Proposition 5.25. *Let G be a $k \times n$ matrix over $\mathbb{F}$ and $W_G(X, Y, T)$ the associated extended weight enumerator. Then the following statements hold:*

(i) $W_G(X, Y, T)$ is invariant under row-equivalence of matrices.
(ii) Let G' be an $l \times n$ matrix with the same row-space as G, then we have $W_G(X, Y, T) = T^{k-l} W_{G'}(X, Y, T)$. In particular, if G is a generator matrix of an $[n, k]$ code C, we have $W_G(X, Y, T) = W_C(X, Y, T)$.
(iii) $W_G(X, Y, T)$ is invariant under permutation of the columns of G.
(iv) $W_G(X, Y, T)$ is invariant under multiplying a column of G with an element of $\mathbb{F}^*$.
(v) If G is the direct sum of G_1 and G_2, i.e. of the form

$$\begin{pmatrix} G_1 & 0 \\ 0 & G_2 \end{pmatrix},$$

then $W_G(X, Y, T) = W_{G_1}(X, Y, T) \cdot W_{G_2}(X, Y, T)$.

Proof. (i) If we multiply G from the left with an invertible $k \times k$ matrix, the $r(J)$ do not change, and therefore (i) holds.

For (ii), we may assume without loss of generality that $k \geq l$. Because G and G' have the same row-space, the ranks $r(G_J)$ and $r(G'_J)$ are the same. So $l(J, G) = k - l + l(J, G')$. Using Proposition 5.24 we have for G

$$W_G(X, Y, T) = \sum_{t=0}^{n} \sum_{|J|=t} T^{l(J,G)} (X - Y)^t Y^{n-t}$$

$$= \sum_{t=0}^{n} \sum_{|J|=t} T^{k-l+l(J,G')} (X - Y)^t Y^{n-t}$$

$$= T^{k-l} \sum_{t=0}^{n} \sum_{|J|=t} T^{l(J,G')} (X - Y)^t Y^{n-t}$$

$$= T^{k-l} W_{G'}(X, Y, T).$$

The last part of (ii) and (iii)–(v) follow directly from the definitions. □

With the use of the extended weight enumerator for general matrices, we can derive a recursive algorithm to determine the extended weight enumerator of a code. Let G be a $k \times n$ matrix with entries in $\mathbb{F}$. Suppose that the j-th column is not the zero vector. Then there exists a matrix row-equivalent to G such that the j-th column is of the form $(1, 0, \ldots, 0)^T$. Such a matrix is called *reduced* at the j-th column. In general, this reduction is not unique.

Let G be a matrix that is reduced at the j-th column a. The matrix $G \setminus a$ is the $k \times (n-1)$ matrix G with the column a removed, and G/a is the $(k-1) \times (n-1)$ matrix G with the column a and the first row removed. We can view $G \setminus a$ as G punctured by a, and G/a as G shortened by a.

For the extended weight enumerators of these matrices, we have the following connection (we omit the (X, Y, T) part for clarity):

Proposition 5.26. *Let G be a $k \times n$ matrix that is reduced at the j-th column a. For the extended weight enumerator of a reduced matrix G holds*

$$W_G = (X - Y)W_{G/a} + YW_{G \setminus a}.$$

Proof. We distinguish between two cases here. First, assume that $G \setminus a$ and G/a have the same rank. Then we can choose a G with all zeros in the first row, except for the 1 in the column a. So G is the direct sum of 1 and G/a. By Proposition 5.25 parts (v) and (ii) we have

$$W_G = (X + (T-1)Y)W_{G/a} \quad \text{and} \quad W_{G \setminus a} = TW_{G/a}.$$

Combining the two gives

$$\begin{aligned} W_G &= (X + (T-1)Y)W_{G/a} \\ &= (X - Y)W_{G/a} + YTW_{G/a} \\ &= (X - Y)W_{G/a} + YW_{G \setminus a}. \end{aligned}$$

For the second case, assume that $G \setminus a$ and G/a do not have the same rank. So $r(G \setminus a) = r(G/a) + 1$. This implies G and $G \setminus a$ do have the same rank. We have that

$$W_G(X, Y, T) = \sum_{t=0}^{n} \sum_{|J|=t} T^{l(J,G)}(X-Y)^t Y^{n-t}$$

by Proposition 5.24. This double sum splits into the sum of two parts by distinguishing between the cases $j \in J$ and $j \notin J$.

Let $j \in J$, $t = |J|$, $J' = J \setminus \{j\}$ and $t' = |J'| = t - 1$. Then

$$l(J', G/a) = k - 1 - r((G/a)_{J'}) = k - r(G_J) = l(J, G).$$

So the first part is equal to

$$\sum_{t=0}^{n} \sum_{\substack{|J|=t \\ j \in J}} T^{l(J,G)}(X-Y)^t Y^{n-t} = \sum_{t'=0}^{n-1} \sum_{|J'|=t'} T^{l(J',G/a)}(X-Y)^{t'+1} Y^{n-1-t'}$$

which is equal to $(X - Y)W_{G/a}$.

Let $j \notin J$. Then $(G \setminus a)_J = G_J$. So $l(J, G \setminus a) = l(J, G)$. Hence the second part is equal to

$$\sum_{t=0}^{n} \sum_{\substack{|J|=t \\ j \notin J}} T^{l(J,G)}(X-Y)^t Y^{n-t} = Y \sum_{t'=0}^{n-1} \sum_{\substack{|J|=t' \\ j \notin J}} T^{l(J,G \setminus a)}(X-Y)^{t'} Y^{n-1-t'}$$

which is equal to $YW_{G \setminus a}$. □

Theorem 5.5. *Let G be a $k \times n$ matrix over $\mathbb{F}$ with $n > k$ of the form $G = (I_k | P)$, where P is a $k \times (n - k)$ matrix over $\mathbb{F}$. Let $A \subseteq [k]$ and write P_A for the matrix formed by the rows of P indexed by A. Let $W_A(X, Y, T) = W_{P_A}(X, Y, T)$. Then the following holds:*

$$W_C(X, Y, T) = \sum_{l=0}^{k} \sum_{|A|=l} Y^l (X - Y)^{k-l} W_A(X, Y, T).$$

Proof. We use the formula of the last proposition recursively. We denote the construction of $G \setminus a$ by G_1 and the construction of G/a by G_2. Repeating this procedure, we get the matrices G_{11}, G_{12}, G_{21} and G_{22}. So we get for the weight enumerator

$$W_G = Y^2 W_{G_{11}} + Y(X-Y)W_{G_{12}} + Y(X-Y)W_{G_{21}} + (X-Y)^2 W_{G_{22}}.$$

Repeating this procedure k times, we get 2^k matrices with $n - k$ columns and $0, \ldots, k$ rows, which form exactly the P_A. In the diagram are the sizes of the matrices of the first two steps: note that only the $k \times n$ matrix on top has to be of full rank. The number of matrices of size $(k - i) \times (n - j)$

is given by the binomial coefficient $\binom{j}{i}$.

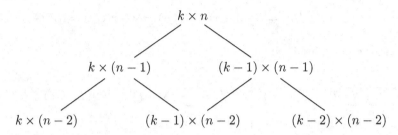

On the last line we have $W_0(X,Y,T) = X^{n-k}$. This proves the formula. □

Example 5.29. Let C be the even weight code of length $n = 6$ over $\mathbb{F}_2$. Then a generator matrix of C is the 5×6 matrix $G = (I_5|P)$ with $P = (1,1,1,1,1,1)^T$. So the matrices P_A are $l \times 1$ matrices with all ones. We have $W_0(X,Y,T) = X$ and $W_l(X,Y,T) = T^{l-1}(X + (T-1)Y)$ by part (ii) of Proposition 5.25. Therefore the weight enumerator of C is equal to

$$W_C(X,Y,T) = W_G(X,Y,T)$$
$$= X(X-Y)^5 + \sum_{l=1}^{5} \binom{5}{l} Y^l (X-Y)^{5-l} T^{l-1}(X + (T-1)Y)$$
$$= X^6 + 15(T-1)X^4Y^2 + 20(T^2 - 3T + 2)X^3Y^3$$
$$+ 15(T^3 - 4T^2 + 6T - 3)X^2Y^4$$
$$+ 6(T^4 - 5T^3 + 10T^2 - 10T + 4)XY^5$$
$$+ (T^5 - 6T^4 + 15T^3 - 20T^2 + 15T - 5)Y^6.$$

For $T = 2$ we get $W_C(X,Y,2) = X^6 + 15X^4Y^2 + 15X^2Y^4 + Y^6$, which we indeed recognize as the weight enumerator of the even weight code that we found in Example 5.20.

5.5.4. *Connections*

There is a connection between the extended weight enumerator and the generalized weight enumerators. We first prove the next proposition.

Proposition 5.27. *Let C be a linear $[n,k]$ code over $\mathbb{F}_q$, and let C^m be the linear subspace consisting of the $m \times n$ matrices over $\mathbb{F}_q$ whose rows are in C. Then there is an isomorphism of $\mathbb{F}_q$-vector spaces between $C \otimes \mathbb{F}_{q^m}$ and C^m.*

Proof. Let α be a primitive m-th root of unity in $\mathbb{F}_{q^m}$. Then we can write an element of $\mathbb{F}_{q^m}$ in a unique way on the basis $(1, \alpha, \alpha^2, \ldots, \alpha^{m-1})$ with coefficients in $\mathbb{F}_q$. If we do this for all the coordinates of a word in $C \otimes \mathbb{F}_{q^m}$, we get an $m \times n$ matrix over $\mathbb{F}_q$. The rows of this matrix are words of C, because C and $C \otimes \mathbb{F}_{q^m}$ have the same generator matrix. This map is clearly injective. There are $(q^m)^k = q^{km}$ words in $C \otimes \mathbb{F}_{q^m}$, and the number of elements of C^m is $(q^k)^m = q^{km}$, so our map is a bijection. It is given by

$$\left(\sum_{i=0}^{m-1} c_{i1}\alpha^i, \sum_{i=0}^{m-1} c_{i2}\alpha^i, \ldots, \sum_{i=0}^{m-1} c_{in}\alpha^i \right) \mapsto$$

$$\begin{pmatrix} c_{01} & c_{02} & c_{03} & \cdots & c_{0n} \\ c_{11} & c_{12} & c_{13} & \cdots & c_{1n} \\ \vdots & \vdots & \vdots & \ddots & \vdots \\ c_{(m-1)1} & c_{(m-1)2} & c_{(m-1)3} & \cdots & c_{(m-1)n} \end{pmatrix}.$$

We see that the map is $\mathbb{F}_q$-linear, so it gives an isomorphism of $\mathbb{F}_q$-vector spaces $C \otimes \mathbb{F}_{q^m} \to C^m$. □

Note that this isomorphism depends on the choice of a primitive element α. The use of this isomorphism for the proof of Theorem 5.6 was suggested by Simonis [21]. We also need the next subresult.

Lemma 5.4. *Let* $\mathbf{c} \in C \otimes \mathbb{F}_{q^m}$ *and* $M \in C^m$ *the corresponding* $m \times n$ *matrix under a given isomorphism. Let* $D \subseteq C$ *be the subcode generated by the rows of* M. *Then* $wt(\mathbf{c}) = wt(D)$.

Proof. If the j-th coordinate c_j of $\mathbf{c}$ is zero, then the j-th column of M consists of only zero's, because the representation of c_j on the basis $(1, \alpha, \alpha^2, \ldots, \alpha^{m-1})$ is unique. On the other hand, if the j-th column of M consists of all zeros, then c_j is also zero. Therefore $\text{wt}(\mathbf{c}) = \text{wt}(D)$. □

Proposition 5.28. *Let* C *be a linear code over* $\mathbb{F}_q$. *Then the weight numerator of an extension code and the generalized weight enumerators are connected via*

$$A_w(q^m) = \sum_{r=0}^{m} [m, r]_q A_w^{(r)}.$$

Proof. We count the number of words in $C \otimes \mathbb{F}_{q^m}$ of weight w in two ways, using the bijection of Proposition 5.27. The first way is just by substituting

$T = q^m$ in $A_w(T)$: this gives the left side of the equation. For the second way, note that every $M \in C^m$ generates a subcode of C whose weight is equal to the weight of the corresponding word in $C \otimes \mathbb{F}_{q^m}$. Fix this weight w and a dimension r: there are $A_w^{(r)}$ subcodes of C of dimension r and weight w. Every such subcode is generated by a $r \times n$ matrix whose rows are words of C. Left multiplication by an $m \times r$ matrix of rank r gives an element of C^m that generates the same subcode of C, and all such elements of C^m are obtained this way. The number of $m \times r$ matrices of rank r is $[m, r]_q$, so summation over all dimensions r gives

$$A_w(q^m) = \sum_{r=0}^{k} [m, r]_q A_w^{(r)}.$$

We can let the summation run to m, because $A_w^{(r)} = 0$ for $r > k$ and $[m, r]_q = 0$ for $r > m$. This proves the given formula. □

This result first appears in [18, Theorem 3.2], although the term "generalized weight enumerator" was yet to be invented. In general, we have the following theorem.

Theorem 5.6. *Let C be a linear code over $\mathbb{F}_q$. Then the extended weight numerator and the generalized weight enumerators are connected via*

$$W_C(X, Y, T) = \sum_{r=0}^{k} \left(\prod_{j=0}^{r-1} (T - q^j) \right) W_C^{(r)}(X, Y).$$

Proof. If we know $A_w^{(r)}$ for all r, we can determine $A_w(q^m)$ for every m. If we have $k+1$ values of m for which $A_w(q^m)$ is known, we can use Lagrange interpolation to find $A_w(T)$, for this is a polynomial in T of degree at most k. In fact, we have

$$A_w(T) = \sum_{r=0}^{k} \left(\prod_{j=0}^{r-1} (T - q^j) \right) A_w^{(r)}.$$

This formula has the right degree and is correct for $T = q^m$ for all integers $m \geq 0$, so we know it must be the correct polynomial. Therefore the theorem follows. □

The converse of the theorem is also true: we can write the generalized weight enumerator in terms of the extended weight enumerator.

Theorem 5.7. *Let C be a linear code over $\mathbb{F}_q$. Then the generalized weight enumerator and the extended weight enumerator are connected via*

$$W_C^{(r)}(X,Y) = \frac{1}{\langle r \rangle_q} \sum_{j=0}^{r} \begin{bmatrix} r \\ j \end{bmatrix}_q (-1)^{r-j} q^{\binom{r-j}{2}} W_C(X,Y,q^j).$$

Proof. We consider the generalized weight enumerator in terms of Proposition 5.24. Then rewriting gives the following:

$$W_C^{(r)}(X,Y) = \sum_{t=0}^{n} B_t^{(r)}(X-Y)^t Y^{n-t}$$

$$= \sum_{t=0}^{n} \sum_{|J|=t} \begin{bmatrix} l(J) \\ r \end{bmatrix}_q (X-Y)^t Y^{n-t}$$

$$= \sum_{t=0}^{n} \sum_{|J|=t} \left(\prod_{j=0}^{r-1} \frac{q^{l(J)} - q^j}{q^r - q^j} \right) (X-Y)^t Y^{n-t}$$

$$= \frac{1}{\prod_{v=0}^{r-1}(q^r - q^v)} \sum_{t=0}^{n} \sum_{|J|=t} \left(\prod_{j=0}^{r-1}(q^{l(J)} - q^j) \right) (X-Y)^t Y^{n-t}$$

$$= \frac{1}{\langle r \rangle_q} \sum_{t=0}^{n} \sum_{|J|=t} \sum_{j=0}^{r} \begin{bmatrix} r \\ j \end{bmatrix}_q (-1)^{r-j} q^{\binom{r-j}{2}} q^{j \cdot l(J)} (X-Y)^t Y^{n-t}$$

$$= \frac{1}{\langle r \rangle_q} \sum_{j=0}^{r} \begin{bmatrix} r \\ j \end{bmatrix}_q (-1)^{r-j} q^{\binom{r-j}{2}} \sum_{t=0}^{n} \sum_{|J|=t} (q^j)^{l(J)} (X-Y)^t Y^{n-t}$$

$$= \frac{1}{\langle r \rangle_q} \sum_{j=0}^{r} \begin{bmatrix} r \\ j \end{bmatrix}_q (-1)^{r-j} q^{\binom{r-j}{2}} W_C(X,Y,q^j).$$

In the fourth step, we use the following identity, which can be proven by induction:

$$\prod_{j=0}^{r-1}(Z - q^j) = \sum_{j=0}^{r} \begin{bmatrix} r \\ j \end{bmatrix}_q (-1)^{r-j} q^{\binom{r-j}{2}} Z^j.$$

See [19, 27–30]. □

5.5.5. *MDS-codes*

We can use the theory in Sections 5.5.1 and 5.5.2 to calculate the weight distribution, generalized weight distribution, and extended weight distribution of a linear $[n,k]$ code C. This is done by determining the values $l(J)$

for each $J \subseteq [n]$. In general, we have to look at the 2^n different subcodes of C to find the $l(J)$, but for the special case of MDS codes we can find the weight distributions much faster.

Proposition 5.29. *Let C be a linear $[n,k]$ MDS code, and let $J \subseteq [n]$. Then we have*

$$l(J) = \begin{cases} 0 & \text{for } t > k \\ k-t & \text{for } t \leq k \end{cases}$$

so for a given t the value of $l(J)$ is independent of the choice of J.

Proof. We know that the dual of an MDS code is also MDS, so $d^\perp = k+1$. Now use $d = n - k + 1$ in Lemma 5.2. $\square$

Now that we know all the $l(J)$ for an MDS code, it is easy to find the weight distribution.

Theorem 5.8. *Let C be an MDS code with parameters $[n,k]$. Then the generalized weight distribution is given by*

$$A_w^{(r)} = \binom{n}{w} \sum_{j=0}^{w-d} (-1)^j \binom{w}{j} \begin{bmatrix} w-d+1-j \\ r \end{bmatrix}_q.$$

The coefficients of the extended weight enumerator for $w > 0$ are given by

$$A_w(T) = \binom{n}{w} \sum_{j=0}^{w-d} (-1)^j \binom{w}{j} (T^{w-d+1-j} - 1).$$

Proof. We will give the construction for the generalized weight enumerator here: the case of the extended weight enumerator goes similar and is left as an exercise. We know from Proposition 5.29 that for an MDS code, $B_t^{(r)}$ depends only on the size of J, so $B_t^{(r)} = \binom{n}{t} \begin{bmatrix} k-t \\ r \end{bmatrix}_q$. Using this in the formula for $A_w^{(r)}$ and substituting $j = t - n + w$, we have

$$\begin{aligned}
A_w^{(r)} &= \sum_{t=n-w}^{n-d_r} (-1)^{n+w+t} \binom{t}{n-w} B_t^{(r)} \\
&= \sum_{t=n-w}^{n-d_r} (-1)^{t-n+w} \binom{t}{n-w} \binom{n}{t} \begin{bmatrix} k-t \\ r \end{bmatrix}_q \\
&= \sum_{j=0}^{w-d_r} (-1)^j \binom{n}{w} \binom{w}{j} \begin{bmatrix} k+w-n-j \\ r \end{bmatrix}_q \\
&= \binom{n}{w} \sum_{j=0}^{w-d_r} (-1)^j \binom{w}{j} \begin{bmatrix} w-d+1-j \\ r \end{bmatrix}_q.
\end{aligned}$$

In the second step, we are using the binomial equivalence

$$\binom{n}{t}\binom{t}{n-w} = \binom{n}{n-w}\binom{n-(n-w)}{t-(n-w)} = \binom{n}{w}\binom{w}{n-t}. \qquad \square$$

So, for all MDS-codes with given parameters $[n, k]$ the extended and generalized weight distributions are the same. But not all such codes are equivalent. We can conclude from this, that the generalized and extended weight enumerators are not enough to distinguish between codes with the same parameters. We illustrate the non-equivalence of two MDS codes by an example.

Example 5.30. Let C be a linear $[n, 3]$ MDS code over $\mathbb{F}_q$. It is possible to write the generator matrix G of C in the following form:

$$\begin{pmatrix} 1 & 1 & \dots & 1 \\ x_1 & x_2 & \dots & x_n \\ y_1 & y_2 & \dots & y_n \end{pmatrix}.$$

Because C is MDS we have $d = n - 2$. We now view the n columns of G as points in the projective plane $\mathbb{P}^2(\mathbb{F}_q)$, say $P_1, \dots, P_n$. The MDS property that every k columns of G are independent is now equivalent to saying that no three points are on a line. To see that these n points do not always determine an equivalent code, consider the following construction. Through the n points there are $\binom{n}{2} = N$ lines, the set $\mathcal{N}$. These lines determine (the generator matrix of) an $[n, 3]$ code $\hat{C}$. The minimum distance of the code $\hat{C}$ is equal to the total number of lines minus the maximum number of lines from $\mathcal{N}$ through an arbitrarily point $P \in \mathbb{P}^2(\mathbb{F}_q)$ by Proposition 5.16. If

$P \notin \{P_1, \ldots, P_n\}$ then the maximum number of lines from $\mathcal{N}$ through P is at most $\frac{1}{2}n$, since no three points of $\mathcal{N}$ lie on a line. If $P = P_i$ for some $i \in [n]$ then P lies on exactly $n - 1$ lines of $\mathcal{N}$, namely the lines $P_i P_j$ for $j \neq i$. Therefore the minimum distance of $\hat{C}$ is $d = N - n + 1$.

We now have constructed an $[n, 3, N - n + 1]$ code $\hat{C}$ from the original code C. Notice that two codes $\hat{C}_1$ and $\hat{C}_2$ are generalized equivalent if C_1 and C_2 are generalized equivalent. The generalized and extended weight enumerators of an MDS code of length n and dimension k are completely determined by the pair (n, k), but this is not generally true for the weight enumerator of $\hat{C}$.

Take for example $n = 6$ and $q = 9$, so $\hat{C}$ is a $[15, 3, 10]$ code. Look at the codes C_1 and C_2 generated by the following matrices respectively, where $\alpha \in \mathbb{F}_9$ is a primitive element:

$$\begin{pmatrix} 1 & 1 & 1 & 1 & 1 & 1 \\ 0 & 1 & 0 & 1 & \alpha^5 & \alpha^6 \\ 0 & 0 & 1 & \alpha^3 & \alpha & \alpha^3 \end{pmatrix} \quad \begin{pmatrix} 1 & 1 & 1 & 1 & 1 & 1 \\ 0 & 1 & 0 & \alpha^7 & \alpha^4 & \alpha^6 \\ 0 & 0 & 1 & \alpha^5 & \alpha & 1 \end{pmatrix}$$

Being both MDS codes, the weight distribution is $(1, 0, 0, 120, 240, 368)$. If we now apply the above construction, we get $\hat{C}_1$ and $\hat{C}_2$ generated by

$$\begin{pmatrix} 1 & 0 & 0 & 1 & 1 & \alpha^4 & \alpha^6 & \alpha^3 & \alpha^7 & \alpha & 1 & \alpha^2 & 1 & \alpha^7 & 1 \\ 0 & 1 & 0 & \alpha^7 & 1 & 0 & 0 & \alpha^4 & 1 & 1 & 0 & \alpha^6 & \alpha & 1 & \alpha^3 \\ 0 & 0 & 1 & 1 & 0 & 1 & 1 & 1 & 0 & 0 & 1 & 1 & 1 & 1 & 1 \end{pmatrix}$$

$$\begin{pmatrix} 1 & 0 & 0 & \alpha^7 & \alpha^2 & \alpha^3 & \alpha & 0 & \alpha^7 & \alpha^7 & \alpha^4 & \alpha^7 & \alpha & 0 & 0 \\ 0 & 1 & 0 & 1 & 0 & \alpha^3 & 0 & \alpha^6 & \alpha^6 & 0 & \alpha^7 & \alpha & \alpha^6 & \alpha^3 & \alpha \\ 0 & 0 & 1 & \alpha^5 & \alpha^5 & \alpha^6 & \alpha^3 & \alpha^7 & \alpha^4 & \alpha^3 & \alpha^5 & \alpha^2 & \alpha^4 & \alpha & \alpha^5 \end{pmatrix}.$$

The weight distribution of $\hat{C}_1$ and $\hat{C}_2$ are, respectively,

$(1, 0, 0, 0, 0, 0, 0, 0, 0, 0, 48, 0, 16, 312, 288, 64)$ and
$(1, 0, 0, 0, 0, 0, 0, 0, 0, 0, 48, 0, 32, 264, 336, 48)$.

So the latter two codes are not generalized equivalent, and therefore not all $[6, 3, 4]$ MDS codes over $\mathbb{F}_9$ are generalized equivalent.

Another example was given in [31, 32] showing that two $[6, 3, 4]$ MDS codes could have distinct covering radii.

5.6. Matroids and codes

Matroids were introduced by Whitney [33], axiomatizing and generalizing the concepts of "independence" in linear algebra and "cycle-free" in graph theory. In the theory of arrangements one uses the notion of a geometric lattice that will be treated in Section 5.7.2. In graph and coding theory one usually refers more to matroids. See [34–38] for basic facts of the theory of matroids.

5.6.1. *Matroids*

A *matroid* M is a pair $(E, \mathcal{I})$ consisting of a finite set E and a collection $\mathcal{I}$ of subsets of E such that the following three conditions hold.

(I.1) $\emptyset \in \mathcal{I}$.
(I.2) If $J \subseteq I$ and $I \in \mathcal{I}$, then $J \in \mathcal{I}$.
(I.3) If $I, J \in \mathcal{I}$ and $|I| < |J|$, then there exists a $j \in (J \setminus I)$ such that $I \cup \{j\} \in \mathcal{I}$.

A subset I of E is called *independent* if $I \in \mathcal{I}$, otherwise it is called *dependent*. Condition (I.2) is called the *independence augmentation axiom*.

If J is a subset of E, then J has a *maximal independent subset*, that is there exists an $I \in \mathcal{I}$ such that $I \subseteq J$ and I is maximal with respect to this property and the inclusion. If I_1 and I_2 are maximal independent subsets of J, then $|I_1| = |I_2|$ by condition (I.3). The *rank* or *dimension* $r(J)$ of a subset J of E is the number of elements of a maximal independent subset of J. An independent set of rank $r(M)$ is called a *basis* of M. The collection of all bases of M is denoted by $\mathcal{B}$.

Let $M_1 = (E_1, \mathcal{I}_1)$ and $M_2 = (E_2, \mathcal{I}_2)$ be matroids. A map $\varphi : E_1 \to E_2$ is called a *morphism of matroids* if $\varphi(I)$ is dependent in M_2 for all I that are dependent in M_1. The map is called an *isomorphism of matroids* if it is a morphism of matroids and there exists a map $\psi : E_2 \to E_1$ such that it is a morphism of matroids and it is the inverse of φ. The matroids are called *isomorphic* if there is an isomorphism of matroids between them.

Example 5.31. Let n and k be nonnegative integers such that $k \leq n$. Let $\mathcal{I}_{n,k} = \{I \subseteq [n] : |I| \leq k\}$. Then $U_{n,k} = ([n], \mathcal{I}_{n,k})$ is a matroid that is called the *uniform matroid* of rank k on n elements. A subset B of $[n]$ is a basis of $U_{n,k}$ if and only if $|B| = k$. The matroid $U_{n,n}$ has no dependent sets and is called *free*.

Let $(E,\mathcal{I})$ be a matroid. An element x in E is called a *loop* if $\{x\}$ is a dependent set. Let x and y in E be two distinct elements that are not loops. Then x and y are called *parallel* if $r(\{x,y\}) = 1$. The matroid is called *simple* if it has no loops and no parallel elements. Now $U_{n,2}$ is up to isomorphism the only simple matroid on n elements of rank two.

Let G be a $k \times n$ matrix with entries in a field $\mathbb{F}$. Let E be the set $[n]$ indexing the columns of G and $\mathcal{I}_G$ be the collection of all subsets I of E such that the submatrix G_I consisting of the columns of G at the positions of I are independent. Then $M_G = (E, \mathcal{I}_G)$ is a matroid. Suppose that $\mathbb{F}$ is a finite field and G_1 and G_2 are generator matrices of a code C, then $(E, \mathcal{I}_{G_1}) = (E, \mathcal{I}_{G_2})$. So the matroid $M_C = (E, \mathcal{I}_C)$ of a code C is well defined by $(E, \mathcal{I}_G)$ for some generator matrix G of C. If C is degenerate, then there is a position i such that $c_i = 0$ for every codeword $\mathbf{c} \in C$ and all such positions correspond one-to-one with loops of M_C. Let C be non-degenerate. Then M_C has no loops, and the positions i and j with $i \neq j$ are parallel in M_C if and only if the i-th column of G is a scalar multiple of the j-th column. The code C is projective if and only if the arrangement $\mathcal{A}_G$ is simple if and only if the matroid M_C is simple. An $[n, k]$ code C is MDS if and only if the matroid M_C is the uniform matroid $U_{n,k}$.

A matroid M is called *realizable* or *representable* over the field $\mathbb{F}$ if there exists a matrix G with entries in $\mathbb{F}$ such that M is isomorphic with M_G.

For more on representable matroids we refer to Tutte [39] and Whittle [40, 41]. Let g_n be the number of isomorphism classes of simple matroids on n points. The values of g_n are determined for $n \leq 8$ by [42] and are given in the following table:

n	1	2	3	4	5	6	7	8
g_n	1	1	2	4	9	26	101	950

Extended tables can be found in [43]. Clearly $g_n \leq 2^{2^n}$. Asymptotically the number g_n is given in [44] and is as follows:

$$\log_2 \log_2 g_n \leq n - \log_2 n + \mathcal{O}(\log_2 \log_2 n),$$

$$\log_2 \log_2 g_n \geq n - \tfrac{3}{2} \log_2 n + \mathcal{O}(\log_2 \log_2 n).$$

A crude upper bound on the number of $k \times n$ matrices with $k \leq n$ and entries in $\mathbb{F}_q$ is given by $(n+1)q^{n^2}$. Hence the vast majority of all matroids on n elements is not representable over a given finite field for $n \to \infty$.

Let $M = (E, \mathcal{I})$ be a matroid. Let $\mathcal{B}$ be the collection of all bases of M. Define $B^\perp = (E \setminus B)$ for $B \in \mathcal{B}$, and $\mathcal{B}^\perp = \{B^\perp : B \in \mathcal{B}\}$. Define $\mathcal{I}^\perp = \{I \subseteq E : I \subseteq B \text{ for some } B \in \mathcal{B}^\perp\}$. Then $(E, \mathcal{I}^\perp)$ is called the *dual matroid* of M and is denoted by $M^\perp$.

The dual matroid is indeed a matroid. Let C be a code over a finite field. Then the matroids $(M_C)^\perp$ and $M_{C^\perp}$ are isomorphic.

Let e be a loop of the matroid M. Then e is not a member of any basis of M. Hence e is in every basis of $M^\perp$. An element of M is called an *isthmus* if it is an element of every basis of M. Hence e is an isthmus of M if and only if e is a loop of $M^\perp$.

Proposition 5.30. *Let $(E, \mathcal{I})$ be a matroid with rank function r. Then the dual matroid has rank function $r^\perp$ given by*

$$r^\perp(J) = |J| - r(E) + r(E \setminus J).$$

Proof. The proof is based on the observation that $r(J) = \max_{B \in \mathcal{B}} |B \cap J|$ and $B \setminus J = B \cap (E \setminus J)$.

$$\begin{aligned}
r^\perp(J) &= \max_{B \in \mathcal{B}^\perp} |B \cap J| \\
&= \max_{B \in \mathcal{B}} |(E \setminus B) \cap J| \\
&= \max_{B \in \mathcal{B}} |J \setminus B| \\
&= |J| - \min_{B \in \mathcal{B}} |J \cap B| \\
&= |J| - (|B| - \max_{B \in \mathcal{B}} |B \setminus J|) \\
&= |J| - r(E) + \max_{B \in \mathcal{B}} |B \cap (E \setminus J)| \\
&= |J| - r(E) + r(E \setminus J).
\end{aligned}$$
$\square$

5.6.2. Graphs, codes and matroids

Graph theory is regarded to start with the paper of Euler [45] with his solution of the problem of the Königbergs bridges. For an introduction to the theory of graphs we refer to [46, 47].

A *graph* Γ is a pair (V, E) where V is a non-empty set and E is a set disjoint from V. The elements of V are called *vertices*, and members of E are called *edges*. Edges are *incident* to one or two vertices, which are called the *ends* of the edge. If an edge is incident with exactly one vertex, then it is called a *loop*. If u and v are vertices that are incident with an edge, then they are called *neighbors* or *adjacent*. Two edges are called *parallel* if they are incident with the same vertices. The graph is called *simple* if it has no loops and no parallel edges.

A graph is called *planar* if there is an injective map $f : V \to \mathbb{R}^2$ from the set of vertices V to the real plane such that for every edge e with ends u and v there is a simple curve in the plane connecting the ends of the edge such that mutually distinct simple curves do not intersect except at the endpoints. More formally: for every edge e with ends u and v there is an injective continuous map $g_e : [0, 1] \to \mathbb{R}^2$ from the unit interval to the plane such that $\{f(u), f(v)\} = \{g_e(0), g_e(1)\}$, and $g_e(0,1) \cap g_{e'}(0,1) = \emptyset$ for all edges e, e' with $e \neq e'$.

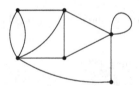

Fig. 5.7. A planar graph

Example 5.32. Consider the next riddle:

> Three newly-build houses have to be connected to the three nearest terminals for gas, water and electricity. For security reasons, the connections are not allowed to cross. How can this be done?

The answer is "not", because the corresponding graph (see Figure 5.9) is not planar. This riddle is very suitable to occupy kids who like puzzles, but make sure to have an easy explainable proof of the improbability. We leave it to the reader to find one.

Let $\Gamma_1 = (V_1, E_1)$ and $\Gamma_2 = (V_2, E_2)$ be graphs. A map $\varphi : V_1 \to V_2$ is called a *morphism of graphs* if $\varphi(v)$ and $\varphi(w)$ are connected in Γ_2 for all

$v, w \in V_1$ that are connected in Γ_1. The map is called an *isomorphism of graphs* if it is a morphism of graphs and there exists a map $\psi : V_2 \to V_1$ such that it is a morphism of graphs and it is the inverse of φ. The graphs are called *isomorphic* if there is an isomorphism of graphs between them.

By deleting loops and parallel edges from a graph Γ one gets a simple graph. There is a choice in the process of deleting parallel edges, but the resulting graphs are all isomorphic. We call this simple graph the *simplification* of the graph and it is denoted by $\bar{\Gamma}$.

Let $\Gamma = (V, E)$ be a graph. Let K be a finite set and $k = |K|$. The elements of K are called *colors*. A *k-coloring* of Γ is a map $\gamma : V \to K$ such that $\gamma(u) \neq \gamma(v)$ for all distinct adjacent vertices u and v in V. So vertex u has color $\gamma(u)$ and all other adjacent vertices have a color distinct from $\gamma(u)$. Let $P_\Gamma(k)$ be the number of k-colorings of Γ. Then P_Γ is called the *chromatic polynomial* of Γ.

If the graph Γ has no edges, then $P_\Gamma(k) = k^v$ where $|V| = v$ and $|K| = k$, since it is equal to the number of all maps from V to K. In particular there is no map from V to an empty set in case V is nonempty. So the number of 0-colorings is zero for every graph.

The number of colorings of graphs was studied by Birkhoff [48], Whitney [49, 50] and Tutte [51–55]. Much research on the chromatic polynomial was motivated by the four-color problem of planar graphs.

Let K_n be the *complete graph* on n vertices in which every pair of two distinct vertices is connected by exactly one edge. Then there is no k coloring if $k < n$. Now let $k \geq n$. Take an enumeration of the vertices. Then there are k possible choices of a color of the first vertex and $k-1$ choices for the second vertex, since the first and second vertices are connected. Now suppose by induction that we have a coloring of the first i vertices, then there are $k - i$ possibilities to color the next vertex, since the $(i + 1)$-th vertex is connected to the first i vertices. Hence

$$P_{K_n}(k) = k(k-1)\cdots(k-n+1).$$

So $P_{K_n}(k)$ is a polynomial in k of degree n.

Proposition 5.31. *Let $\Gamma = (V, E)$ be a graph. Then $P_\Gamma(k)$ is a polynomial in k.*

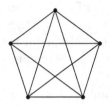

Fig. 5.8. The complete graph K_5

Proof. See [48]. Let $\gamma : V \to K$ be a k-coloring of Γ with exactly i colors. Let σ be a permutation of K. Then the composition of maps $\sigma \circ \gamma$ is also k-coloring of Γ with exactly i colors. Two such colorings are called equivalent. Then $k(k-1)\cdots(k-i+1)$ is the number of colorings in the equivalence class of a given k-coloring of Γ with exactly i colors. Let m_i be the number of equivalence classes of colorings with exactly i colors of the set K. Let $v = |V|$. Then $P_\Gamma(k)$ is equal to

$$m_1 k + m_2 k(k-1) + \ldots + m_i k(k-1)\cdots(k-i+1) + \ldots + m_v k(k-1)\cdots(k-v+1).$$

Therefore $P_\Gamma(k)$ is a polynomial in k. □

A graph $\Gamma = (V, E)$ is called *bipartite* if V is the disjoint union of two nonempty sets M and N such that the ends of an edge are in M and in N. Hence no two points in M are adjacent and no two points in N are adjacent. Let m and n be integers such that $1 \leq m \leq n$. The *complete bipartite graph* $K_{m,n}$ is the graph on a set of vertices V that is the disjoint union of two sets M and N with $|M| = m$ and $|N| = n$, and such that every vertex in M is connected with every vertex in N by a unique edge.

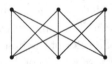

Fig. 5.9. The complete bipartite graph $K_{3,3}$

Another tool to show that $P_\Gamma(k)$ is a polynomial is by deletion-contraction of graphs, a process which is similar to the puncturing and shortening of codes from Section 5.5.3.

Let $\Gamma = (V, E)$ be a graph. Let e be an edge that is incident to the vertices u and v. Then the *deletion* $\Gamma \setminus e$ is the graph with vertices V and edges $E \setminus \{e\}$. The *contraction* Γ/e is the graph obtained by identifying u and v and deleting e. Formally this is defined as follows. Let $\tilde{u} = \tilde{v} = \{u, v\}$, and $\tilde{w} = \{w\}$ if $w \neq u$ and $w \neq v$. Let $\tilde{V} = \{\tilde{w} : w \in V\}$. Then Γ/e is the graph $(\tilde{V}, E \setminus \{e\})$, where an edge $f \neq e$ is incident with $\tilde{w}$ in Γ/e if f is incident with w in Γ.

Notice that the number of k-colorings of Γ does not change by deleting loops and a parallel edge. Hence the chromatic polynomial Γ and its simplification $\bar{\Gamma}$ are the same.

The following proposition is due to Foster. See the concluding note in [50].

Proposition 5.32. *Let* $\Gamma = (V, E)$ *be a simple graph. Let e be an edge of* Γ. *Then the following* deletion-contraction *formula holds:*

$$P_\Gamma(k) = P_{\Gamma \setminus e}(k) - P_{\Gamma/e}(k)$$

for all positive integers k.

Proof. Let u and v be the vertices of e. Then $u \neq v$, since the graph is simple. Let γ be a k-coloring of $\Gamma \setminus e$. Then γ is also a coloring of Γ if and only if $\gamma(u) \neq \gamma(v)$. If $\gamma(u) = \gamma(v)$, then consider the induced map $\tilde{\gamma}$ on $\tilde{V}$ defined by $\tilde{\gamma}(\tilde{u}) = \gamma(u)$ and $\tilde{\gamma}(\tilde{w}) = \gamma(w)$ if $w \neq u$ and $w \neq v$. The map $\tilde{\gamma}$ gives a k-coloring of Γ/e. Conversely, every k-coloring of Γ/e gives a k-coloring γ of $\Gamma \setminus e$ such that $\gamma(u) = \gamma(v)$. Therefore

$$P_{\Gamma \setminus e}(k) = P_\Gamma(k) + P_{\Gamma/e}(k).$$

This follows also from a more general deletion-contraction formula for matroids that will be treated in Section 5.6.4 and Proposition 5.8.1. □

Let $\Gamma = (V, E)$ be a graph. Suppose that $V' \subseteq V$ and $E' \subseteq E$ and all the endpoints of e' in E' are in V'. Then $\Gamma' = (V', E')$ is a graph and it is called a *subgraph* of Γ.

Two vertices u to v are *connected* by a *path* from u to v if there is a t-tuple of mutually distinct vertices $(v_1, \ldots, v_t)$ with $u = v_1$ and $v = v_t$, and a $(t-1)$-tuple of mutually distinct edges $(e_1, \ldots, e_{t-1})$ such that e_i is incident with v_i and v_{i+1} for all $1 \leq i < t$. If moreover e_t is an edge that is incident with u and v and distinct from e_i for all $i < t$, then $(e_1, \ldots, e_{t-1}, e_t)$ is called a *cycle*. The length of the smallest cycle is called the *girth* of the graph and is denoted by $\gamma(\Gamma)$.

The graph is called *connected* if every two vertices are connected by a path. A maximal connected subgraph of Γ is called a *connected component* of Γ. The vertex set V of Γ is a disjoint union of subsets V_i and set of edges E is a disjoint union of subsets E_i such that $\Gamma_i = (V_i, E_i)$ is a connected component of Γ. The number of connected components of Γ is denoted by $c(\Gamma)$.

An edge of a graph is called an *isthmus* if the number of components of the graph increases by deleting the edge. If the graph is connected, then deleting an isthmus gives a graph that is no longer connected. Therefore an isthmus is also called a *bridge*. An edge is an isthmus if and only if it is not an edge of a cycle. Therefore an edge that is an isthmus is also called an *acyclic edge*.

Let $\Gamma = (V, E)$ be a finite graph. Suppose that V consists of m elements enumerated by $v_1, \ldots, v_m$. Suppose that E consists of n elements enumerated by $e_1, \ldots, e_n$. The *incidence matrix* $I(\Gamma)$ is an $m \times n$ matrix with entries a_{ij} defined by

$$a_{ij} = \begin{cases} 1 & \text{if } e_j \text{ is incident with } v_i \text{ and } v_k \text{ for some } i < k, \\ -1 & \text{if } e_j \text{ is incident with } v_i \text{ and } v_k \text{ for some } i > k, \\ 0 & \text{otherwise.} \end{cases}$$

Suppose moreover that Γ is simple. Then $\mathcal{A}_\Gamma$ is the arrangement $(H_1, \ldots, H_n)$ of hyperplanes where $H_j = X_i - X_k$ if e_j is incident with v_i and v_k with $i < k$. An arrangement $\mathcal{A}$ is called *graphic* if $\mathcal{A}$ is isomorphic with $\mathcal{A}_\Gamma$ for some graph Γ.

The *graph code* of Γ over $\mathbb{F}_q$ is the $\mathbb{F}_q$-linear code that is generated by the rows of the incidence matrix $I(\Gamma)$. The *cycle code* C_Γ of Γ is the dual of the graph code of Γ.

Let Γ be a finite graph without loops. Then the arrangement $\mathcal{A}_\Gamma$ is isomorphic with $\mathcal{A}_{C_\Gamma}$.

Proposition 5.33. *Let Γ be a finite graph. Then C_Γ is a code with parameters $[n, k, d]$, where $n = |E|$, $k = |E| - |V| + c(\Gamma)$ and $d = \gamma(\Gamma)$.*

Proof. See [46, Prop. 4.3] □

Let $M = (E, \mathcal{I})$ be a matroid. A subset C of E is called a *circuit* if it is dependent and all its proper subsets are independent. A circuit of the dual matroid $M^\perp$ is called a *cocircuit* of M.

Proposition 5.34. *Let $\mathcal{C}$ be the collection of circuits of a matroid. Then*

(C.1) $\emptyset \notin \mathcal{C}$.
(C.2) *If $C_1, C_2 \in \mathcal{C}$ and $C_1 \subseteq C_2$, then $C_1 = C_2$.*
(C.3) *If $C_1, C_2 \in \mathcal{C}$ and $C_1 \neq C_2$ and $x \in C_1 \cap C_2$, then there exists a $C_3 \in \mathcal{C}$ such that $C_3 \subseteq (C_1 \cup C_2) \setminus \{x\}$.*

Proof. See [35, Lemma 1.1.3]. □

Condition (C.3) is called the *circuit elimination axiom*. The converse of Proposition 5.34 holds.

Proposition 5.35. *Let $\mathcal{C}$ be a collection of subsets of a finite set E that satisfies conditions (C.1), (C.2) and (C.3). Let $\mathcal{I}$ be the collection of all subsets of E that contain no member of $\mathcal{C}$. Then $(E, \mathcal{I})$ is a matroid with $\mathcal{C}$ as its collection of circuits.*

Proof. See [35, Theorem 1.1.4]. □

Proposition 5.36. *Let $\Gamma = (V, E)$ be a finite graph. Let $\mathcal{C}$ the collection of all subsets $\{e_1, \ldots, e_t\}$ such that $(e_1, \ldots, e_t)$ is a cycle in Γ. Then $\mathcal{C}$ is the collection of circuits of a matroid M_Γ on E. This matroid is called the* cycle matroid *of Γ.*

Proof. See [35, Proposition 1.1.7]. □

Loops in Γ correspond one-to-one to loops in M_Γ. Two edges that are no loops, are parallel in Γ if and only if they are parallel in M_Γ. So Γ is simple if and only if M_Γ is simple. Let e in E. Then e is an isthmus in the graph Γ if and only if e is an isthmus in the matroid M_Γ.

A matroid M is called *graphic* if M is isomorphic with M_Γ for some graph Γ, and it is called *cographic* if $M^\perp$ is graphic. If Γ is a planar graph, then the matroid M_Γ is graphic by definition but it is also cographic.

Let Γ be a finite graph with incidence matrix $I(\Gamma)$. The rows of $I(\Gamma)$ generate the code C_Γ over a field $\mathbb{F}$. Suppose that $\mathbb{F}$ is the binary field. Look at all the columns indexed by the edges of a cycle of Γ. Since every vertex in a cycle is incident with exactly two edges, the sum of these columns is zero and therefore they are dependent. Removing a column gives an independent set of vectors. Hence the circuits of the matroid M_{C_Γ} coincide with the cycles in Γ. Therefore M_Γ is isomorphic with M_{C_Γ}. One can generalize this argument for any field. Hence graphic matroids are representable over any field.

The matroid of the binary Hamming $[7,4,3]$ code is not graphic and not cographic. Clearly the matroids M_{K_5} and $M_{K_{3,3}}$ are graphic by definition, but both are not cographic. Tutte [56] found a classification for graphic matroids.

5.6.3. The weight enumerator and the Tutte polynomial

See [1, 26, 57–63] for references of this section.

Definition 5.3. Let $M = (E, \mathcal{I})$ be a matroid. Then the *Whitney rank generating function* $R_M(X, Y)$ is defined by

$$R_M(X,Y) = \sum_{J \subseteq E} X^{r(E)-r(J)} Y^{|J|-r(J)}$$

and the *Tutte-Whitney* or *Tutte polynomial* by

$$t_M(X,Y) = \sum_{J \subseteq E} (X-1)^{r(E)-r(J)} (Y-1)^{|J|-r(J)} .$$

In other words,

$$t_M(X,Y) = R_M(X-1, Y-1).$$

Whitney [50] defined the coefficients m_{ij} of the polynomial $R_M(X, Y)$ such that

$$R_M(X,Y) = \sum_{i=0}^{r(M)} \sum_{j=0}^{|M|} m_{ij} X^i Y^j,$$

but he did not define the polynomial $R_M(X, Y)$ as such. It is clear that these coefficients are nonnegative, since they count the number of elements of certain sets. The coefficients of the Tutte polynomial are also nonnegative, but this is not a trivial fact, it follows from the counting of certain *internal* and *external* bases of a matroid. See [64].

As we have seen, we can interpret a linear $[n, k]$ code C over $\mathbb{F}_q$ as a matroid via the columns of a generator matrix G.

Proposition 5.37. *Let C be an $[n,k]$ code over $\mathbb{F}_q$. Then the Tutte polynomial t_C associated with the matroid M_C of the code C is*

$$t_C(X,Y) = \sum_{t=0}^{n} \sum_{|J|=t} (X-1)^{l(J)} (Y-1)^{l(J)-(k-t)} .$$

Proof. This follows from $l(J) = k - r(J)$ by Lemma 5.1 and $r(M) = k$. $\square$

This formula and Proposition 5.24 suggest the next connection between the weight enumerator and the Tutte polynomial. Greene [26] was the first to notice this connection.

Theorem 5.9. *Let C be an $[n,k]$ code over $\mathbb{F}_q$ with generator matrix G. Then the following holds for the Tutte polynomial and the extended weight enumerator:*

$$W_C(X,Y,T) = (X-Y)^k Y^{n-k} \, t_C\left(\frac{X+(T-1)Y}{X-Y}, \frac{X}{Y}\right) .$$

Proof. By using Proposition 5.37 about the Tutte polynomial, rewriting, and Proposition 5.24 we get

$$(X-Y)^k Y^{n-k} \, t_C\left(\frac{X+(T-1)Y}{X-Y}, \frac{X}{Y}\right)$$

$$= (X-Y)^k Y^{n-k} \sum_{t=0}^{n} \sum_{|J|=t} \left(\frac{TY}{X-Y}\right)^{l(J)} \left(\frac{X-Y}{Y}\right)^{l(J)-(k-t)}$$

$$= (X-Y)^k Y^{n-k} \sum_{t=0}^{n} \sum_{|J|=t} T^{l(J)} Y^{k-t} (X-Y)^{-(k-t)}$$

$$= \sum_{t=0}^{n} \sum_{|J|=t} T^{l(J)} (X-Y)^t Y^{n-t}$$

$$= W_C(X,Y,T).$$

$\square$

We use the extended weight enumerator here, because extending a code does not change the generator matrix and therefore not the matroid G. The converse of this theorem is also true: the Tutte polynomial is completely defined by the extended weight enumerator.

Theorem 5.10. *Let C be an $[n,k]$ code over $\mathbb{F}_q$. Then the following holds for the extended weight enumerator and the Tutte polynomial:*

$$t_C(X,Y) = Y^n (Y-1)^{-k} W_C(1, Y^{-1}, (X-1)(Y-1)) .$$

Proof. The proof of this theorem goes analogous to the proof of the previous theorem.

$$Y^n(Y-1)^{-k}W_C(1, Y^{-1}, (X-1)(Y-1))$$

$$= Y^n(Y-1)^{-k} \sum_{t=0}^{n} \sum_{|J|=t} ((X-1)(Y-1))^{l(J)} (1-Y^{-1})^t Y^{-(n-t)}$$

$$= \sum_{t=0}^{n} \sum_{|J|=t} (X-1)^{l(J)}(Y-1)^{l(J)} Y^{-t}(Y-1)^t Y^{-(n-k)} Y^n (Y-1)^{-k}$$

$$= \sum_{t=0}^{n} \sum_{|J|=t} (X-1)^{l(J)}(Y-1)^{l(J)-(k-t)}$$

$$= t_C(X, Y).$$

□

We see that the Tutte polynomial depends on two variables, while the extended weight enumerator depends on three variables. This is no problem, because the weight enumerator is given in its homogeneous form here: we can view the extended weight enumerator as a polynomial in two variables via $W_C(Z, T) = W_C(1, Z, T)$.

Greene [26] already showed that the Tutte polynomial determines the weight enumerator, but not the other way round. By using the extended weight enumerator, we get a two-way equivalence and the proof reduces to rewriting.

We can also give expressions for the generalized weight enumerator in terms of the Tutte polynomial, and the other way round. The first formula was found by Britz [61] and independently by Jurrius [1].

Theorem 5.11. *For the generalized weight enumerator of an $[n, k]$ code C and the associated Tutte polynomial we have that $W_C^{(r)}(X, Y)$ is equal to*

$$\frac{1}{\langle r \rangle_q} \sum_{j=0}^{r} \begin{bmatrix} r \\ j \end{bmatrix}_q (-1)^{r-j} q^{\binom{j}{2}} (X-Y)^k Y^{n-k} \; t_C\left(\frac{X+(q^j-1)Y}{X-Y}, \frac{X}{Y}\right).$$

And, conversely,

$$t_C(X, Y) = Y^n(Y-1)^{-k} \sum_{r=0}^{k} \left(\prod_{j=0}^{r-1}((X-1)(Y-1) - q^j)\right) W_C^{(r)}(1, Y^{-1}).$$

Proof. For the first formula, use Theorems 5.7 and 5.9. Use Theorems 5.6 and 5.10 for the second formula. □

5.6.4. Deletion and contraction of matroids

Let $M = (E, \mathcal{I})$ be a matroid of rank k. Let e be an element of E. Then the *deletion* $M \setminus e$ is the matroid on the set $E \setminus \{e\}$ with independent sets of the form $I \setminus \{e\}$ where I is independent in M. The *contraction* M/e is the matroid on the set $E \setminus \{e\}$ with independent sets of the form $I \setminus \{e\}$ where I is independent in M and $e \in I$.

Let C be a code with reduced generator matrix G at position e. So $a = (1, 0, \ldots, 0)^T$ is the column of G at position e. Then $M \setminus e = M_{G \setminus a}$ and $M/e = M_{G/a}$. A puncturing-shortening formula for the extended weight enumerator is given in Proposition 5.26. By virtue of the fact that the extended weight enumerator and the Tutte polynomial of a code determine each other by Theorems 5.9 and 5.10, one expects that an analogous generalization for the Tutte polynomial of matroids holds.

Proposition 5.38. *Let $M = (E, \mathcal{I})$ be a matroid. Let $e \in E$ that is not a loop and not an isthmus. Then the following deletion-contraction formula holds:*

$$t_M(X, Y) = t_{M \setminus e}(X, Y) + t_{M/e}(X, Y).$$

Proof. See [25, 53, 65, 66]. □

Let M be a graphic matroid. So $M = M_\Gamma$ for some finite graph Γ. Let e be an edge of Γ, then $M \setminus e = M_{\Gamma \setminus e}$ and $M/e = M_{\Gamma/e}$.

5.6.5. MacWilliams type property for duality

For both codes and matroids we defined the dual structure. These objects obviously completely define their dual. But how about the various polynomials associated to a code and a matroid? We know from Example 5.30 that the weight enumerator is a less strong invariant for a code than the code itself: this means there are non-equivalent codes with the same weight enumerator. So it is a priori not clear that the weight enumerator of a code completely defines the weight enumerator of its dual code. We already saw that there is in fact such a relation, namely the MacWilliams identity in Theorem 5.2. We will give a proof of this relation by considering the more general question for the extended weight enumerator. We will prove the MacWilliams identities using the Tutte polynomial. We do this because of the following simple and very useful relation between the Tutte polynomial of a matroid and its dual.

Theorem 5.12. *Let $t_M(X,Y)$ be the Tutte polynomial of a matroid M, and let $M^\perp$ be the dual matroid. Then*

$$t_M(X,Y) = t_{M^\perp}(Y,X).$$

Proof. Let M be a matroid on the set E. Then $M^\perp$ is a matroid on the same set. In Proposition 5.30 we proved $r^\perp(J) = |J| - r(E) + r(E \setminus J)$. In particular, we have $r^\perp(E) + r(E) = |E|$. Substituting this relation into the definition of the Tutte polynomial for the dual code, gives

$$\begin{aligned} t_{M^\perp}(X,Y) &= \sum_{J \subseteq E} (X-1)^{r^\perp(E) - r^\perp(J)} (Y-1)^{|J| - r^\perp(J)} \\ &= \sum_{J \subseteq E} (X-1)^{r^\perp(E) - |J| - r(E\setminus J) + r(E)} (Y-1)^{r(E) - r(E\setminus J)} \\ &= \sum_{J \subseteq E} (X-1)^{|E\setminus J| - r(E\setminus J)} (Y-1)^{r(E) - r(E\setminus J)} \\ &= t_M(Y,X). \end{aligned}$$

In the last step, we use that the summation over all $J \subseteq E$ is the same as a summation over all $E \setminus J \subseteq E$. This proves the theorem. $\square$

If we consider a code as a matroid, then the dual matroid is the dual code. Therefore we can use the above theorem to prove the MacWilliams relations. Greene [26] was the first to use this idea, see also Brylawsky and Oxley [67].

Theorem 5.13 (MacWilliams). *Let C be a code and let $C^\perp$ be its dual. Then the extended weight enumerator of C completely determines the extended weight enumerator of $C^\perp$ and vice versa, via the following formula:*

$$W_{C^\perp}(X,Y,T) = T^{-k} W_C(X + (T-1)Y, X - Y, T).$$

Proof. Let G be the matroid associated to the code. Using the previous theorem and the relation between the weight enumerator and the Tutte polynomial, we find

$$T^{-k}W_C(X+(T-1)Y,X-Y,T)$$
$$=T^{-k}(TY)^k(X-Y)^{n-k}\,t_C\left(\frac{X}{Y},\frac{X+(T-1)Y}{X-Y}\right)$$
$$=Y^k(X-Y)^{n-k}\,t_{C^\perp}\left(\frac{X+(T-1)Y}{X-Y},\frac{X}{Y}\right)$$
$$=W_{C^\perp}(X,Y,T).$$

Notice in the last step that $\dim C^\perp = n-k$, and $n-(n-k)=k$. □

We can use the relations in Theorems 5.6 and 5.7 to prove the MacWilliams identities for the generalized weight enumerator.

Theorem 5.14. *Let C be a code and let $C^\perp$ be its dual. Then the generalized weight enumerators of C completely determine the generalized weight enumerators of $C^\perp$ and vice versa, via the following formula:*

$$W_{C^\perp}^{(r)}(X,Y) = \sum_{j=0}^{r}\sum_{l=0}^{j}(-1)^{r-j}\frac{q^{\binom{r-j}{2}}-j(r-j)-l(j-l)-jk}{\langle r-j\rangle_q\langle j-l\rangle_q}W_C^{(l)}(X+(q^j-1)Y,X-Y).$$

Proof. We write the generalized weight enumerator in terms of the extended weight enumerator, use the MacWilliams identities for the extended weight enumerator, and convert back to the generalized weight enumerator.

$$W_{C^\perp}^{(r)}(X,Y) = \frac{1}{\langle r\rangle_q}\sum_{j=0}^{r}\begin{bmatrix}r\\j\end{bmatrix}_q(-1)^{r-j}q^{\binom{r-j}{2}}\,W_{C^\perp}(X,Y,q^j)$$
$$=\sum_{j=0}^{r}(-1)^{r-j}\frac{q^{\binom{r-j}{2}}-j(r-j)}{\langle j\rangle_q\langle r-j\rangle_q}q^{-jk}W_c(X+(q^j-1)Y,X-Y,q^j)$$
$$=\sum_{j=0}^{r}(-1)^{r-j}\frac{q^{\binom{r-j}{2}}-j(r-j)-jk}{\langle j\rangle_q\langle r-j\rangle_q}$$
$$\times\sum_{l=0}^{j}\frac{\langle j\rangle_q}{q^{l(j-l)}\langle j-l\rangle_q}W_C^{(l)}(X+(q^j-1)Y,X-Y)$$
$$=\sum_{j=0}^{r}\sum_{l=0}^{j}(-1)^{r-j}\frac{q^{\binom{r-j}{2}}-j(r-j)-l(j-l)-jk}{\langle r-j\rangle_q\langle j-l\rangle_q}$$
$$\times W_C^{(l)}(X+(q^j-1)Y,X-Y).$$

□

This theorem was proved by Kløve [68], although the proof uses only half of the relations between the generalized weight enumerator and the extended weight enumerator. Using both makes the proof much shorter.

5.7. Posets and lattices

In this section we consider the theory of posets and lattices and the Möbius function. Geometric lattices are defined and its connection with matroids is given. See [30, 69–72].

5.7.1. Posets, the Möbius function and lattices

Let L be a set and $\leq$ a relation on L such that:

(PO.1) $x \leq x$, for all x in L *(reflexive)*.
(PO.2) If $x \leq y$ and $y \leq x$, then $x = y$, for all $x, y \in L$ *(anti-symmetric)*.
(PO.3) If $x \leq y$ and $y \leq z$, then $x \leq z$, for all x, y and z in L *(transitive)*.

The pair $(L, \leq)$, or just L, is called a *poset* with *partial order* $\leq$ on the set L. Define $x < y$ if $x \leq y$ and $x \neq y$. The elements x and y in L are *comparable* if $x \leq y$ or $y \leq x$. A poset L is called a *linear order* if every two elements are comparable. Define $L_x = \{y \in L : x \leq y\}$ and $L^x = \{y \in L : y \leq x\}$ and *the interval* between x and y by $[x, y] = \{z \in L : x \leq z \leq y\}$. Notice that $[x, y] = L_x \cap L^y$.

Let $(L, \leq)$ be a poset. A *chain of length r from x to y in L* is a sequence of elements $x_0, x_1, \ldots, x_r$ in L such that

$$x = x_0 < x_1 < \cdots < x_r = y.$$

Let $r \geq 0$ be an integer. Let $x, y \in L$. Then $c_r(x, y)$ denotes the *number of chains* of length r from x to y. Now $c_r(x, y)$ is finite if L is finite. The poset is called *locally finite* if $c_r(x, y)$ is finite for all $x, y \in L$ and every integer $r \geq 0$.

Proposition 5.39. *Let L be a locally finite poset. Let $x \leq y$ in L. Then*

(N.1) $c_0(x, y) = 0$ *if x and y are not comparable.*
(N.2) $c_0(x, x) = 1$, $c_r(x, x) = 0$ *for all $r > 0$ and $c_0(x, y) = 0$ if $x < y$.*
(N.3) $c_{r+1}(x, y) = \sum_{x \leq z < y} c_r(x, z) = \sum_{x < z \leq y} c_r(z, y).$

Proof. Statements (N.1) and (N.2) are trivial. Let $z < y$ and $x = x_0 < x_1 < \cdots < x_r = z$ a chain of length r from x to z, then $x = x_0 < x_1 < \cdots < x_r < x_{r+1} = y$ is a chain of length $r + 1$ from x to y, and every chain of length $r + 1$ from x to y is obtained uniquely in this way. Hence $c_{r+1}(x, y) = \sum_{x \leq z < y} c_r(x, z)$. The last equality is proved similarly. $\square$

Definition 5.4. The *Möbius function* of L, denoted by μ_L or μ is defined by

$$\mu(x, y) = \sum_{r=0}^{\infty} (-1)^r c_r(x, y).$$

Proposition 5.40. *Let L be a locally finite poset. Then for all $x, y \in L$:*

(M.1) $\mu(x, y) = 0$ *if x and y are not comparable.*
(M.2) $\mu(x, x) = 1$.
(M.3) *If $x < y$, then $\sum_{x \leq z \leq y} \mu(x, z) = \sum_{x \leq z \leq y} \mu(z, y) = 0$.*
(M.4) *If $x < y$, then $\mu(x, y) = -\sum_{x \leq z < y} \mu(x, z) = -\sum_{x < z \leq y} \mu(z, y)$.*

Proof. (M.1) and (M.2) follow from (N.1) and (N.2), respectively, of Proposition 5.39. (M.3) is clearly equivalent with (M.4). If $x < y$, then $c_0(x, y) = 0$. So

$$\mu(x, y) = \sum_{r=1}^{\infty} (-1)^r c_r(x, y)$$

$$= \sum_{r=0}^{\infty} (-1)^{r+1} c_{r+1}(x, y)$$

$$= -\sum_{r=0}^{\infty} (-1)^r \sum_{x \leq z < y} c_r(x, z)$$

$$= -\sum_{x \leq z < y} \sum_{r=0}^{\infty} (-1)^r c_r(x, z)$$

$$= -\sum_{x \leq z < y} \mu(x, z).$$

The first and last equalities use the definition of μ. The second equality starts counting at $r = 0$ instead of $r = 1$, the third uses (N.3) of Proposition 5.39 and in the fourth the order of summation is interchanged. $\square$

Remark 5.5. (M.2) and (M.4) of Proposition 5.40 can be used as an alternative way to compute $\mu(x, y)$ by induction.

Let L be a poset. If L has an element 0_L such that 0_L is the unique minimal element of L, then 0_L is called the *minimum* of L. Similarly 1_L is called the *maximum* of L if 1_L is the unique maximal element of L. If $x, y \in L$ and $x \leq y$, then the interval $[x, y]$ has x as minimum and y as maximum. Suppose that L has 0_L and 1_L as minimum and maximum, also denoted by 0 and 1, respectively. Then $0 \leq x \leq 1$ for all $x \in L$. Define $\mu(x) = \mu(0, x)$ and $\mu(L) = \mu(0, 1)$ if L is finite.

Let L be a locally finite poset with a minimum element. Let A be an abelian group and $f: L \to A$ a map from L to A. The *sum function* $\hat{f}$ of f is defined by

$$\hat{f}(x) = \sum_{y \leq x} f(y).$$

Define similarly the sum function $\check{f}$ of f by $\check{f}(x) = \sum_{x \leq y} f(y)$ if L is a locally finite poset with a maximum element.

A poset L is locally finite if and only if $[x, y]$ is finite for all $x \leq y$ in L. So $[0, x]$ is finite if L is a locally finite poset with minimum element 0. Hence the sum function $\hat{f}(x)$ is well-defined, since it is a finite sum of $f(y)$ in A with y in $[0, x]$. In the same way $\check{f}(x)$ is well-defined, since $[x, 1]$ is finite.

Theorem 5.15 (Möbius inversion formula). *Let L be a locally finite poset with a minimum element. Then*

$$f(x) = \sum_{y \leq x} \mu(y, x) \hat{f}(y).$$

Similarly $f(x) = \sum_{x \leq y} \mu(x, y) \check{f}(y)$ if L is a locally finite poset with a maximum element.

Proof. Let x be an element of L. Then

$$\sum_{y \leq x} \mu(y, x) \hat{f}(y) = \sum_{y \leq x} \sum_{z \leq y} \mu(y, x) f(z)$$
$$= \sum_{z \leq x} f(z) \sum_{z \leq y \leq x} \mu(y, x)$$
$$= f(x)\mu(x, x) + \sum_{z < x} f(z) \sum_{z \leq y \leq x} \mu(y, x)$$
$$= f(x).$$

The first equality uses the definition of $\hat{f}(y)$. In the second equality the order of summation is interchanged. In the third equality the first summation is split in the parts $z = x$ and $z < x$, respectively. Finally $\mu(x,x) = 1$ and the second summation is zero for all $z < x$, by Proposition 5.40.
The proof of the second equality is similar. □

Example 5.33. Let $f(x) = 1$ if $x = 0$ and $f(x) = 0$ otherwise. Then the sum function $\hat{f}(x) = \sum_{y \leq x} f(y)$ is constant 1 for all x. The Möbius inversion formula gives that

$$\sum_{y \leq x} \mu(x) = \begin{cases} 1 & \text{if } x = 0, \\ 0 & \text{if } x > 0, \end{cases}$$

which is a special case of Proposition 5.40.

Remark 5.6. Let $(L, \leq)$ be a poset. Let $\leq_R$ be the *reverse relation* on L defined by $x \leq_R y$ if and only if $y \leq x$. Then $(L, \leq_R)$ is a poset. Suppose that $(L, \leq)$ is locally finite with Möbius function μ. Then the number of chains of length r from x to y in $(L, \leq_R)$ is the same as the number of chains of length r from y to x in $(L, \leq)$. Hence $(L, \leq_R)$ is locally finite with Möbius function μ_R such that $\mu_R(x,y) = \mu(y,x)$. If $(L, \leq)$ has minimum 0_L or maximum 1_L, then $(L, \leq_R)$ has minimum 1_L or maximum 0_L, respectively.

Definition 5.5. Let L be a poset. Let $x, y \in L$. Then y is called a *cover* of x if $x < y$, and there is no z such that $x < z < y$. The *Hasse diagram* of L is a directed graph that has the elements of L as vertices, and there is a directed edge from y to x if and only if y is a cover of x.

Example 5.34. Let $L = \mathbb{Z}$ be the set of integers with the usual linear order. The Hasse diagram of this poset looks as follows:

$$\ldots \longrightarrow n+1 \longrightarrow n \longrightarrow n-1 \longrightarrow \ldots \longrightarrow 1 \longrightarrow 0 \longrightarrow -1 \longrightarrow \ldots$$

Let $x, y \in L$ and $x \leq y$. Then $c_0(x,x) = 1$, $c_0(x,y) = 0$ if $x < y$, and $c_r(x,y) = \binom{y-x-1}{r-1}$ for all $r \geq 1$. So L infinite and locally finite. Furthermore $\mu(x,x) = 1$, $\mu(x, x+1) = -1$ and $\mu(x,y) = 0$ if $y > x+1$.

Let L be a poset. Let $x, y \in L$. Then x and y have a *least upper bound* if there is a $z \in L$ such that $x \leq z$ and $y \leq z$, and if $x \leq w$ and $y \leq w$, then $z \leq w$ for all $w \in L$. If x and y have a least upper bound, then such an element is unique and it is called the *join* of x and y and denoted by $x \vee y$. Similarly the *greatest lower bound* of x and y is defined. If it exists,

then it is unique and it is called the *meet* of x and y and denoted by $x \wedge y$.
A poset L is called a *lattice* if $x \vee y$ and $x \wedge y$ exist for all $x, y \in L$.

Remark 5.7. Let $(L, \leq)$ be a finite poset with maximum 1 such that $x \wedge y$ exists for all $x, y \in L$. The collection $\{z : x \leq z, y \leq z\}$ is finite and not empty, since it contains 1. The meet of all the elements in this collection is well defined and is given by

$$x \vee y = \bigwedge \{z : x \leq z, y \leq z\}.$$

Hence L is a lattice. Similarly L is a lattice if L is a finite poset with minimum 0 such that $x \vee y$ exists for all $x, y \in L$, since $x \wedge y = \bigvee \{z : z \leq x, z \leq y\}$.

Example 5.35. Let L be the collection of all finite subsets of a given set $\mathcal{X}$. Let $\leq$ be defined by the inclusion, that means $I \leq J$ if and only if $I \subseteq J$. Then $0_L = \emptyset$, and L has a maximum if and only if $\mathcal{X}$ is finite in which case $1_L = \mathcal{X}$. For $\mathcal{X} = \{a, b, c, d\}$ the Hasse diagram of the poset is given in Figure 5.10.

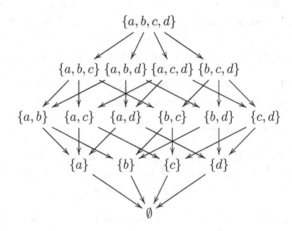

Fig. 5.10. The Hasse diagram of the poset of all subsets of $\{a, b, c, d\}$

Let $I, J \in L$ and $I \leq J$. Then $|I| \leq |J| < \infty$. Let $m = |J| - |I|$. Then

$$c_r(I, J) = \sum_{m_1 < m_2 < \cdots < m_{r-1} < m} \binom{m_2}{m_1}\binom{m_3}{m_2}\cdots\binom{m}{m_{r-1}}.$$

Hence L is locally finite. L is finite if and only if $\mathcal{X}$ is finite. Furthermore $I \vee J = I \cup J$ and $I \wedge J = I \cap J$. So L is a lattice. Using Remark 5.5 we see that $\mu(I, J) = (-1)^{|J|-|I|}$ if $I \leq J$. This is much easier than computing $\mu(I, J)$ by means of Definition 5.4.

Example 5.36. Let $\mathcal{X} = [n]$. Let k be an integer between 0 and n. Let $L_k = \{\mathcal{X}\}$ and L_i be the collection of all subsets of $\mathcal{X}$ of size i for all $i < k$. Let the partial order be given by the inclusion. Then L is a poset and $\mu(I, J) = (-1)^{|J|-|I|}$ if $I \leq J$ and $|J| < k$ as in Example 5.35, and $\mu(I, \mathcal{X}) = -\sum_{I \leq J < \mathcal{X}} (-1)^{|J|-|I|}$ for all $I < \mathcal{X}$ by Proposition 5.40.

Example 5.37. Now suppose again that $\mathcal{X} = [n]$. Let L be the poset of subsets of $\mathcal{X}$. Let $A_1, \ldots, A_n$ be a collection of subsets of a finite set A. Define for a subset J of $\mathcal{X}$

$$A_J = \bigcap_{j \in J} A_j \quad \text{and} \quad f(J) = \left| A_J \setminus \left(\bigcup_{I < J} A_I \right) \right|.$$

Then A_J is the disjoint union of the subsets $A_I \setminus (\bigcup_{K < I} A_K)$ for all $I \leq J$. Hence the sum function is equal to

$$\hat{f}(J) = \sum_{I \leq J} f(I) = \sum_{I \leq J} |A_I \setminus \left(\bigcup_{K < I} A_K \right) | = |A_J|.$$

Möbius inversion gives that

$$\left| A_J \setminus \left(\bigcup_{I < J} A_I \right) \right| = \sum_{I \leq J} (-1)^{|J|-|I|} |A_I|,$$

which is called the *principle of inclusion/exclusion*.

Example 5.38. A variant of the principle of inclusion/exclusion is given as follows. Let $H_1, \ldots, H_n$ be a collection of subsets of a finite set H. Let L be the poset of all intersections of the H_j with the reverse inclusion as partial order. Then H is the minimum of L and $H_1 \cap \cdots \cap H_n$ is the maximum of L. Let $x \in L$. Define

$$f(x) = \left| x \setminus \left(\bigcup_{x < y} y \right) \right|.$$

Then
$$\tilde{f}(x) = \sum_{x \leq y} f(y) = \sum_{x \leq y} |y \setminus \left(\bigcup_{y<z} z \right)| = |x|.$$

Hence
$$|x \setminus \left(\bigcup_{x<y} y \right)| = \sum_{x \leq y} \mu(x,y)|y|.$$

Example 5.39. Let $L = \mathbb{N}$ be the set of positive integers with the divisibility relation as partial order. Then $0_L = 1$ is the minimum of L, it is locally finite and it has no maximum. Now $m \vee n = \mathrm{lcm}(m,n)$ and $m \wedge n = \gcd(m,n)$. Hence L is a lattice. By Remark 5.5 we see that

$$\mu(n) = \begin{cases} 1 & \text{if } n = 1; \\ (-1)^r & \text{if } n \text{ is the product of } r \text{ mutually distinct primes}; \\ 0 & \text{if } n \text{ is divisible by the square of a prime}. \end{cases}$$

Hence $\mu(n)$ is the classical Möbius function. Furthermore, $\mu(d,n) = \mu(\frac{n}{d})$ if $d|n$. Let
$$\varphi(n) = |\{i \in \mathbb{N} : \gcd(i,n) = 1\}|$$
be *Euler's φ function*. Define
$$V_d = \{i \in [n] : \gcd(i,n) = \tfrac{n}{d}\}$$
for $d|n$. Then
$$\{ i \cdot \tfrac{n}{d} : i \in [d] , \gcd(i,d) = 1 \} = V_d$$
so $|V_d| = \varphi(d)$. Now $[n]$ is the disjoint union of the subsets V_d with $d|n$. Hence the sum function of $\varphi(n)$ is given by
$$\hat{\varphi}(n) = \sum_{d|n} \varphi(d) = n.$$
Therefore by Möbius inversion
$$\varphi(n) = \sum_{d|n} \mu(d)\frac{n}{d}.$$

Example 5.40. Consider the poset L of Example 5.39 with the divisibility as partial order. Let $\mathrm{Irr}_q(n)$ be the number of irreducible monic polynomials over $\mathbb{F}_q$ of degree n. Define $f(d) = d \cdot \mathrm{Irr}_q(d)$. Then the sum function

$\hat{f}(n) = \sum_{d|n} f(d)$ is equal to q^n. See [73, Corollary 3.21]. The Möbius inversion formula of Theorem 5.15 implies that

$$\text{Irr}_q(n) = \frac{1}{n} \sum_{d|n} \mu\left(\frac{n}{d}\right) q^d.$$

Let $(L_1, \leq_1)$ and $(L_2, \leq_2)$ be posets. A map $\varphi : L_1 \to L_2$ is called *monotone* if $\varphi(x) \leq_2 \varphi(y)$ for all $x \leq_1 y$ in L_1. The map φ is called *strictly monotone* if $\varphi(x) <_2 \varphi(y)$ for all $x <_1 y$ in L_1. The map is called an *isomorphism of posets* if it is strictly monotone and there exists a strictly monotone map $\psi : L_2 \to L_1$ that is the inverse of φ. The posets are called *isomorphic* if there is an isomorphism of posets between them.

If $\varphi : L_1 \to L_2$ is an isomorphism between locally finite posets with a minimum, then $\mu_2(\varphi(x), \varphi(y)) = \mu_1(x, y)$ for all x, y in L_1. If $(L_1, \leq_1)$ and $(L_2, \leq_2)$ are isomorphic posets and L_1 is a lattice, then L_2 is also a lattice.

Example 5.41. Let n be a positive integer that is the product of r mutually distinct primes $p_1, \ldots, p_r$. Let L_1 be the set of all positive integers that divide n with divisibility as partial order $\leq_1$ as in Example 5.39. Let L_2 be the collection of all subsets of $[r]$ with the inclusion as partial order $\leq_2$ as in Example 5.35. Define the maps $\varphi : L_1 \to L_2$ and $\psi : L_2 \to L_1$ by $\varphi(d) = \{i : p_i \text{ divides } n\}$ and $\psi(x) = \prod_{i \in x} p_i$. Then φ and ψ are strictly monotone and they are inverses of each other. Hence L_1 and L_2 are isomorphic lattices.

5.7.2. Geometric lattices

Let $(L, \leq)$ be a lattice without infinite chains. Then L has a minimum and a maximum. Let L be a lattice with minimum 0. An *atom* is an element $a \in L$ that is a cover of 0. A lattice is called *atomic* if for every $x > 0$ in L there exist atoms $a_1, \ldots, a_r$ such that $x = a_1 \vee \cdots \vee a_r$. The minimum length of a chain from 0 to x is called the *rank* of x and is denoted by $r_L(x)$ or $r(x)$ for short. A lattice is called *semimodular* if for all mutually distinct $x, y \in L$, $x \vee y$ covers x and y if there exists a z such that x and y cover z. A lattice is called *modular* if $x \vee (y \wedge z) = (x \vee y) \wedge z$ for all $x, y, z \in L$ such that $x \leq z$. A lattice L is called a *geometric lattice* if it is atomic and semimodular and has no infinite chains. If L is a geometric lattice L, then it has a minimum and a maximum and $r(1)$ is called the rank of L and is denoted by $r(L)$.

Example 5.42. Let L be the collection of all finite subsets of a given set

$\mathcal{X}$ as in Example 5.35. The atoms are the singleton sets, that is subsets consisting of exactly one element of $\mathcal{X}$. Every $x \in L$ is the finite union of its singleton subsets. So L is atomic and $r(x) = |x|$. Now y covers x if and only if there is an element Q not in x such that $y = x \cup \{Q\}$. If $x \neq y$ and x and y both cover z, then there is an element P not in z such that $x = z \cup \{P\}$, and there is an element Q not in z such that $y = z \cup \{Q\}$. Now $P \neq Q$, since $x \neq y$. Hence $x \vee y = z \cup \{P, Q\}$ covers x and y. Hence L is semimodular. In fact L is modular. L is locally finite. L is a geometric lattice if and only if $\mathcal{X}$ is finite.

Example 5.43. Let L be the set of positive integers with the divisibility relation as in Example 5.39. The atoms of L are the primes. But L is not atomic, since a square is not the join of finitely many elements. L is semimodular. The interval $[1, n]$ in L is a geometric lattice if and only if n is square free. If n is square free and $m \leq n$, then $r(m) = r$ if and only if m is the product of r mutually distinct primes.

Let L be a geometric lattice. Let $x, y \in L$ and $x \leq y$. The chain $x = y_0 < y_1 < \cdots < y_s = y$ from x to y is called an *extension* of the chain $x = x_0 < x_1 < \cdots < x_r = y$ if $\{x_0, x_1, \ldots, x_r\}$ is a subset of $\{y_0, y_1, \ldots, y_s\}$. A chain from x to y is called *maximal* if there is no extension to a longer chain from x to y.

Proposition 5.41. *Let L be a geometric lattice. Then for all $x, y \in L$:*

(GL.1) *If $x < y$, then $r(x) < r(y)$* (strictly monotone)
(GL.2) *$r(x \vee y) + r(x \wedge y) \leq r(x) + r(y)$* (semimodular inequality)
(GL.3) *If $x \leq y$, then every chain from x to y can be extended to a maximal chain with the same end points, and all such maximal chains have the same length $r(y) - r(x)$* (Jordan-Hölder property).

Proof. See [30, Prop. 3.3.2] and [72, Prop. 3.7]. □

Let L be an atomic lattice. Then L is semimodular if and only if the semimodular inequality (GL.2) holds for all $x, y \in L$. And L is modular if and only if the *modular equality* holds for all $x, y \in L$:

$$r(x \vee y) + r(x \wedge y) = r(x) + r(y).$$

Then the *Hasse diagram* of L is a graph that has the elements of L as vertices. If $x, y \in L$, $x < y$ and $r(y) = r(x) + 1$, then x and y are connected by an edge. So only elements between two consecutive levels L_j and L_{j+1} are connected by an edge. The Hasse diagram of L considered as a poset as in Definition 5.5 is the directed graph with an arrow from y to x if $x, y \in L$, $x < y$ and $r(y) = r(x) + 1$.

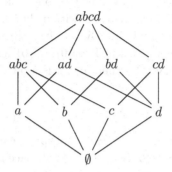

Fig. 5.11. The Hasse diagram of the geometric lattice in Example 5.48

Let L be a geometric lattice. Then L_x is a geometric lattice with x as minimum element and of rank $r_L(1) - r_L(x)$, and $\mu_{L_x}(y) = \mu(x, y)$ and $r_{L_x}(y) = r_L(y) - r_L(x)$ for all $x \in L$ and $y \in L_x$. Similar remarks hold for L^x and $[x, y]$.

Example 5.44. Let L be the collection of all linear subspaces of a given finite dimensional vector space V over a field $\mathbb{F}$ with the inclusion as partial order. Then $0_L = \{0\}$ is the minimum and $1_L = V$ is the maximum of L. The partial order L is locally finite if and only if L is finite if and only if the field $\mathbb{F}$ is finite. Let x and y be linear subspaces of V. Then $x \cap y$ the intersection of x and y is the largest linear subspace that is contained in x and y. So $x \wedge y = x \cap y$. The sum $x + y$ of x and y is by definition the set of elements $a + b$ with a in x and b in y. Then $x + y$ is the smallest linear subspace containing both x and y. Hence $x \vee y = x + y$. So L is a lattice. The atoms are the one dimensional linear subspaces. Let x be a subspace of dimension r over $\mathbb{F}$. So x is generated by a basis $\mathbf{g}_1, \ldots, \mathbf{g}_r$. Let a_i be the one dimensional subspace generated by $\mathbf{g}_i$. Then $x = a_1 \vee \cdots \vee a_r$. Hence L is atomic and $r(x) = \dim(x)$. Moreover L is modular, since

$$\dim(x \cap y) + \dim(x + y) = \dim(x) + \dim(y)$$

for all $x, y \in L$. Furthermore L has no infinite chains, since V is finite dimensional. Therefore L is a modular geometric lattice.

Example 5.45. Let $\mathbb{F}$ be a field. Let $\mathcal{V} = (\mathbf{v}_1, \ldots, \mathbf{v}_n)$ be an n-tuple of nonzero vectors in $\mathbb{F}^k$. Let $L = L(\mathcal{V})$ be the collection of all linear subspaces of $\mathbb{F}^k$ that are generated by subsets of $\mathcal{V}$ with inclusion as partial order. So L is finite and a fortiori locally finite. By definition $\{0\}$ is the linear subspace space generated by the empty set. Then $0_L = \{0\}$ and 1_L is the subspace generated by all $\mathbf{v}_1, \ldots, \mathbf{v}_n$. Furthermore L is a lattice with $x \vee y = x + y$ and
$$x \wedge y = \bigvee \{z : z \leq x, z \leq y\}$$
by Remark 5.7. Let a_j be the linear subspace generated by $\mathbf{v}_j$. Then $a_1, \ldots, a_n$ are the atoms of L. Let x be the subspace generated by $\{\mathbf{v}_j : j \in J\}$. Then $x = \bigvee_{j \in J} a_j$. If x has dimension r, then there exists a subset I of J such that $|I| = r$ and $x = \bigvee_{i \in I} a_i$. Hence L is atomic and $r(x) = \dim(x)$. Now $x \wedge y \subseteq x \cap y$, so
$$r(x \vee y) + r(x \wedge y) \leq \dim(x + y) + \dim(x \cap y) = r(x) + r(y).$$
Hence the semimodular inequality holds and L is a geometric lattice. In most cases L is not modular.

Example 5.46. Let $\mathbb{F}$ be a field. Let $\mathcal{A} = (H_1, \ldots, H_n)$ be an arrangement over $\mathbb{F}$ of hyperplanes in the vector space $V = \mathbb{F}^k$. Let $L = L(\mathcal{A})$ be the collection of all nonempty intersections of elements of $\mathcal{A}$. By definition $\mathbb{F}^k$ is the empty intersection. Define the partial order $\leq$ by
$$x \leq y \text{ if and only if } y \subseteq x.$$
Then V is the minimum element and $\{0\}$ is the maximum element. Furthermore
$$x \vee y = x \cap y \text{ if } x \cap y \neq \emptyset, \text{ and } x \wedge y = \bigcap \{z : x \cup y \subseteq z\}.$$
Suppose that $\mathcal{A}$ is a central arrangement. Then $x \cap y$ is nonempty for all $x, y \in L$. So $x \vee y$ and $x \wedge y$ exist for all $x, y \in L$, and L is a lattice. Let $\mathbf{v}_j = (v_{1j}, \ldots, v_{kj})$ be a nonzero vector such that $\sum_{i=1}^k v_{ij} X_i = 0$ is a homogeneous equation of H_j. Let $\mathcal{V} = (\mathbf{v}_1, \ldots, \mathbf{v}_n)$. Consider the map $\varphi : L(\mathcal{V}) \to L(\mathcal{A})$ defined by
$$\varphi(x) = \bigcap_{j \in J} H_j \text{ if } x \text{ is the subspace generated by } \{\mathbf{v}_j : j \in J\}.$$

Now $x \subset y$ if and only if $\varphi(y) \subset \varphi(x)$ for all $x, y \in L(\mathcal{V})$. So φ is a strictly monotone map. Furthermore φ is a bijection and its inverse map is also strictly monotone. Hence $L(\mathcal{V})$ and $L(\mathcal{A})$ are isomorphic lattices. Therefore $L(\mathcal{A})$ is also a geometric lattice.

Example 5.47. Let G be a generator matrix of the simplex code $\mathcal{S}_r(q)$ of Example 5.22. Let $\mathcal{A} = \mathcal{A}_G$ be the arrangement of the matrix G. Then the projective hyperplanes of the arrangement of $\mathcal{A}$ are all the $(q^r - 1)/(q - 1)$ hyperplanes of $\mathbb{P}^{r-1}(\mathbb{F}_q)$. The geometric lattice $L(\mathcal{A})$ consists of all possible intersections of these hyperplanes, so they are all projective subspaces of $\mathbb{P}^{r-1}(\mathbb{F}_q)$ with the reverse inclusion as partial order. This geometric order is self dual, that means that it is isomorphic under map that sends points to hyperplanes with the inclusion and the reverse inclusion as partial orders. In this way we see that the geometric lattice of the simplex code $\mathcal{S}_r(q)$ is isomorphic with the geometric lattice of all linear subspaces of a given vector space V of Example 5.44 with $\mathbb{F} = \mathbb{F}_q$ and $V = \mathbb{F}_q^r$.

Example 5.48. Consider the following matrix over a field $\mathbb{F}$.

$$\begin{pmatrix} 1 & 1 & 0 & 0 \\ 0 & 0 & 0 & 1 \\ 0 & 1 & 1 & 0 \end{pmatrix}.$$

Denote the columns of the matrix by a, b, c, d respectively. We can interpret the columns either as vectors in $\mathbb{F}^3$, or as the coefficients of hyperplanes in $\mathbb{F}^3$. The corresponding lattices will be isomorphic by Examples 5.45 and 5.46. The corresponding Hasse diagram is given in Figure 5.11.

5.7.3. Geometric lattices and matroids

The notion of a geometric lattice is "*cryptomorphic*", that is almost equivalent to the concept of a matroid. See [35, 63, 69, 72, 74].

Proposition 5.42. *Let L be a finite geometric lattice. Let $M(L)$ be the set of all atoms of L. Let $\mathcal{I}(L)$ be the collection of all subsets I of $M(L)$ such that $r(a_1 \vee \cdots \vee a_r) = r$ if $I = \{a_1, \ldots, a_r\}$ is a collection of r atoms of L. Then $(M(L), \mathcal{I}(L))$ is a matroid.*

Proof. The proof is left as an exercise. $\square$

Let C be a projective code with generator matrix G. Then $\mathcal{A}_G$ is an essential simple arrangement with geometric lattice $L(\mathcal{A}_G)$. Furthermore the matroids $M(L(\mathcal{A}_G))$ and M_C are isomorphic.

Let $M = (E, \mathcal{I})$ be a matroid. A *k-flat* of M is a maximal subset of E of rank k. Let $L(M)$ be the collection of all flats of M, it is called the *lattice of flats* of M. Let J be a subset of E. Then the *closure* $\bar{J}$ is by definition the intersection of all flats that contain J. Flats of size $k-1$ are sometimes called *hyperplanes*, but in this chapter we will avoid this terminology.

The whole set E is a k-flat with $k = r(M)$. If F_1 and F_2 are flats, then $F_1 \cap F_2$ is also a flat. Consider $L(M)$ with the inclusion as partial order. Then E is the maximum of $L(M)$ and $F_1 \cap F_2 = F_1 \wedge F_2$ for all F_1 and F_2 in $L(M)$. Hence $L(M)$ is indeed a lattice by Remark 5.7. Let J be a subset of E, then $\bar{J}$ is a flat, since it is a nonempty, finite intersection of flats. So $\bar{\emptyset}$ is the minimum of $L(M)$.

An element x in E is a loop if and only if $\bar{x} = \bar{\emptyset}$. If $x, y \in E$ are not loops, then x and y are parallel if and only if $\bar{x} = \bar{y}$. Let $\bar{E} = \{\bar{x} : x \in M, \bar{x} \neq \bar{\emptyset}\}$. Let $\bar{\mathcal{I}} = \{\bar{I} : I \in \mathcal{I}, \bar{\emptyset} \notin \bar{I}\}$. Then $\bar{M} = (\bar{E}, \bar{\mathcal{I}})$ is a simple matroid.

Let G be a generator matrix of a code C. The *simplified matrix* $\bar{G}$ is the matrix obtained from G by deleting all zero columns from G and all columns that are a scalar multiple of a previous column. The *simplified code* $\bar{C}$ of C is the code with generator matrix $\bar{G}$.

Remark 5.8. Let G be a generator matrix of a code C. The definition of the simplified code $\bar{C}$ by means of $\bar{G}$ does not depend on the choice of the generator matrix G of C. The matroids $\bar{M}_G$ and $M_{\bar{G}}$ are isomorphic.

Let J be a subset of $[n]$. Then the closure $\bar{J}$ is equal to the complement in $[n]$ of the support of $C(J)$ and $C(J) = C(\bar{J})$.

Proposition 5.43. *Let M be a matroid. Then $L(M)$ with the inclusion as partial order is a geometric lattice and $L(M)$ is isomorphic with $L(\bar{M})$.*

Proof. See [75] and [72, Theorem 3.8]. □

Example 5.49. The geometric lattice of the matroid $U_{n,k}$ is isomorphic with the lattice consisting of $[n]$ and all its subsets of size at most $k-1$ of Example 5.36.

5.8. The characteristic polynomial

The characteristic polynomial of geometric lattices is defined. This is generalized in two variable polynomials in two ways: the coboundary polynomial and the Möbius polynomial. For simple matroids and codes the coboundary polynomial is equivalent to the Tutte polynomial and the extended weight enumerator. The Möbius polynomial contains information on the number of minimal subcodes and codewords. The coboundary and Möbius polynomial do not determine each other. This will be shown by examples of three dimensional codes.

5.8.1. The characteristic and coboundary polynomial

Let L be a finite geometric lattice. The *characteristic polynomial* $\chi_L(T)$ and the *Poincaré polynomial* $\pi_L(T)$ of L are defined by:

$$\chi_L(T) = \sum_{x \in L} \mu_L(x) T^{r(L)-r(x)} \text{ and } \pi_L(T) = \sum_{x \in L} \mu_L(x)(-T)^{r(x)}.$$

The *two variable characteristic polynomial* or *coboundary polynomial* is defined by

$$\chi_L(S,T) = \sum_{x \in L} \sum_{x \leq y \in L} \mu(x,y) S^{a(x)} T^{r(L)-r(y)},$$

where $a(x)$ is the number of atoms a in L such that $a \leq x$.

Now $\chi_L(1) = 0$ if and only if L consists of one element. Furthermore $\chi_L(T) = T^{r(L)} \pi_L(-T^{-1})$ and $\chi_L(0,T) = \chi_L(T)$.

Remark 5.9. Let n be the number of atoms of L. Then the following relation holds for the coboundary polynomial in terms of characteristic polynomials:

$$\chi_L(S,T) = \sum_{i=0}^{n} S^i \chi_i(T) \text{ with } \chi_i(T) = \sum_{\substack{x \in L \\ a(x)=i}} \chi_{L_x}(T).$$

$\chi_i(T)$ is called the *i-defect* polynomial. See [63, 76].

Example 5.50. Let L be the lattice of all subsets of a given finite set of r elements as in Examples 5.35 and 5.42. Then $r(x) = a(x)$ and $\mu(x,y) = (-1)^{a(y)-a(x)}$ if $x \leq y$. Hence

$$\chi_L(T) = \sum_{j=0}^{r} \binom{r}{j}(-1)^j T^{r-j} = (T-1)^r \text{ and } \chi_i(T) = \binom{r}{i}(T-1)^{r-i}.$$

Therefore $\chi_L(S,T) = (S+T-1)^r$.

Example 5.51. Let L be the lattice of the simplex code, that is of all linear subspaces of a given vector space of dimension r over the finite field $\mathbb{F}_q$ as in Example 5.47. Then the number of atoms of L is $n = \frac{q^r-1}{q-1}$ and $r(x)$ is the dimension of x over $\mathbb{F}_q$. The number of subspaces of dimension i is counted in Remark 5.4. It is left as an exercise to show that

$$\mu(x,y) = (-1)^i q^{(j-i)(j-i-1)/2}$$

if $r(x) = i$, $r(y) = j$ and $x \leq y$, and

$$\chi_L(T) = \sum_{i=0}^{r} \begin{bmatrix} r \\ i \end{bmatrix}_q (-1)^i q^{\binom{i}{2}} T^{r-i}$$
$$= (T-1)(T-q) \cdots (T-q^{r-1}).$$

See [19] and the proof of Theorem 5.7. More generally if $0 \leq i \leq \frac{q^r-1}{q-1}$ and $0 \leq j \leq r$, then

$$\chi_i(T) = \begin{cases} 0 & \text{if } \frac{q^{j-1}-1}{q-1} < i < \frac{q^j-1}{q-1}, \\ 1 & \text{if } i = \frac{q^r-1}{q-1}, \\ \begin{bmatrix} r \\ j \end{bmatrix}_q \prod_{l=0}^{r-j-1}(T-q^l) & \text{if } i = \frac{q^j-1}{q-1}, j < r. \end{cases}$$

See [77, Prop. 3.3].

Proposition 5.44 (Rota's Crosscut Theorem). *Let L be a finite geometric lattice. Let $M(L) = (E, \mathcal{I})$ be the matroid associated with L. Then*

$$\chi_L(T) = \sum_{J \subseteq E} (-1)^{|J|} T^{r(L)-r(J)}.$$

Proof. See [71] and [78, Theorem 3.1]. □

As a consequence of Proposition 5.44 we have the following way to describe the characteristic polynomial of L in terms of the Tutte polynomial of $M(L)$:

$$\chi_L(T) = (-1)^{r(L)} t_{M(L)}(1-T, 0).$$

Theorem 5.16. *The two-variable characteristic or coboundary polynomial $\chi_L(S,T)$ of a finite geometric lattice L is related to the Whitney rank generating function of $M(L)$ by the formula*

$$\chi_L(S,T) = (S-1)^{r(L)} R_{M(L)}\left(\frac{T}{S-1}, S-1\right).$$

Proof. See [79, p. 605] and [76, Theorem II]. □

Remark 5.10. Because of Theorem 5.16 we have the following relations between $t_{M(L)}(X,Y)$ and $\chi_L(S,T)$:

$$\chi_L(S,T) = (S-1)^{r(L)} t_{M(L)}\left(\frac{S+T-1}{S-1}, S\right)$$

and, vice versa,

$$t_{M(L)}(X,Y) = (Y-1)^{-r(L)} \chi_L(Y, (X-1)(Y-1)).$$

Therefore the polynomials $\chi_L(S,T)$ and $t_{M(L)}(X,Y)$ completely determine each other.

Starting with an arbitrary matroid M one has the associated geometric lattice $L(M)$, but $M(L(M))$ is isomorphic with M if and only if M is simple by Proposition 5.43. Therefore $t_M(X,Y)$ and $\chi_{L(M)}(S,T)$ completely determine each other if M is simple, but $t_M(X,Y)$ is a stronger invariant than $\chi_{L(M)}(S,T)$ if M is not simple. We will see a counterexample in Example 5.59.

Remark 5.11. The relation between $t_{M(L)}(X,Y)$ and $\chi_L(S,T)$ shows great similarity with the formula in Theorem 5.10. Combining the relations we find that for projective codes

$$\chi_i(T) = A_{n-i}(T)$$

for all $0 \leq i \leq n$.

Let $\Gamma = (V, E)$ be a finite simple graph. Let χ_Γ be the characteristic polynomial of the geometric lattice $L(M_\Gamma)$. Then for all positive integers k, $P_\Gamma(k) = \chi_\Gamma(k)$. So the chromatic polynomial of a graph is the prime example of a characteristic polynomial and the two-variable characteristic polynomial of a graph is also called the *dichromatic* polynomial of the graph. See [51, 53, 65].

Let γ be a coloring of Γ. Then an edge is called *bad* if it joins two vertices with the same color. The i-defect polynomial $\chi_i(T)$ counts up to a factor of T the number of ways of coloring Γ with i bad edges. See [63, §6.3.F].

5.8.2. The Möbius polynomial and Whitney numbers

Let L be a finite geometric lattice. The *two-variable Möbius polynomial* $\mu_L(S,T)$ in S and T is defined by

$$\mu_L(S,T) = \sum_{x \in L} \sum_{x \leq y \in L} \mu(x,y) S^{r(x)} T^{r(L)-r(y)}.$$

Now $\mu(L) = \chi_L(0)$ and $\mu_L(0,T) = \chi_L(0,T) = \chi_L(T)$.

Remark 5.12. Let r be the rank of L. Then the following relation holds for the Möbius polynomial in terms of characteristic polynomials

$$\mu_L(S,T) = \sum_{i=0}^{r} S^i \mu_i(T) \quad \text{with} \quad \mu_i(T) = \sum_{x \in L_i} \chi_{L_x}(T).$$

The Möbius polynomial was introduced by Zaslavsky [80, Section 1] for hyperplane arrangements and for signed graph colorings in [81, Section 2] where it is called the Whitney polynomial. See also [57].

Example 5.52. In Examples 5.42 and 5.50 we considered the lattice L of all subsets of a given finite set of r elements. Since $r(x) = a(x)$ for all $x \in L$, the Möbius polynomial of L is equal to the coboundary polynomial of L, so $\mu_L(S,T) = (S+T-1)^r$.

Example 5.53. Let L be the lattice of all linear subspaces of a given vector space of dimension r over the finite field $\mathbb{F}_q$ as in Example 5.44. In Example 5.51 we calculated the characteristic polynomial of this lattice. In the same way, we find that

$$\mu_i(T) = \begin{bmatrix} r \\ i \end{bmatrix}_q (T-1)(T-q) \cdots (T-q^{r-i-1}).$$

Remark 5.13. Let L be a geometric lattice. Then

$$\sum_{i=0}^{r(L)} \mu_i(T) = \mu_L(1,T)$$

$$= \sum_{y \in L} \sum_{0 \leq x \leq y} \mu(x,y) T^{r(L)-r(y)}$$

$$= T^{r(L)}$$

since $\sum_{0 \leq x \leq y} \mu(x,y) = 0$ for all $0 < y$ in L by Proposition 5.40. Similarly $\sum_{i=0}^{n} \chi_i(T) = \chi_L(1,T) = T^{r(L)}$. Also $\sum_{w=0}^{n} A_w(T) = T^k$ for the extended weights of a code of dimension k by Propositions 5.21 and 5.22 for $t=0$.

Example 5.54. Let L be the lattice consisting of $[n]$ and all its subsets of size at most $k-1$ as in Example 5.36, which is also the lattice of the uniform matroid $U_{n,k}$ and the lattice of an MDS code with parameters $[n, k, n-k+1]$. Then $\mu_i(T)$ and $\chi_i(T)$ are both equal to $\binom{n}{i}(T-1)^{n-i}$ for all $i < k$ as in Example 5.50, and $\chi_i(T) = 0$ for all $k \leq i < n$, since $a(1_L) = n$, $r(1_L) = k$ and $a(x) = r(x)$ for all x in L_i and $i < k$. Remark 5.13 implies

$$\mu_k(T) = T^k - \sum_{i<k} \binom{n}{i}(T-1)^{n-i} \text{ and } \chi_n(T) = T^k - \sum_{i<k} \binom{n}{i}(T-1)^{n-i}.$$

Let L be a finite geometric lattice. The *Whitney numbers* w_i and W_i of *the first and second kind*, respectively, are defined by

$$w_i = \sum_{x \in L_i} \mu(x) \text{ and } W_i = |L_i|.$$

The *doubly indexed Whitney numbers* w_{ij} and W_{ij} of *the first and second kind*, respectively, are defined by

$$w_{ij} = \sum_{x \in L_i} \sum_{y \in L_j} \mu(x,y) \text{ and } W_{ij} = |\{(x,y) : x \in L_i, y \in L_j, x \leq y\}|.$$

In particular $w_j = w_{0j}$ and $W_j = W_{0j}$. See [82] and [63, §6.6.D] and [35, Chapter 14] and [30, §3.11].

Remark 5.14. We have that

$$\chi_L(T) = \sum_{i=0}^{r(L)} w_i T^{r(L)-i} \text{ and } \mu_L(S,T) = \sum_{i=0}^{r(L)} \sum_{j=0}^{r(L)} w_{ij} S^i T^{r(L)-j}.$$

Hence the (doubly indexed) Whitney numbers of the first kind are determined by $\mu_L(S,T)$. The leading coefficient of

$$\mu_i(T) = \sum_{x \in L_i} \sum_{x \leq y} \mu(x,y) T^{r(L_x) - r_{L_x}(y)}$$

is equal to $\sum_{x \in L_i} \mu(x,x) = |L_i| = W_i$. Hence the Whitney numbers of the second kind W_i are also determined by $\mu_L(S,T)$. We will see in Example 5.60 that the Whitney numbers are not determined by $\chi_L(S,T)$. Finally, let $r = r(L)$. Then

$$\mu_{r-1}(T) = (T-1) \cdot W_{r-1}.$$

5.8.3. *Minimal codewords and subcodes*

A *minimal codeword* of a code C is a codeword whose support does not properly contain the support of another codeword.

The zero word is a minimal codeword. Notice that a nonzero scalar multiple of a minimal codeword is again a minimal codeword. Nonzero minimal codewords play a role in minimum distance decoding [11, 58, 83] and secret sharing schemes and access structures [84, 85]. We can generalize this notion to subcodes instead of words.

A *minimal subcode of dimension* r of a code C is an r-dimensional subcode whose support is not properly contained in the support of another r-dimensional subcode.

A minimal codeword generates a minimal subcode of dimension one, and all the elements of a minimal subcode of dimension one are minimal codewords. A codeword of minimal weight is a nonzero minimal codeword, but the converse is not always the case.

In Example 5.60 we will see two codes that have the same Tutte polynomial, but a different number of minimal codewords. Hence the number of minimal codewords and subcodes is not determined by the Tutte polynomial. However, the number of minimal codewords and the number of minimal subcodes of a given dimension are given by the Möbius polynomial.

Theorem 5.17. *Let C be a code of dimension k. Let $0 \leq r \leq k$. Then the number of minimal subcodes of dimension r is equal to W_{k-r}, the $(r-k)$-th Whitney number of the second kind, and it is determined by the Möbius polynomial.*

Proof. Let D be a subcode of C of dimension r. Let J be the complement in $[n]$ of the support of D. If $\mathbf{d} \in D$ and $d_j \neq 0$, then $j \in \text{supp}(D)$ and $j \notin J$. Hence $D \subseteq C(J)$. Now suppose moreover that D is a minimal subcode of C. Without loss of generality we may assume that D is systematic at the first r positions. So D has a generator matrix of the form $(I_r|A)$. Let $\mathbf{d}_i$ be the i-th row of this matrix. Let $\mathbf{c} \in C(J)$. If $\mathbf{c} - \sum_{i=1}^{r} c_i \mathbf{d}_i$ is not the zero word, then the subcode D' of C generated by $\mathbf{c}, \mathbf{d}_2, \ldots, \mathbf{d}_r$ has dimension r and its support is contained in $\text{supp}(D) \setminus \{1\}$ and $1 \in \text{supp}(D)$. This contradicts the minimality of D. Hence $\mathbf{c} - \sum_{i=1}^{r} c_i \mathbf{d}_i = 0$ and $\mathbf{c} \in D$. Therefore $D = C(J)$.

To find a minimal subcode of dimension r, we fix $l(J) = r$ and minimize the support of $C(J)$ with respect to inclusion. Because J is contained in the complement in $[n]$ of the support of $C(J)$, this is equivalent to maximize J with respect to inclusion. In matroid terms this means we are maximizing J for $r(J) = k - l(J) = k - r$. This means $J = \bar{J}$ is a flat of rank $k - r$ by Remark 5.8. The flats of a matroid are the elements in the geometric lattice $L = L(M)$. The number of $(k-r)$-dimensional elements in $L(M)$ is equal to $|L_{k-r}|$, which is equal to the Whitney number of the second kind W_{k-r} and thus equal to the leading coefficient of $\mu_{k-r}(T)$ by Remark 5.14. Hence the Möbius polynomial determines all the numbers of minimal subcodes of dimension r for $0 \leq r \leq k$. □

Note that the flats of dimension $k - r$ in a matroid are exactly the hyperplanes in the $(r-1)$-th truncated matroid $T^{r-1}(M)$. This gives another proof of the result of Britz [61, Theorem 3] that the minimal supports of dimension r are the cocircuits of the $(r-1)$-th truncated matroid. For $r = 1$, this gives the well-known equivalence between nonzero minimal codewords and cocircuits. See [35, Theorem 9.2.4] and [39, 1.21].

The number of minimal subcodes of dimension r does not change after extending the code under a finite field extension, since this number is determined by the Möbius polynomial of the lattice of the code, and this lattice does not change under a finite field extension.

5.8.4. *The characteristic polynomial of an arrangement*

Let $\mathcal{X}$ be an *affine variety* in $\mathbb{A}^k$ defined over $\mathbb{F}_q$, that is the zeroset of a collection of polynomials in $\mathbb{F}_q[X_1, \ldots, X_k]$. Then $\mathcal{X}(\mathbb{F}_{q^m})$ is the set of all points $\mathcal{X}$ with coordinates in $\mathbb{F}_{q^m}$, also called the set of $\mathbb{F}_{q^m}$-*rational points* of $\mathcal{X}$. Note the similarity with extension codes.

A central arrangement $\mathcal{A}$ gives rise to a geometric lattice $L(\mathcal{A})$ and characteristic polynomial $\chi_{L(\mathcal{A})}$ that will be denoted by $\chi_{\mathcal{A}}$. Similarly $\pi_{\mathcal{A}}$ denotes the Poincaré polynomial of $\mathcal{A}$. If $\mathcal{A}$ is an arrangement over the real numbers, then $\pi_{\mathcal{A}}(1)$ counts the number of connected components of the complement of the arrangement. See [80]. Something similar can be said about arrangements over finite fields.

Proposition 5.45. *Let q be a prime power, and let $\mathcal{A} = (H_1, \ldots, H_n)$ be a simple and central arrangement in $\mathbb{F}_q^k$. Then*

$$\chi_{\mathcal{A}}(q^m) = |\mathbb{F}_{q^m}^k \setminus (H_1 \cup \cdots \cup H_n)|.$$

Proof. See [57, Theorem 2.2] and [86, Proposition 3.2] and [74, Sect. 16] and [70, Theorem 2.69].

Let $A = \mathbb{F}_{q^m}^k$ and $A_j = H_j(\mathbb{F}_{q^m})$. Let L be the poset of all intersections of the A_j with the reverse inclusion as partial order. The principle of inclusion/exclusion as formulated in Example 5.38 gives that

$$|\mathbb{F}_{q^m}^k \setminus (H_1 \cup \cdots \cup H_n)| = \sum_{x \in L} \mu(x)|x| = \sum_{x \in L} \mu(x) q^{m \dim(x)}.$$

The expression on the right-hand side is equal to $\chi_{\mathcal{A}}(q^m)$, since L is isomorphic with the geometric lattice $L(\mathcal{A})$ of the arrangement $\mathcal{A} = (H_1, \ldots, H_n)$ with rank function $r = r_L$, so $\dim(x) = r(L) - r(x)$. □

Let $\mathcal{A} = (H_1, \ldots, H_n)$ be an arrangement in $\mathbb{F}^k$ over the field $\mathbb{F}$. Let $H = H_i$. Then the *deletion* $\mathcal{A} \setminus H$ is the arrangement in $\mathbb{F}^k$ obtained from $(H_1, \ldots, H_n)$ by deleting all the H_j such that $H_j = H$.

Let $x = \cap_{i \in I} H_i$ be an intersection of hyperplanes of $\mathcal{A}$. Let l be the dimension of x. The *restriction* $\mathcal{A}_x$ is the arrangement in $\mathbb{F}^l$ of all hyperplanes $x \cap H_j$ in x such that $x \cap H_j \neq \emptyset$ and $x \cap H_j \neq x$, for a chosen isomorphism of x with $\mathbb{F}^l$.

Proposition 5.46 (Deletion-restriction formula). *Let $\mathcal{A} = (H_1, \ldots, H_n)$ be a simple and central arrangement in $\mathbb{F}^k$ over the field $\mathbb{F}$. Let $H = H_i$. Then*

$$\chi_{\mathcal{A}}(T) = \chi_{\mathcal{A} \setminus H}(T) - \chi_{\mathcal{A}_H}(T).$$

Proof. Note the similarity of this theorem with the corresponding Proposition 5.32 for graphs. A proof can be given using the deletion-contraction formula for matroids in Proposition 5.38, Remark 5.11 that gives the relation between the two-variable characteristic and the Tutte polynomial and the fact that $\chi_{\mathcal{A}}(T) = \chi_{\mathcal{A}}(0, T)$. Another proof for an arbitrary field can be found in Orlik [70, Theorem 2.56]. Here a proof of the special case of a central arrangement over the finite field $\mathbb{F}_q$ will be given by a counting argument. Without loss of generality we may assume that $H = H_1$. Denote $H_j(\mathbb{F}_{q^m})$ by H_j and $\mathbb{F}_{q^m}^k$ by V. Then we have the following disjoint union:

$$V \setminus (H_2 \cup \cdots \cup H_n) = (V \setminus (H_1 \cup H_2 \cup \cdots \cup H_n)) \cup (H_1 \setminus (H_2 \cup \cdots \cup H_n)).$$

The number of elements of the left-hand side is equal to $\chi_{\mathcal{A}\setminus H}(q^m)$, and the number of elements of the two sets on the right-hand side are equal to $\chi_{\mathcal{A}}(q^m)$ and $\chi_{\mathcal{A}_H}(q^m)$, respectively by Proposition 5.45. Hence

$$\chi_{\mathcal{A}\setminus H}(q^m) = \chi_{\mathcal{A}}(q^m) + \chi_{\mathcal{A}_H}(q^m)$$

for all positive integers m, since the union is disjoint. Therefore the identity of the polynomial holds. □

Let $\mathcal{A} = (H_1, \ldots, H_n)$ be a central simple arrangement over the field $\mathbb{F}$ in $\mathbb{F}^k$. Let $J \subseteq [n]$. Define $H_J = \cap_{j \in J} H_j$. Consider the decreasing sequence

$$\mathcal{N}_k \subset \mathcal{N}_{k-1} \subset \cdots \subset \mathcal{N}_1 \subset \mathcal{N}_0$$

of algebraic subsets of the affine space $\mathbb{A}^k$, defined by

$$\mathcal{N}_i = \bigcup_{\substack{J \subseteq [n] \\ r(H_J) = i}} H_J.$$

Define $\mathcal{M}_i = (\mathcal{N}_i \setminus \mathcal{N}_{i+1})$.

Note that $\mathcal{N}_0 = \mathbb{A}^k$, $\mathcal{N}_1 = \cup_{j=1}^n H_j$, $\mathcal{N}_k = \{0\}$ and $\mathcal{N}_{k+1} = \emptyset$. Furthermore, $\mathcal{N}_i$ is a union of linear subspaces of $\mathbb{A}^k$ all of dimension $k - i$. Remember from Remark 5.3 that H_J is isomorphic with $C(J)$ in case $\mathcal{A}$ is the arrangement of the generator matrix G of the code C.

Proposition 5.47. *Let $\mathcal{A} = (H_1, \ldots, H_n)$ be a central simple arrangement over the field $\mathbb{F}$ in $\mathbb{F}^k$. Let $z(\mathbf{x}) = \{j \in [n] : \mathbf{x} \in H_j\}$ and $r(\mathbf{x}) = r(H_{z(\mathbf{x})})$ the rank of $\mathbf{x}$ for $\mathbf{x} \in \mathbb{A}^k$. Then*

$$\mathcal{N}_i = \{\mathbf{x} \in \mathbb{A}^k : r(\mathbf{x}) \geq i\} \quad \text{and} \quad \mathcal{M}_i = \{\mathbf{x} \in \mathbb{A}^k : r(\mathbf{x}) = i\}.$$

Proof. Let $\mathbf{x} \in \mathbb{A}^k$ and $\mathbf{c} = \mathbf{x}G$. Let $\mathbf{x} \in \mathcal{N}_i$. Then there exists a $J \subseteq [n]$ such that $r(H_J) = i$ and $\mathbf{x} \in H_J$. So $c_j = 0$ for all $j \in J$. So $J \subseteq z(\mathbf{x})$. Hence $H_{z(\mathbf{x})} \subseteq H_J$. Therefore $r(\mathbf{x}) = r(H_{z(\mathbf{x})}) \geq r(H_J) = i$. The converse implication is proved similarly.

The statement about $\mathcal{M}_i$ is a direct consequence of the one about $\mathcal{N}_i$. □

Proposition 5.48. *Let $\mathcal{A}$ be a central simple arrangement over $\mathbb{F}_q$. Let $L = L(\mathcal{A})$ be the geometric lattice of $\mathcal{A}$. Then*

$$\mu_i(q^m) = |\mathcal{M}_i(\mathbb{F}_{q^m})|.$$

Proof. See also [57, Theorem 6.3]. Remember from Remark 5.12 that $\mu_i(T) = \sum_{r(x)=i} \chi_{L_x}(T)$. Let $L = L(\mathcal{A})$ and $x \in L$. Then $L(\mathcal{A}_x) = L_x$. Let $\cup \mathcal{A}_x$ be the union of the hyperplanes of $\mathcal{A}_x$. Then $|(x \setminus (\cup \mathcal{A}_x))(\mathbb{F}_{q^m})| = \chi_{L_x}(q^m)$ by Proposition 5.45. Now $\mathcal{M}_i$ is the disjoint union of complements of the arrangements of $\mathcal{A}_x$ for all $x \in L$ such that $r(x) = i$ by Proposition 5.47. Hence

$$|\mathcal{M}_i(\mathbb{F}_{q^m})| = \sum_{\substack{x \in L \\ r(x)=i}} |(x \setminus (\cup \mathcal{A}_x))(\mathbb{F}_{q^m})|$$

$$= \sum_{\substack{x \in L \\ r(x)=i}} \chi_{L_x}(q^m).$$

$\square$

5.8.5. The characteristic polynomial of a code

Proposition 5.49. *Let C be a nondegenerate $\mathbb{F}_q$-linear code. Then*

$$A_n(T) = \chi_C(T).$$

Proof. The short proof is given by $\chi_C(T) = \chi_C(0, T) = \chi_0(T) = A_n(T)$. The geometric interpretation is as follows.

The elements in $\mathbb{F}_{q^m}^k \setminus (H_1 \cup \cdots \cup H_n)$ correspond one-to-one to codewords of weight n in $C \otimes \mathbb{F}_{q^m}$ by Proposition 5.16 and because the arrangements corresponding to C and $C \otimes \mathbb{F}_{q^m}$ are the same. So $A_n(q^m) = \chi_C(q^m)$ for all positive integers m by Proposition 5.45. Hence $A_n(T) = \chi_C(T)$. $\square$

Let G be a generator matrix of an $[n, k]$ code C over $\mathbb{F}_q$. Define

$$\mathcal{Y}_i = \{\mathbf{x} \in \mathbb{A}^k : \mathrm{wt}(\mathbf{x}G) \leq n - i\} \quad \text{and} \quad \mathcal{X}_i = \{\mathbf{x} \in \mathbb{A}^k : \mathrm{wt}(\mathbf{x}G) = n - i\}.$$

The $\mathcal{Y}_i$ form a decreasing sequence

$$\mathcal{Y}_n \subseteq \mathcal{Y}_{n-1} \subseteq \cdots \subseteq \mathcal{Y}_1 \subseteq \mathcal{Y}_0$$

of algebraic subsets of $\mathbb{A}^k$, and $\mathcal{X}_i = (\mathcal{Y}_i \setminus \mathcal{Y}_{i+1})$. Suppose that G has no zero column. Let $\mathcal{A}_G$ be the arrangement of G. Then

$$\mathcal{X}_i = \{\mathbf{x} \in \mathbb{A}^k : \mathbf{x} \text{ is in exactly } i \text{ hyperplanes of } \mathcal{A}_G\}.$$

Proposition 5.50. *Let C be a projective code of length n. Then*

$$\chi_i(q^m) = |\mathcal{X}_i(\mathbb{F}_{q^m})| = A_{n-i}(q^m).$$

Proof. Every $\mathbf{x} \in \mathbb{F}_{q^m}^k$ corresponds one-to-one to codeword in $C \otimes \mathbb{F}_{q^m}$ via the map $\mathbf{x} \mapsto \mathbf{x}G$. So $|\mathcal{X}_i(\mathbb{F}_{q^m})| = A_{n-i}(q^m)$. Also, $A_{n-i}(q^m) = \chi_i(q^m)$ for all i, by Remark 5.11. See also [87, Theorem 3.3]. $\square$

The similarity between Proposition 5.50 and Proposition 5.16 gives a geometric argument for the relation between $\chi_C(S,T)$ and $W_C(X,Y,T)$ in Remark 5.11.

Remark 5.15. Another way to define $\mathcal{X}_i$ is the collection of all points $P \in \mathbb{A}^k$ such that P is on exactly i distinct hyperplanes of the arrangement $\mathcal{A}_G$. Denote the arrangement of hyperplanes in $\mathbb{P}^{k-1}$ also by $\mathcal{A}_G$, and let $\bar{P}$ be the point in $\mathbb{P}^{k-1}$ corresponding to $P \in \mathbb{A}^k$. Define

$$\bar{\mathcal{X}}_i = \{\bar{P} \in \mathbb{P}^{k-1} : \bar{P} \text{ is on exactly } i \text{ hyperplanes of } \mathcal{A}_G\}.$$

For all $i < n$ the polynomial $\chi_i(T)$ is divisible by $T-1$. Define $\bar{\chi}_i(T) = \chi_i(T)/(T-1)$. Then $\bar{\chi}_i(q^m) = |\bar{\mathcal{X}}_i(\mathbb{F}_{q^m})|$ for all $i < n$ by Proposition 5.50.

Theorem 5.18. *Let G be a generator matrix of a nondegenerate code C. Let $\mathcal{A}_G$ be the associated central arrangement. Let $d^\perp = d(C^\perp)$. Then $\mathcal{N}_i \subseteq \mathcal{Y}_i$ for all i, equality holds for all $i < d^\perp$. Also, $\mathcal{M}_i = \mathcal{X}_i$ for all $i < d^\perp - 1$. If furthermore C is projective, then*

$$\mu_i(T) = \chi_i(T) = A_{n-i}(T) \text{ for all } i < d^\perp - 1.$$

Proof. Let $\mathbf{x} \in \mathcal{N}_i$. Then $\mathbf{x} \in H_J$ for some $J \subseteq [n]$ such that $r(H_J) = i$. So $|J| \geq i$ and $\mathrm{wt}(\mathbf{x}G) \leq n-i$ by Proposition 5.16. Hence $\mathbf{x} \in \mathcal{Y}_i$. Therefore $\mathcal{N}_i \subseteq \mathcal{Y}_i$.

Let $i < d^\perp$ and $\mathbf{x} \in \mathcal{Y}_i$. Then $\mathrm{wt}(\mathbf{x}G) \leq n-i$. Let J be the complement of $\mathrm{supp}(\mathbf{x}G)$ in $[n]$. Then $|J| \geq i$. Take a subset I of J such that $|I| = i$. Then $\mathbf{x} \in H_I$ and $r(I) = |I| = i$ by Lemma 5.2, since $i < d^\perp$. Hence $\mathbf{x} \in \mathcal{N}_i$. Therefore $\mathcal{Y}_i \subseteq \mathcal{N}_i$. So $\mathcal{Y}_i = \mathcal{N}_i$ for all $i < d^\perp$, and $\mathcal{M}_i = \mathcal{X}_i$ for all $i < d^\perp - 1$.

The code is nondegenerate. So $d(C^\perp) \geq 2$. Suppose furthermore that C is projective. Then $\mu_i(T) = \chi_i(T) = A_{n-i}(T)$ for all $i < d^\perp - 1$, by Remark 5.11 and Propositions 5.50 and 5.48. □

The extended and generalized weight enumerators are determined by the pair (n,k) for an $[n,k]$ MDS code by Theorem 5.8. If C is an $[n,k]$ code, then $d(C^\perp)$ is at most $k+1$. Furthermore $d(C^\perp) = k+1$ if and only if C is MDS if and only if $C^\perp$ is MDS. An $[n,k,d]$ code is called *almost MDS* if $d = n - k$. So $d(C^\perp) = k$ if and only if $C^\perp$ is almost MDS. If C is almost MDS, then $C^\perp$ is not necessarily almost MDS. The code C is called *near MDS* if both C and $C^\perp$ are almost MDS. See [88].

Proposition 5.51. *Let C be an $[n,k,d]$ code such that $C^\perp$ is MDS or almost MDS and $k \geq 3$. Then both $\chi_C(S,T)$ and $W_C(X,Y,T)$ determine*

$\mu_C(S,T)$. In particular

$$\mu_i(T) = \chi_i(T) = A_{n-i}(T) \text{ for all } i < k-1,$$

$$\mu_{k-1}(T) = \sum_{i=k-1}^{n-1} \chi_i(T) = \sum_{i=k-1}^{n-1} A_{n-i}(T),$$

and $\mu_k(T) = 1$.

Proof. Let C be a code such that $d(C^\perp) \geq k \geq 3$. Then C is projective and $\mu_i(T) = \chi_i(T) = A_{n-i}(T)$ for all $i < k-1$ by Theorem 5.18. Furthermore, $\mu_k(T) = \chi_n(T) = A_0(T) = 1$.

Finally let $L = L(C)$. Then $\sum_{i=0}^{k} \mu_i(T) = T^k$, $\sum_{i=0}^{n} \chi_i(T) = T^k$ and $\sum_{i=0}^{n} A_i(T) = T^k$ by Remark 5.13. Hence the formula for $\mu_{k-1}(T)$ holds. Therefore $\mu_C(S,T)$ is determined both by $W_C(X,Y,T)$ and $\chi_C(S,T)$. □

Projective codes of dimension 3 are examples of codes C such that $C^\perp$ is almost MDS. In the following we will give explicit formulas for $\mu_C(S,T)$ for such codes.

Let C be a projective code of length n and dimension 3 over $\mathbb{F}_q$ with generator matrix G. The arrangement $\mathcal{A}_G = (H_1, \ldots, H_n)$ of planes in $\mathbb{F}_q^3$ is simple and essential, and the corresponding arrangement of lines in $\mathbb{P}^2(\mathbb{F}_q)$ is also denoted by $\mathcal{A}_G$. We defined in Remark 5.15 that

$$\bar{\mathcal{X}}_i(\mathbb{F}_{q^m}) = \{\bar{P} \in \mathbb{P}^2(\mathbb{F}_{q^m}) : \bar{P} \text{ is on exactly } i \text{ lines of } \mathcal{A}_G\}$$

and $\bar{\chi}_i(q^m) = |\bar{\mathcal{X}}_i(\mathbb{F}_{q^m})|$ for all $i < n$.

Notice that for projective codes of dimension 3 we have $\bar{\mathcal{X}}_i(\mathbb{F}_{q^m}) = \bar{\mathcal{X}}_i(\mathbb{F}_q)$ for all positive integers m and $2 \leq i < n$. Abbreviate in this case $\bar{\chi}_i(q^m) = \bar{\chi}_i$ for $2 \leq i < n$.

Proposition 5.52. *Let C be a projective code of length n and dimension 3 over $\mathbb{F}_q$. Then*

$$\begin{cases} \mu_0(T) = (T-1)\left(T^2 - (n-1)T + \sum_{i=2}^{n-1}(i-1)\bar{\chi}_i - n + 1\right), \\ \mu_1(T) = (T-1)\left(nT + n - \sum_{i=2}^{n-1} i\bar{\chi}_i\right), \\ \mu_2(T) = (T-1)\left(\sum_{i=2}^{n-1} \bar{\chi}_i\right), \\ \mu_3(T) = 1. \end{cases}$$

Proof. A more general statement and proof is possible for $[n, k]$ codes C such that $d(C^\perp) \geq k$, using Proposition 5.51, the fact that $B_t(T) = T^{k-t} - 1$ for all $t < d(C^\perp)$ by Lemma 5.2 and the expression of $B_t(T)$ in terms of $A_w(T)$ by Proposition 5.17. We will give a second geometric proof for the special case of projective codes of dimension 3.

By Lagrange interpolation it is enough to show this proposition with $T = q^m$ for all m. Notice that $\mu_i(q^m)$ is the number of elements of $\mathcal{M}_i(\mathbb{F}_{q^m})$ by Proposition 5.48. Let $\bar{P}$ be the corresponding point in $\mathbb{P}^2(\mathbb{F}_{q^m})$ for $P \in \mathbb{F}_{q^m}^3$ and $P \neq 0$. Abbreviate $\mathcal{M}_i(\mathbb{F}_{q^m})$ by $\mathcal{M}_i$. Define $\bar{\mathcal{M}}_i = \{\bar{P} : P \in \mathcal{M}_i\}$. Then $|\mathcal{M}_i| = (q^m - 1)|\bar{\mathcal{M}}_i|$ for all $i < 3$.

If $\bar{P} \in \bar{\mathcal{M}}_2$, then $\bar{P} \in H_j \cap H_k$ for some $j \neq k$. Hence $\bar{P} \in \bar{\mathcal{X}}_i(\mathbb{F}_q)$ for some $i \geq 2$, since the code is projective. So $\bar{\mathcal{M}}_2$ is the disjoint union of the $\bar{\mathcal{X}}_i(\mathbb{F}_q)$ for $2 \leq i < n$. Therefore $|\bar{\mathcal{M}}_2| = \sum_{i=2}^{n-1} \bar{\chi}_i$.

$\bar{P} \in \bar{\mathcal{M}}_1$ if and only if $\bar{P}$ is on exactly one line H_j. There are n lines, and every line has $q^m + 1$ points that are defined over $\mathbb{F}_{q^m}$. If $i \geq 2$, then every $\bar{P} \in \bar{\mathcal{X}}_i(\mathbb{F}_q)$ is on i lines H_j. Hence $|\bar{\mathcal{M}}_1| = n(q^m + 1) - \sum_{i=2}^{n-1} i\bar{\chi}_i$.

Finally, $\mathbb{P}^2(\mathbb{F}_{q^m})$ is the disjoint union of $\bar{\mathcal{M}}_0$, $\bar{\mathcal{M}}_0$ and $\bar{\mathcal{M}}_2$. The numbers $|\bar{\mathcal{M}}_2|$ and $|\bar{\mathcal{M}}_1|$ are already computed, and $|\mathbb{P}^2(\mathbb{F}_{q^m})| = q^{2m} + q^m + 1$. From this we derive the number of elements of $\bar{\mathcal{M}}_0$. □

Note that $\mu_i(T)$ is divisible by $T - 1$ for all $0 \leq i < k$. Define $\bar{\mu}_i = \mu_i(T)/(T - 1)$. Define similarly $\bar{A}_w = A_w(T)/(T - 1)$ for all $0 < w \leq n$.

5.8.6. *Examples and counterexamples*

Example 5.55. Consider the matrix G given by

$$G = \begin{pmatrix} 1 & 0 & 0 & 0 & 1 & 1 & 1 \\ 0 & 1 & 0 & 1 & 0 & 1 & 1 \\ 0 & 0 & 1 & 1 & 1 & 0 & 1 \end{pmatrix}.$$

Let C be the code over $\mathbb{F}_q$ with generator matrix G. For $q = 2$, this is the simplex code $\mathcal{S}_2(2)$. The columns of G represent also the coefficients of the lines of $\mathcal{A}_G$. The projective picture of $\mathcal{A}_G$ is given in Figure 5.12.

If q is odd, then there are 3 points on two lines and 6 points on three lines, so $\bar{\chi}_2 = 3$ and $\bar{\chi}_3 = 6$. The number of points that are on one line is equal to the number of points on each of the seven lines, minus the points we already counted, with multiplicity: $7(T+1) - 3 \cdot 2 - 6 \cdot 3 = 7T - 17$. There are no points on more than three lines, so $\bar{\chi}_i = 0$ for $i > 3$. We calculate $\bar{\chi}_0$ via $\bar{\chi}_0 + \bar{\chi}_1 + \bar{\chi}_2 + \bar{\chi}_3 = T^2 + T + 1$.

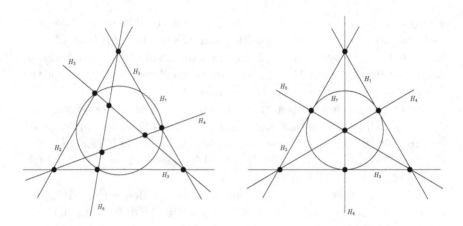

Fig. 5.12. The arrangement of G for q odd and q even

If q is even, we can do the same kind of calculation. The values of $\bar{\mu}_i$ can be calculated using Proposition 5.52, but they follow more directly from Proposition 5.51. The results are in the next table:

	i	0	1	2	3	4	5	6	7
q odd	$\bar{\chi}_i$	T^2-6T+9	$7T-17$	3	6	0	0	0	
	$\bar{A}_i$		0	0	0	6	3	$7T-17$	T^2-6T+9
	$\bar{\mu}_i$	T^2-6T+9	$7T-17$	9	1				
q even	$\bar{\chi}_i$	T^2-6T+8	$7T-14$	0	7	0	0	0	
	$\bar{A}_i$		0	0	0	7	0	$7T-14$	T^2-6T+8
	$\bar{\mu}_i$	T^2-6T+8	$7T-14$	7	1				

Notice that there is a codeword of weight 7 in case q is even and $q > 4$ or q is odd and $q > 3$, since $\bar{A}_7 = (T-2)(T-4)$ or $\bar{A}_7 = (T-3)^2$, respectively.

Example 5.56. Let G be a $3 \times n$ generator matrix of an MDS code. As mentioned in Example 5.30, the lines of the arrangement $\mathcal{A}_G$ are in general position. That means that every two distinct lines meet in one point, and every three mutually distinct lines have an empty intersection. So $\bar{\chi}_2 = \binom{n}{2}$ and $\bar{\chi}_i = 0$ for all $i > 2$. By Proposition 5.52 we have $\bar{\mu}_2 = \binom{n}{2}$ and $\bar{\mu}_1 = nT + 2n - n^2$ and $\bar{\mu}_0 = T^2 - (n-1)T + \binom{n-1}{2}$. By Proposition 5.48 we find $A_i = 0$ for $0 < i < n-2$, $\bar{A}_{n-2} = \bar{\chi}_2$ and $\bar{A}_{n-1} = \bar{\chi}_1 = \bar{\mu}_1$ and $\bar{A}_n = \bar{\chi}_0 = \bar{\mu}_0$. The values found for the extended weight enumerator are in agreement with Theorem 5.8.

Example 5.57. Let a and b be positive integers such that $2 < a < b$. Let $n = a + b$. Let G be a $3 \times n$ generator matrix of a nondegenerate code. Suppose that there are two points P and Q in the projective plane over $\mathbb{F}_q$ such that the $a + b$ lines of the projective arrangement of $\mathcal{A}_G$ consists of a distinct lines incident with P, and b distinct lines incident with Q and there is no line incident with P and Q. Then $\bar{\chi}_2 = \bar{A}_{n-2} = ab$, $\bar{\chi}_a = \bar{A}_b = 1$ and $\bar{\chi}_b = \bar{A}_a = 1$. Hence $\bar{\mu}_2(T) = ab + 2$. Furthermore

$$\bar{\mu}_1 = \bar{A}_{n-1} = (a+b)T - 2ab,$$

$$\bar{\mu}_0 = \bar{A}_n = T^2 - (a+b-1)T + ab - 1$$

and $\bar{A}_i = 0$ for all $i \notin \{a, b, n-2, n-1, n\}$.

Example 5.58. Let a, b and c be positive integers such that $2 < a < b < c$. Let $n = a + b + c$. Let G be a $3 \times n$ generator matrix of a nondegenerate code $C(a, b, c)$. Suppose that there are three points P, Q and R in the projective plane over $\mathbb{F}_q$ such that the lines of the projective arrangement of $\mathcal{A}_G$ consist of a distinct lines incident with P and not with Q and R, b distinct lines incident with Q and not with P and R, and c distinct lines incident with R and not with P and Q. The a lines through P intersect the b lines through Q in ab points. Similarly statements hold for the lines through P and R intersecting in ac points, and the lines through Q and R intersecting in bc points. Suppose that all these $ab + bc + ac$ intersection points are mutually distinct, so every intersection point lies on exactly two lines of the arrangement. If q is large enough, then such a configuration exists.

The number of points on two lines of the arrangement is $\bar{\chi}_2 = ab+bc+ca$. Since P is the unique point on exactly a lines of the arrangement, we have $\bar{\chi}_a = 1$. Similarly $\bar{\chi}_b = \bar{\chi}_c = 1$. Finally $\bar{\chi}_i = 0$ for all $2 \leq i < n$ and $i \notin \{2, a, b, c\}$. Propositions 5.51 and 5.52 imply that $\bar{A}_{n-a} = \bar{A}_{n-b} = \bar{A}_{n-c} = 1$ and $\bar{A}_{n-2} = ab + bc + ca$ and $\bar{\mu}_2 = ab + bc + ca + 3$. Furthermore

$$\bar{\mu}_1 = \bar{\chi}_1 = \bar{A}_{n-1} = nT - 2(ab + bc + ca),$$

$$\bar{\mu}_0 = \bar{\chi}_0 = \bar{A}_n = T^2 - (n-1)T + ab + bc + ca - 2$$

and $\bar{A}_i(T) = 0$ for all $i \notin \{0, n-c, n-b, n-a, n-2, n-1, n\}$.

Therefore $W_{C(a,b,c)}(X,Y,T) = W_{C(a',b',c')}(X,Y,T)$ if and only if $(a,b,c) = (a',b',c')$, and $\mu_{C(a,b,c)}(S,T) = \mu_{C(a',b',c')}(S,T)$ if and only if $a + b + c = a' + b' + c'$ and $ab + bc + ca = a'b' + b'c' + c'a'$. In particular let $C_1 = C(3, 9, 14)$ and $C_2 = C(5, 6, 15)$. Then C_1 and C_2 are two projective codes with the same Möbius polynomial $\mu_C(S,T)$ but distinct extended weight enumerators and coboundary polynomial $\chi_C(S,T)$.

Now $d(C(a,b,c)) = n - c$. Hence $d(C_1) = 12$ and $d(C_2) = 11$. Therefore $\mu_C(S,T)$ does not determine the minimum distance although it gives the number of minimal codewords.

Example 5.59. Consider the codes C_3 and C_4 over $\mathbb{F}_q$ with $q > 2$ with generator matrices G_3 and G_4 given by

$$G_3 = \begin{pmatrix} 1\ 1\ 0\ 0\ 1\ 0\ 0 \\ 0\ 1\ 1\ 1\ 0\ 1\ 0 \\ -1\ 0\ 1\ 1\ 0\ 0\ 1 \end{pmatrix} \quad \text{and} \quad G_4 = \begin{pmatrix} 1\ 1\ 0\ 0\ 1\ 0\ 0 \\ 0\ 1\ 1\ 1\ 0\ 1\ 0 \\ 0\ 1\ 1\ a\ 0\ 0\ 1 \end{pmatrix}$$

where $a \in \mathbb{F}_q \setminus \{0,1\}$. It was shown in Brylawsky [63, Exercise 6.96] that the duals of these codes have the same Tutte polynomial. So the codes C_3 and C_4 have the same Tutte polynomial

$$t_C(X,Y) = 2X + 2Y + 3X^2 + 5XY + 4Y^2 + X^3 + X^2Y + 2XY^2 + 3Y^3 + Y^4.$$

Hence C_3 and C_4 have the extended weight enumerator given by

$$X^7 + (2T-2)X^4Y^3 + (3T-3)X^3Y^4 + (T^2-T)X^2Y^5$$
$$+ (5T^2 - 15T + 10)XY^6 + (T^3 - 6T^2 + 11T - 6)Y^7.$$

The codes C_3 and C_4 are not projective and their simplifications $\bar{C}_3$ and $\bar{C}_4$, respectively, have generator matrices given by

$$\bar{G}_3 = \begin{pmatrix} 1\ 1\ 0\ 1\ 0\ 0 \\ 0\ 1\ 1\ 0\ 1\ 0 \\ -1\ 0\ 1\ 0\ 0\ 1 \end{pmatrix} \quad \text{and} \quad \bar{G}_4 = \begin{pmatrix} 1\ 1\ 0\ 0\ 0\ 0 \\ 0\ 1\ 1\ 1\ 1\ 0 \\ 0\ 1\ 1\ a\ 0\ 1 \end{pmatrix}$$

where $a \in \mathbb{F}_q \setminus \{0,1\}$.

From the arrangement $\mathcal{A}(\bar{C}_3)$ and $\mathcal{A}(\bar{C}_4)$ in Figure 5.13 we deduce the $\bar{\chi}_i$ that are given in the following table.

code \ i	0	1	2	3	4	5
C_3	$T^2 - 5T + 6$	$6T - 12$	3	4	0	0
C_4	$T^2 - 5T + 6$	$6T - 13$	6	1	1	0

Therefore $t_{C_3}(X,Y) = t_{C_4}(X,Y)$ but $\chi_{C_3}(S,T) \neq \chi_{C_4}(S,T)$ and $t_{\bar{C}_3}(X,Y) \neq t_{\bar{C}_4}(X,Y)$.

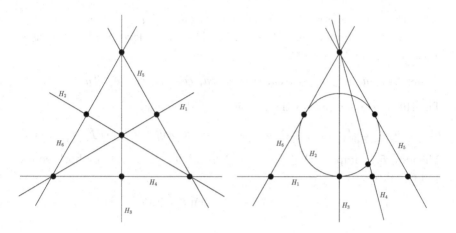

Fig. 5.13. The arrangements of $\bar{G}_3$ and $\bar{G}_4$

Example 5.60. Let $C_5 = C_3^\perp$ and $C_6 = C_4^\perp$. Their generator matrices are

$$G_5 = \begin{pmatrix} 1 & 0 & 0 & 0 & 1 & 0 & -1 \\ 0 & 1 & 0 & 0 & 1 & 1 & 0 \\ 0 & 0 & 1 & 0 & 0 & 1 & 1 \\ 0 & 0 & 0 & 1 & 0 & 1 & 1 \end{pmatrix} \quad \text{and} \quad G_6 = \begin{pmatrix} 1 & 0 & 0 & 0 & 1 & 0 & 0 \\ 0 & 1 & 0 & 0 & 1 & 1 & 1 \\ 0 & 0 & 1 & 0 & 0 & 1 & 1 \\ 0 & 0 & 0 & 1 & 0 & 1 & a \end{pmatrix}$$

where $a \in \mathbb{F}_q \setminus \{0,1\}$. Then C_5 and C_6 have the same Tutte polynomial $t_{C^\perp}(X,Y) = t_C(Y,X)$ as given by Example 5.59:

$$2X + 2Y + 4X^2 + 5XY + 3Y^2 + 3X^3 + 2X^2Y + XY^2 + Y^3 + 3X^4.$$

Hence C_5 and C_6 have the same extended weight enumerator given by

$$X^7 + (T-1)X^5Y^2$$
$$+ (6T-6)X^4Y^3 + (2T^2 - T - 1)X^3Y^4 + (15T^2 - 43T + 28)X^2Y^5$$
$$+ (7T^3 - 36T^2 + 60T - 31)XY^6 + (T^4 - 7T^3 + 19T^2 - 23T + 10)Y^7.$$

The geometric lattice $L(C_5)$ has atoms a, b, c, d, e, f, g corresponding to the first, second, etc. column of G_5. The second level of $L(C_5)$ consists of the following 17 elements:

abe, ac, ad, af, ag, bc, bd, bf, bg, cd, ce, cf, cg, de, df, dg, efg.

The third level consists of the following 12 elements:

$abce$, $abde$, $abefg$, $acdg$, acf, adf, $bcdf$, bcg, bdg, cde, $cefg$, $defg$.

Similarly, the geometric lattice $L(C_6)$ has atoms a, b, c, d, e, f, g corresponding to the first, second, etc. column of G_6. The second level of $L(C_6)$ consists of the following 17 elements:

$abe, ac, ad, af, ag, bc, bd, bf, bg, cd, ce, cf, cg, de, dfg, ef, eg.$

The third level consists of the following 13 elements:

$abce, abde, abef, abeg, acd, acf, acg, adfg, bcdfg, cde, cef, ceg, defg.$

Theorem 5.18 implies that $\mu_0(T)$ and $\mu_1(T)$ are the same for both codes and equal to

$$\mu_0(T) = \chi_0(T) = A_7(T) = (T-1)(T-2)(T^2 - 4T + 5)$$

$$\mu_1(T) = \chi_1(T) = A_6(T) = (T-1)(7T^2 - 29T + 31).$$

The polynomials $\mu_3(T)$ and $\mu_2(T)$ are given in the following table using Remarks 5.14 and 5.13.

	C_5	C_6
$\mu_2(T)$	$17T^2 - 49T + 32$	$17T^2 - 50T + 33$
$\mu_3(T)$	$12T - 12$	$13T - 13$

This example shows that for projective codes the Möbius polynomial $\mu_C(S, T)$ is not determined by the coboundary polynomial $\chi_C(S, T)$.

5.9. Overview of polynomial relations

We have established relations between the generalized weight enumerators for $0 \leq r \leq k$, the extended weight enumerator and the Tutte polynomial. We summarize this in Figure 5.14.

We see that the Tutte polynomial, the extended weight enumerator and the collection of generalized weight enumerators all contain the same amount of information about a code, because they completely define each other. The original weight enumerator $W_C(X, Y)$ contains less information and therefore does not determine $W_C(X, Y, T)$ or $\{W_C^{(r)}(X, Y)\}_{r=0}^k$. See Simonis [21].

One may wonder if the method of generalizing and extending the weight enumerator can be continued, creating the generalized extended weight enumerator, in order to get a stronger invariant. The answer is no: the generalized extended weight enumerator can be defined, but does not contain

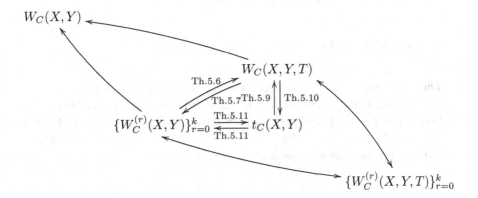

Fig. 5.14. Relations between the weight enumerator and Tutte polynomial

more information than the three underlying polynomials.

Now $t_C(X,Y)$, $R_{M_C}(X,Y)$ and $\chi_C(S,T)$ determine each other on the class of projective codes by Theorem 5.16. This is summarized in Figure 5.15. The dotted arrows only apply if the matroid is simple or, equivalently, if the code is projective.

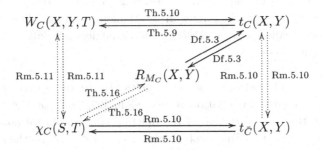

Fig. 5.15. Relations between the weight enumerator, characteristic, and Tutte polynomial

The polynomials $\chi_C(S,T)$ and $\mu_C(S,T)$ do not determine each other by Examples 5.58 and 5.60.

5.10. Further reading and open problems

5.10.1. Multivariate and other polynomials

The *multivariate* Tutte or *polychromatic* polynomial of a graph and a matroid is considered in [89–92] and is related to the partition function of the Potts-model in statistical mechanics [93, 94]. The multivariate weight enumerator of a code is considered in [95]. The characteristic and multivariate Tutte polynomial of arrangements are studied in [57, 87, 96].

The *tree polynomial* of a graph is generalized to the *basis polynomial* of a matroid [97]. The characteristic polynomial of a graph is the characteristic polynomial $\det(\lambda I - A)$ of the adjacency matrix A of the graph [46] and is distinct from the chromatic polynomial of the graph and from the characteristic polynomial of the geometric lattice of the graph. The *spectrum* of a graph is the set of eigenvalues of the characteristic polynomial of the graph.

Gray gave an example of two non-isomorphic graphs that have the same Tutte polynomial. This result was generalized in [54] on codichromatic graphs and in [89] on copolychromatic graphs.

Every polynomial in one variable with coefficients in a field $\mathbb{F}$ factorizes in linear factors over the algebraic closure $\bar{\mathbb{F}}$ of $\mathbb{F}$. In Examples 5.50 and 5.51 we see that $\chi_L(T)$ factorizes in linear factors over $\mathbb{Z}$. This is always the case for so-called *super solvable* geometric lattices and lattices from *free* central arrangements. See [70].

The theory of matroid complexes gives rise to the *spectrum polynomial* [98]. A recurrence relation is proved in [99, 100] for the spectrum polynomial that is a variation of the deletion-contraction formula for the Tutte polynomial. The Tutte polynomial does not determine the spectrum polynomial. The converse problem is an open question. The multivariate spectrum polynomial is considered in [101].

The theory of knots and links and their Kauffman, Jones and Homfly polynomials have connections with graph theory and the Tutte polynomial. See [102–104].

5.10.2. *The coset leader weight enumerator*

Let C be a linear code of length n over $\mathbb{F}_q$. Let $\mathbf{y} \in \mathbb{F}_q^n$. The weight of the coset $\mathbf{y} + C$ is defined by

$$\mathrm{wt}(\mathbf{y} + C) = \min\{\mathrm{wt}(\mathbf{y} + \mathbf{c}) : \mathbf{c} \in C\}.$$

A *coset leader* is a choice of an element $\mathbf{y} \in \mathbb{F}_q^n$ of minimal weight in its coset, that is $\mathrm{wt}(\mathbf{y}) = \mathrm{wt}(\mathbf{y} + C)$. Let α_i be the number of cosets of C that are of weight i. Let λ_i be the number of $\mathbf{y}$ in $\mathbb{F}_q^n$ that are of minimal weight i in its coset. Then $\alpha_C(X,Y)$, the *coset leader weight enumerator* of C and $\lambda_C(X,Y)$, the *list weight enumerator* of C are polynomials defined by

$$\alpha_C(X,Y) = \sum_{i=0}^n \alpha_i X^{n-i} Y^i \quad \text{and} \quad \lambda_C(X,Y) = \sum_{i=0}^n \lambda_i X^{n-i} Y^i.$$

See [8, 105]. The *covering radius* $\rho(C)$ of C is the maximal i such that $\alpha_i(C) \neq 0$. We have $\alpha_i = \lambda_i = \binom{n}{i}(q-1)^i$ for all $i \leq (d-1)/2$, where d is the minimum distance of C. The coset leader weight enumerator gives a formula for the *error probability*, that is the probability that the output of the decoder is the wrong codeword. In this decoding scheme the decoder uses the chosen coset leader as the error vector as explained in Section 5.3.4 and [8, Chap. 1 §5]. The list weight enumerator is of interest in case the decoder has as output the list of all nearest codewords [106, 107]. The coset leader weight enumerator is also used in *steganography* to compute the average of changed symbols [108, 109].

The covering radius is determined by the coset leader weight enumerator of a code. The covering radius of a binary code is in general not determined by the Tutte polynomial of the code by [32]. Hence the Tutte polynomial and the extended weight enumerator of a code do not determine the coset leader weight enumerator.

Consider the functions $\alpha_i(T)$ and $\lambda_i(T)$ such that $\alpha_i(q^m)$ and $\lambda_i(q^m)$ are equal to the number of cosets of weight i and the number of elements in $\mathbb{F}_{q^m}^n$ of minimal weight i in its coset, respectively, with respect to the extended coded $C \otimes \mathbb{F}_{q^m}$. Define the *extended coset leader weight enumerator* and the *extended list weight enumerator* [3], respectively, by:

$$\alpha_C(X,Y,T) = \sum_{i=0}^n \alpha_i(T) X^{n-i} Y^i \quad \text{and} \quad \lambda_C(X,Y,T) = \sum_{i=0}^n \lambda_i(T) X^{n-i} Y^i.$$

In [105, Theorem 2.1] it is shown that the function $\alpha_i(T)$ is determined by finitely many data for all extensions of $\mathbb{F}_q$. In fact, the $\alpha_i(T)$ are polynomials in the variable T. There are well-defined nonnegative integers F_{ij} such that

$$\alpha_C(X,Y,T) = 1 + \sum_{i=1}^{n-k}\sum_{j=1}^{n-k} F_{ij}(T-1)(T-q)\cdots(T-q^{j-1})X^{n-i}Y^i.$$

This is similar to the expression of the extended weight enumerator in terms of the generalized weight enumerator as given in Proposition 5.28.

5.10.3. Graph codes

Graph codes were studied in [110] and used in [111] to show that decoding linear codes is hard, even if preprocessing is allowed. *Sparse graph* codes, *Gallager* or *Low-density parity check* codes and *Tanner graph* codes play an important role in the research of coding theory at this moment. See [112, 113].

5.10.4. The reconstruction problem

The reconstruction problem is whether a structure can be reconstructed from certain substructures. The original *vertex reconstruction problem* of Ulam and Kelly is whether a graph with at least three vertices can be reconstructed from the collection of its vertex deleted subgraphs. The *edge reconstruction problem* of a graph $\Gamma = (V, E)$ with at least four edges is whether this graph can be reconstructed from the collection of its edge deleted subgraphs $\Gamma \setminus e$. Both reconstruction problems are still open. See [114] for a survey. One can formulate a corresponding reconstruction problem for matroids. Let $M = (E, \mathcal{I})$ and $N = (E, \mathcal{J})$ be matroids on the same set E. Are M and N isomorphic if $M \setminus e$ and $N \setminus e$ are isomorphic for all e in E? In [115] a counterexample is given for this reconstruction problem for matroids. One can reconstruct the Tutte polynomial of M if one knows the Tutte polynomial of $M \setminus e$ for all e in E, see [116]. See a similar result for the polychromatic polynomial of graphs in [117].

5.10.5. Questions concerning the Möbius polynomial

Is it true that the Möbius polynomial of M is determined by the collection of Möbius polynomials of all $M \setminus e$ with e in E?

The doubly indexed Whitney numbers of the first kind and the Whitney numbers of the second kind are determined by the Möbius polynomial as was shown in Remark 5.14 and Theorem 5.17. Are the doubly indexed Whitney numbers of the second kind determined by the Möbius polynomial?

The geometric lattice of a matroid M is equal to the geometric lattice of its simplification $\bar{M}$ by Proposition 5.43. So information is lost by this process. The dual of a simple matroid is not necessarily simple. Similarly the dual of a projective code is not necessarily projective. Now suppose that both C and its dual are projective. Is there a MacWilliams type of formula for the Möbius polynomial? In other words: Is $\mu_C(S,T)$ determined by $\mu_{C^\perp}(S,T)$? A similar question could be asked for matroids M such that M and $M^\perp$ are simple.

We have seen in Example 5.58 that the Tutte polynomial and the coboundary polynomial are not determined by the Möbius polynomial of a projective code. Is $\chi_C(S,T)$ determined by the polynomials $\mu_C(S,T)$ and/or $\mu_{C^\perp}(S,T)$ if C and $C^\perp$ are projective?

5.10.6. *Monomial conjectures*

A sequence of real numbers $(v_0, v_1, \ldots, v_r)$ is called *unimodal* if

$$v_i \geq \min\{v_{i-1}, v_{i+1}\} \text{ for all } 0 < i < r.$$

The sequence is called *logarithmically concave* or *log-concave* if

$$v_i^2 \geq v_{i-1} v_{i+1} \text{ for all } 0 < i < r.$$

The Whitney numbers of the first kind are alternating in sign. That is

$$w_i^+ := (-1)^i w_i > 0 \text{ for all } i.$$

It was conjectured by Rota [118] that the Whitney numbers w_i^+ are unimodal. See [119, Problem 12]. Welsh [36] conjectured that the Whitney numbers w_i^+ are log-concave by generalizing a conjecture of Read [120] on graphs. It was shown that the following weaker version of the unimodal property is true for a matroid M of rank r:

$$w_i^+ < w_j^+ \text{ for all } 0 \leq i \leq r/2 \text{ and } i < j \leq r - i.$$

See [121, Corollary 8.4.2].

5.10.7. Complexity issues

The computation of the minimum distance and the weight enumerator of a code are NP hard problems [11, 13]. The computation of the coefficients of the Tutte polynomial of planar graphs is #P hard, but also the evaluation at a specific point (x, y) is #P-hard except for 9 points and two special curves [104, 122–124].

5.10.8. The zeta function

The counting of rational points over field extensions $\mathbb{F}_{q^m}$ is computed by the zeta function. Let $\mathcal{X}$ be an *affine variety* in $\mathbb{A}^k$ defined over $\mathbb{F}_q$, that is the zeroset of a collection of polynomials in $\mathbb{F}_q[X_1, \ldots, X_k]$. Then $\mathcal{X}(\mathbb{F}_{q^m})$ is the set of all points $\mathcal{X}$ with coordinates in $\mathbb{F}_{q^m}$, also called the set of $\mathbb{F}_{q^m}$-*rational points* of $\mathcal{X}$. The *zeta function* $Z_{\mathcal{X}}(T)$ of $\mathcal{X}$ is the formal power series in T defined by

$$Z_{\mathcal{X}}(T) = \exp\left(\sum_{m=1}^{\infty} \frac{|\mathcal{X}(\mathbb{F}_{q^m})|}{r} T^r\right).$$

Theorem 5.19. *Let $\mathcal{A}$ be a central simple arrangement in $\mathbb{F}_q^k$. Let*

$$\chi_{\mathcal{A}}(T) = \sum_{j=0}^{k} c_j T^j$$

be the characteristic polynomial of $\mathcal{A}$. Let $\mathcal{M} = \mathbb{A}^k \setminus (H_1 \cup \cdots \cup H_n)$ be the complement of the arrangement. Then the zeta function of $\mathcal{M}$ is given by:

$$Z_{\mathcal{M}}(T) = \prod_{j=0}^{k} (1 - q^j T)^{-c_j}.$$

Proof. See [86, Theorem 3.6]. □

The numbers $|c_j|$ can be interpreted as the Betti numbers of the cohomology of the complement of the arrangement over the algebraic closure of the finite field, which is analogous to the situation over the complex numbers [70, 125].

The (two-variable) zeta function of a code as defined by Duursma [24, 126, 127] is motivated by algebraic geometry codes on curves and the zeta function of the curve. It is related to the extended and generalized weight enumerator of the code and not to the zeta function of the arrangement of the code.

References

[1] R. Jurrius. Classifying polynomials of linear codes. Master's thesis, Leiden University, (2008).

[2] R. Jurrius and R. Pellikaan. Extended and generalized weight enumerators. In eds. T. Helleseth and Ø. Ytrehus, *Proc. Int. Workshop on Coding and Cryptography WCC-2009*, pp. 76–91. Selmer Center, Bergen, (2009).

[3] R. Jurrius and R. Pellikaan. The extended coset leader weight enumerator. In eds. F. Willems and T. Tjalkens, *Proc. 30th Symposium 2009 on Information Theory in the Benelux*, pp. 217–224. WIC, Eindhoven, (2009).

[4] R. Jurrius and R. Pellikaan. Codes, arrangemens and weight enumerators. Soria Summer School on Computational Mathematics (S3CM): Applied Computational Algebraic Geomertric Modelling, (2009).

[5] E. Berlekamp, *Algebraic coding theory* (Aegon Park Press, Laguna Hills, 1984).

[6] R. Blahut, *Theory and practice of error control codes* (Addison-Wesley, Reading, 1983).

[7] J. v. Lint, *Introduction to coding theory. Third edition, Graduate Texts in Math. vol. 86* (Springer, Berlin, 1999).

[8] F. MacWilliams and N. Sloane, *The theory of error-correcting codes* (North-Holland Mathematical Library, Amsterdam, 1977).

[9] R. Hamming, Error detecting and error correcting codes, *Bell System Technical Journal* **29**, 147–160, (1950).

[10] A. Shannon, A mathematical theory of communication, *Bell System Technical Journal* **27**, 379–423 and 623–656, (1948).

[11] A. Barg. Complexity issues in coding theory. In eds. V. Pless and W. Huffman, *Handbook of coding theory, vol. 1*, pp. 649–754. North-Holland, Amsterdam, (1998).

[12] E. Berlekamp, R. McEliece, and H. van Tilborg, On the inherent intractability of certain coding problems, *IEEE Transactions on Information Theory* **24**, 384–386, (1978).

[13] A. Vardy, The intractability of computing the minimum distance of a code, *IEEE Transactions on Information Theory* **43**, 1757–1766, (1997).

[14] T. Kløve, *Codes for error detection* (Series on Coding Theory and Cryptology, vol. 2. World Scientific Publishing Co. Pte. Ltd., Hackensack, 2007).

[15] G. Katsman and M. Tsfasman, Spectra of algebraic-geometric codes, *Problemy Peredachi Informatsii* **23**, 19–34, (1987).

[16] M. Tsfasman and S. Vlăduț, *Algebraic-geometric codes* (Kluwer Academic Publishers, Dordrecht, 1991).

[17] M. Tsfasman and S. Vlăduț, Geometric approach to higher weights, *IEEE Transactions on Information Theory* **41**, 1564–1588, (1995).

[18] T. Helleseth, T. Kløve, and J. Mykkeltveit, The weight distribution of irreducible cyclic codes with block lengths $n_1((q^l-1)/n)$, *Discrete Mathematics* **18**, 179–211, (1977).

[19] T. Kløve, The weight distribution of linear codes over $GF(q^l)$ having generator matrix over $GF(q)$, *Discrete Mathematics* **23**, 159–168, (1978).

[20] V. Wei, Generalized Hamming weights for linear codes, *IEEE Transactions on Information Theory* **37**, 1412–1418, (1991).
[21] J. Simonis, The effective length of subcodes, *AAECC* **5**, 371–377, (1993).
[22] L. Ozarev and A. Wyner, Wire-tap channel II, *AT&T Bell Labs Technical Journal.* **63**, 2135–2157, (1984).
[23] G. Forney, Dimension/length profiles and trellis complexity of linear block codes, *IEEE Transactions on Information Theory* **40**, 1741–1752, (1994).
[24] I. Duursma. Combinatorics of the two-variable zeta function. In eds. G. Mullen, A. Poli, and H. Stichtenoth, *International Conference on Finite Fields and Applications*, vol. 2948, Lecture Notes in Computer Science, pp. 109–136. Springer, (2003). ISBN 3-540-21324-4.
[25] T. Brylawski, A decomposition for combinatorial geometries, *Tans. Am. Math. Soc.* **171**, 235–282, (1972).
[26] C. Greene, Weight enumeration and the geometry of linear codes, *Studies in Applied Mathematics* **55**, 119–128, (1976).
[27] M. Aigner, *Combinatorial theory* (Springer, New York, 1979).
[28] T. Britz, Extensions of the critical theorem, *Discrete Mathematics* **305**, 55–73, (2005).
[29] J. van Lint and R. M. Wilson, *A course in combinatorics* (Cambridge University Press, Cambridge, 1992).
[30] R. Stanley, *Enumerative combinatorics, vol. 1* (Cambridge University Press, Cambridge, 1997).
[31] A. Skorobogatov. Linear codes, strata of grassmannians, and the problems of segre. In eds. H. Stichtenoth and M. Tfsafsman, *Coding Theory and Algebraic Geometry, Lecture Notes Math. vol 1518*, pp. 210–223. Springer-Verlag, Berlin, (1992).
[32] T. Britz and C. Rutherford, Covering radii are not matroid invariants, *Discrete Mathematics* **296**, 117–120, (2005).
[33] H. Whitney, On the abstract properties of linear dependence, *American Journal of Mathematics* **57**, 509–533, (1935).
[34] J. Kung, *A source book in matroid theory* (Birkhäuser, Boston, 1986).
[35] J. Oxley, *Matroid theory* (Oxford University Press, Oxford, 1992).
[36] D. Welsh, *Matroid theory* (Academic Press, London, 1976).
[37] N. White, *Theory of matroids* (Encyclopedia of Mathmatics and its Applications, vol. 26, Cambridge University Press, Cambridge, 1986).
[38] N. White, *Matroid applications* (Encyclopedia of Mathmatics and its Applications, vol. 40, Cambridge University Press, Cambridge, 1992).
[39] W. Tutte, Lectures on matroids, *Journal of Research of the National Bureau of Standards, Sect. B* **69**, 1–47, (1965).
[40] G. Whittle, A charactrization of the matroids representable over GF(3) and the rationals, *Journal of Combinatorial Theory, Ser. B* **65**(2), 222–261, (1995).
[41] G. Whittle, On matroids representable over GF(3) and other fields, *Transactions of the American Mathematical Society* **349**(2), 579–603, (1997).
[42] J. Blackburn, N. Crapo, and D. Higgs, A catalogue of combinatorial geometries, *Mathematics of Computation* **27**, 155–166, (1973).

[43] W. Dukes, on the number of matroids on a fintie set, *Séminaire Lotharingien de Combinatoire* **51**, Art. B51g, 12 pp., (2004).
[44] D. Knuth, The asymptotic number of geometries, *Journal of Combinatorial Theory, Ser. A* **16**, 398–400, (1974).
[45] L. Euler, Solutio problematis ad geometriam situs pertinentis, *Commentarii Academiae Scientiarum Imperialis Petropolitanae* **8**, 128–140, (1736).
[46] N. Biggs, *Algebraic graph theory* (Cambridge University Press, Cambridge, 1993).
[47] R. Wilson and J. Watkins, *Graphs; An introductory approach* (J. Wiley & Sons, New York, 1990).
[48] G. Birkhoff, On the number of ways of coloring a map, *Proc. Edinburgh Math. Soc.* **2**, 83–91, (1930).
[49] H. Whitney, A logical expansion in mathematics, *Bulletin of the American Mathematical Society* **38**, 572–579, (1932).
[50] H. Whitney, The coloring of graphs, *Annals of Mathematics* **33**, 688–718, (1932).
[51] W. Tutte, A contribution to the theory of chromatic polynomials, *Canadian Journal of Mathematics* **6**, 80–91, (1954).
[52] W. Tutte, On the algebraic theory of graph coloring, *Journal of Combinatorial Theory* **1**, 15–50, (1966).
[53] W. Tutte, On dichromatic polynomials, *Journal of Combinatorial Theory* **2**, 301–320, (1967).
[54] W. Tutte, Cochromatic graphs, *Journal of Combinatorial Theory* **16**, 168–174, (1974).
[55] W. Tutte, Graphs-polynomials, *Advances in Applied Mathematics* **32**, 5–9, (2004).
[56] W. Tutte, Matroids and graphs, *Transactions of the American Mathematical Society* **90**, 527–552, (1959).
[57] C. Athanasiadis, Characteristic polynomials of subspace arrangements and finite fields, *Advances in Mathematics* **122**, 193–233, (1996).
[58] A. Barg, The matroid of supports of a linear code, *AAECC* **8**, 165–172, (1997).
[59] T. Britz. *Relations, matroids and codes*. PhD thesis, Univ. Aarhus, (2002).
[60] T. Britz, MacWilliams identities and matroid polynomials, *The Electronic Journal of Combinatorics* **9**, R19, (2002).
[61] T. Britz, Higher support matroids, *Discrete Mathematics* **307**, 2300–2308, (2007).
[62] T. Britz and K. Shiromoto, A MacWillimas type identity for matroids, *Discrete Mathematics* **308**, 4551–4559, (2008).
[63] T. Brylawski and J. Oxley. The Tutte polynomial and its applications. In ed. N. White, *Matroid Applications*, pp. 173–226. Cambridge University Press, Cambridge, (1992).
[64] G. Etienne and M. Las Vergnas, Computing the Tutte polynomial of a hyperplane arrangement, *Advances in Applied Mathematics* **32**(1), 198–211, (2004).
[65] W. Tutte, A ring in graph theory, *Proc. Cambridge Philos. Soc.* **43**, 26–40,

(1947).
[66] W. Tutte. *An algebraic theory of graphs.* PhD thesis, Univ. Cambridge, (1948).
[67] T. Brylawski and J. Oxley, Several identities for the characteristic polynomial of a combinatorial geometry, *Discrete Mathematics* **31**(2), 161–170, (1980).
[68] T. Kløve, Support weight distribution of linear codes, *Discrete Matematics* **106/107**, 311–316, (1992).
[69] P. Cartier, Les arrangements d'hyperplans: un chapitre de géométrie combinatoire, *Seminaire N. Bourbaki* **561**, 1–22, (1981).
[70] P. Orlik and H. Terao, *Arrangements of hyperplanes.* vol. 300, (Springer-Verlag, Berlin, 1992).
[71] G.-C. Rota, On the foundations of combinatorial theory I: Theory of möbius functions, *Zeit. für Wahrsch.* **2**, 340–368, (1964).
[72] R. Stanley. An introduction to hyperplane arrangements. In *Geometric combinatorics, IAS/Park City Math. Ser.*, *13*, pp. 389–496. Amer. Math. Soc., Providence, RI, (2007).
[73] R. Lidl and H. Niederreiter, *Introduction to finite fields and their applications* (Cambridge University Press, Cambridge, 1994).
[74] H. Crapo and G.-C. Rota, *On the foundations of combinatorial theory: Combinatorial geometries* (MIT Press, Cambridge MA, 1970).
[75] G. Birkhoff, Abstract linear dependence and lattices, *Amer. Journ. Math.* **56**, 800–804, (1935).
[76] H. Crapo, The Tutte polynomial, *Aequationes Math.* **3**, 211–229, (1969).
[77] E. Mphako, Tutte polynomials of perfect matroid designs, *Combinatorics, Probability and Computing* **9**, 363–367, (2000).
[78] A. Blass and B. Sagan, Möbius functions of lattices, *Advances in Mathematics* **129**, 94–123, (1997).
[79] H. Crapo, Möbius inversion in lattices, *Archiv der Mathematik* **19**, 595–607, (1968).
[80] T. Zaslavsky, *Facing up to arrangements: Face-count fomulas for partitions of space by hyperplanes* (Mem. Amer. Math. Soc. vol. 1, No. 154, Amer. Math. Soc., 1975).
[81] T. Zaslavsky, Signed graph colouring, *Discrete. Math.* **39**, 215–228, (1982).
[82] C. Greene and T. Zaslavsky, On the interpretation of Whitney numbers through arrangements of hyperplanes, zonotopes, non-radon partitions and orientations of graphs, *Transactions of the American Mathematical Society* **280**, 97–126, (1983).
[83] A. Ashikhmin and A. Barg, Minimal vectors in linear codes, *IEEE Transactions on Information Theory* **44**(5), 2010–2017, (1998).
[84] J. Massey. Minimal codewords and secret sharing. In *In Proc. Sixth Joint Swedish-Russian Workshop on Information theory, Molle, Sweden*, pp. 276–279, (1993).
[85] D. Stinson, *Cryptography, theory and practice* (CRC Press, Boca Raton, 1995).
[86] A. Björner and T. Ekedahl, Subarrangments over finite fields: Chomological

and enumerative aspects, *Advances in Mathematics* **129**, 159–187, (1997).
[87] F. Ardila, Computing the tutte polynomial of a hyperplane arrangement, *Pacific J. Math.* **230**(5), 1–26, (2007).
[88] M. de Boer, Almost MDS codes, *Designs, Codes and Cryptography* **9**, 143–155, (1996).
[89] T. Brylawski, Intersection theory for graphs, *J. Comb. Theory, Ser. B* **30** (2), 233–246, (1981).
[90] J. Kung. Twelve views of matroid theory. In eds. K. K. S. Hong, J.H. Kwak and F. Roush, *Combinnatorial and Computational Mathematics*, pp. 56–96. World Scientific, River Edge, (2001).
[91] A. Sokal. The multivariate Tutte polynomial (alias Potts model) for graphs and matroids. In *Surveys in combinatorics 2005, London Math. Soc. Lecture Note Ser. vol. 327*, pp. 173–226. Cambridge University Press, Cambridge, (2005).
[92] G. Farr. Tutte-whitney polynomials: Some history and generalizations. In eds. G. Grimmett and C. MacDiarmid, *Combinatorics, Complexity and Chance: A Tribute to D. Welsh*, pp. 28–52. Oxford Univ. Press, Oxford, (2007).
[93] C. Fortuin and P. Kasteleyn, On the random cluster-model. I. Introduction and realtion to other models, *Physica* **57**, 536–564, (1972).
[94] P. Kasteleyn and C. Fortuin, Phase transitions in lattice systems with random local properties, *J. Phys. Soc. Japan* **26**, 11–14, (1969).
[95] T. Britz and K. Shiromoto, Designs from subcode support of linear codes, *Designs, Codes and Cryptography* **46**, 175–189, (2008).
[96] D. Welsh and G. Whittle, Arrangements, channel assignments and associated polynomials, *Advances in Applied Mathematics* **23**, 375–406, (1999).
[97] J. Kung, Preface: Old and new perspectives on the Tutte polynomial, *Annals of Combinatorics* **12**, 133–137, (2008).
[98] V. Kook, W. Reiner and D. Stanton, Combinatorial Laplacians on matroid complexes, *Journal of the American Mathematical Society* **13**, 129–148, (2000).
[99] W. Kook, Recurrence relations for the spectrum polynomial of a matroid, *Discrete Applied Mathematics* **143**, 312–317, (2004).
[100] A. Duval, A common recursion for Laplacians of matroids and shifted simplicial complexes, *Documneta Mathematica* **10**, 583–618, (2005).
[101] G. Denham, The combinatorial Laplacian of the Tutte complex, *J. Algebra* **242**(1), 160–175, (2001).
[102] M. Aigner and J. Seidel, Knoten, Spin modelle und Grahen, *Jber. Dt. Math-Verein* **97**, 75–96, (1995).
[103] L. Kaufmann, *On knots* (Ann. Math. Stud. 115, Princeton Univ. Press, Princeton, 1987).
[104] D. Welsh, *Complexity: Knots, colourings and counting* (London Mathematical Society Lecture Note Series vol. 186, Cambridge University Press, Cambridge, 1993).
[105] T. Helleseth, The weight distribution of the coset leaders of some classes of codes with related parity-check matrices, *Discrete Mathematics* **28**,

161–171, (1979).
[106] J. Justesen and T. Høholdt, Bounds on list decoding of MDS codes, *IEEE Transactions on Information Theory* **47**, 1604–1609, (2001).
[107] M. Sudan, Decoding of reed-solomon codes beyond the error-correction bound, *J. Complexity* **13**, 180–193, (1997).
[108] M. Munuera, Steganography and error-correcting codes, *Signal Processing* **87**, 1528–1533, (2007).
[109] M. Munuera. Steganography from a coding theory point of view. In ed. E. Martínez-Moro, *Algebraic geometry modeling in information theory cryptography*. World Scientific, River Edge, (2011).
[110] S. Hakami and H. Frank, Cut-set matrices and linear codes, *IEEE Transactions on Information Theory* **11**, 457–458, (1965).
[111] J. Bruck and M. Naor, The hardness of decoding linear codes with preprocessing, *IEEE Transactions on Information Theory* **36**(2), 381–385, (1990).
[112] D. MacKay, *Information theory, inference and learning algorithms* (Cambridge University Press, Cambridge, 2003).
[113] T. Richardson and R. Urbanke, *Modern coding theory* (Cambridge University Press, Cambridge, 2008).
[114] F. Harary. A survey of the reconstructing conjecture. In *Lecture Notes in Mathematics*, vol. *406*, pp. 18–28, (1974).
[115] T. Brylawski, On the nonreconstructibility of combinatorial geometries, *J. Comb. Theory, Ser. B* **19**(1), 72–76, (1975).
[116] T. Brylawski. Reconstructing combinatorial geometries. In *Lecture Notes in Mathematics*, vol. *406*, pp. 226–235, (1974).
[117] T. Brylawski, Hyperplane reconstruction of the tutte polynomial of a geometric lattice, *Discrete Mathematics* **35**(1-3), 25–38, (1981).
[118] G.-C. Rota. Combinatorial theory, old and new. In *Proc. Int. Congress Math. 1970 (Nice)*, vol. 3, pp. 229–233, Paris, (1971). Gauthier-Villars.
[119] D. Welsh. Combinatorial problems in matroid theory. In ed. D. Welsh, *Combinatorial mathematics and its applications*, pp. 291–306. Academic Press, London, (1972).
[120] R. Read, An introduction to chromatic polynomials, *Journal of Combinatorial Theory, Series A* **4**, 52–71, (1968).
[121] M. Aigner. Whitney numbers. In ed. N. White, *Combinatorial geometries, Encyclopedia Math. Appl.* vol. *29*, pp. 139–160. Cambridge Univ. Press, Cambridge, (1987).
[122] J. Jaeger, D. Vertigan, and D. Welsh, On the computational complexity of the Jones and Tutte polynomials, *Math. Proc. Camb. Phil. Soc.* **108**, 35–53, (1990).
[123] D. Welsh, The computational complexity of knot and matroid polynmials, *Discrete Mathematics* **124**, 251–269, (1994).
[124] P. K. A. Björklund, T. Husfeldt and M. Koivisto. Computing the Tutte polynomial in vertex-exponential time. In *FOCS*, pp. 677–686. IEEE Computer Society, (2008).
[125] P. Orlik and L. Solomon, Combinatorics and topology of complements of hyperplanes, *Invent. Math.* **56**, 167–189, (1980).

[126] I. Duursma, Weight distributions of geometric Goppa codes, *Transactions of the American Mathematical Society* **351**, 3609–3639, (1999).

[127] I. Duursma, From weight enumerators to zeta functions, *Discrete Applied Mathematics* **111**(1-2), 55–73, (2001).